T0135035

Data Science and Big Data Computing

Zaigham Mahmood
Editor

Data Science and Big Data Computing

Frameworks and Methodologies

Editor
Zaigham Mahmood
Department of Computing and Mathematics
University of Derby
Derby, UK

Business Management and Informatics Unit
North West University
Potchefstroom, South Africa

ISBN 978-3-319-81139-0 ISBN 978-3-319-31861-5 (eBook)
DOI 10.1007/978-3-319-31861-5

Printed on acid-free paper

This Springer imprint is published by Springer Nature
The registered company is Springer International Publishing AG Switzerland

To
Rehana Zaigham Mahmood:
For her Love and Support

Preface

Overview

Huge volumes of data are being generated by commercial enterprises, scientific domains and general public. According to a recent report by IBM, we create 2.5 quintillion bytes of data every day. According to another recent research, data production will be 44 times greater in 2020 than it was in 2009.

Data being a vital organisational resource, its management and analysis is becoming increasingly important: not just for business organisations but also for other domains including education, health, manufacturing and many other sectors of our daily life. This data, due to its volume, variety and velocity, often referred to as *Big Data*, is no longer restricted to sensory outputs and classical databases; it also includes highly unstructured data in the form of textual documents, webpages, photos, spatial and multimedia data, graphical information, social media comments and public opinions. Since Big Data is characterised by massive sample sizes, high-dimensionality and intrinsic heterogeneity, and since noise accumulation, spurious correlation and incidental endogeneity are common features of such datasets, traditional approaches to data management, visualisation and analytics are no longer satisfactorily applicable. There is therefore an urgent need for newer tools, better frameworks and workable methodologies for such data to be appropriately categorised, logically segmented, efficiently analysed and securely managed. This requirement has resulted in an emerging new discipline of *Data Science* that is now gaining much attention with researchers and practitioners in the field of *Data Analytics*.

Although the terms Big Data and Data Science are often used interchangeably, the two concepts have fundamentally different roles to play. Whereas Big Data refers to collection and management of large amounts of varied data from diverse sources, Data Science looks to creating models and providing tools, techniques and scientific approaches to capture the underlying patterns and trends embedded in these datasets, mainly for the purposes of strategic decision making.

In this context, this book, *Data Science and Big Data Computing: Frameworks and Methodologies*, aims to capture the state of the art and present discussions and guidance on the current advances and trends in Data Science and Big Data Analytics. In this reference text, 36 researchers and practitioners from around the world have presented latest research developments, frameworks and methodologies, current trends, state of the art reports, case studies and suggestions for further understanding and development of the Data Science paradigm and Big Data Computing.

Objectives

The aim of this volume is to present the current research and future trends in the development and use of methodologies, frameworks and the latest technologies relating to Data Science, Big Data and Data Analytics. The key objectives include:

- Capturing the state of the art research and practice relating to Data Science and Big Data
- Analysing the relevant theoretical frameworks, practical approaches and methodologies currently in use
- Discussing the latest advances, current trends and future directions in the subject areas relating to Big Data
- Providing guidance and best practices with respect to employing frameworks and methodologies for efficient and effective Data Analytics
- In general, advancing the understanding of the emerging new methodologies relevant to Data Science, Big Data and Data Analytics

Organisation

There are 13 chapters in *Data Science and Big Data Computing: Frameworks and Methodologies*. These are organised in three parts, as follows:

- Part I: *Data Science Applications and Scenarios*. This section has a focus on Big Data (BD) applications. There are four chapters. The first chapter presents a framework for fast data applications, while the second contribution suggests a technique for complex event processing for BD applications. The third chapter focuses on agglomerative approaches for partitioning of networks in BD scenarios; and the fourth chapter presents a BD perspective for identifying minimum-sized influential vertices from large-scale weighted graphs.
- Part II: *Big Data Modelling and Frameworks*. This part of the book also comprises four chapters. The first chapter presents a unified approach to data modelling and management, whereas the second contribution presents a distributed computing perspective on interfacing physical and cyber worlds. The third chapter discusses machine learning in the context of Big Data, and the final

contribution in this section presents an analytics-driven approach to identifying duplicate records in large data repositories.
- Part III: *Big Data Tools and Analytics*: There are five chapters in this section that focus on frameworks, strategies and data analytics technologies. The first two chapters present Apache and other enabling technologies and tools for data mining. The third contribution suggests a framework for data extraction and knowledge discovery. The fourth contribution presents a case study for adaptive decision making; and the final chapter focuses on social impact and social media analysis relating to Big Data.

Target Audiences

The current volume is a reference text aimed at supporting a number of potential audiences, including the following:

- *Data scientists, software architect and business analysts* who wish to adopt the newer approaches to Data Analytics to gain business intelligence to support business managers' strategic decision making
- *Students and lecturers* who have an interest in further enhancing the knowledge of Data Science; and technologies, mechanisms and frameworks relevant to Big Data and Data Analytics
- *Researchers* in this field who need to have up-to-date knowledge of the current practices, mechanisms and frameworks relevant to Data Science and Big Data to further develop the same

Derby, UK
Potchefstroom, South Africa

Zaigham Mahmood

Acknowledgements

The editor acknowledges the help and support of the following colleagues during the review and editing phases of this text:

- Dr. Ashiq Anjum, University of Derby, Derby, UK
- Anupam Biswas, Indian Institute of Technology (BHU) Varanasi, India
- Dr. Alfredo Cuzzocrea, CAR-CNR & Univ. of Calabria, Rende (CS), Italy
- Dr. Emre Erturk, Eastern Institute of Technology, New Zealand
- Prof. Jing He, Kennesaw State University, Kennesaw, GA, USA
- Josip Lorincz, FESB-Split, University of Split, Croatia
- Dr. N Maheswari, School CS & Eng, Chennai, Tamil Nadu, India
- Aleksandar Milić, University of Belgrade, Serbia,
- Prof. Sulata Mitra, Indian Institute of Eng Science and Tech, Shibpur, India
- Prof. Saswati Mukherjee, Anna University, Chennai, India
- Dr. S Parthasarathy, Thiagarajar College of Eng, Tamil Nadu, India
- Daniel Pop, Institute e-Austria Timisoara, West Univ. of Timisoara, Romania
- Dr. Pethuru Raj, IBM Cloud Center of Excellence, Bangalore, India
- Dr. Muthu Ramachandran, Leeds Beckett University, Leeds, UK
- Dr. Lucio Agostinho Rocha, State University of Campinas, Brazil
- Dr. Saqib Saeed, Bahria University, Islamabad, Pakistan
- Prof. Claudio Sartori, University of Bologna, Bologna, Italy
- Dr. Mahmood Shah, University of Central Lancashire, Preston, UK
- Amro Najjar, École Nationale Supérieure des Mines de Saint Étienne, France
- Dr. Fareeha Zafar, GC University, Lahore, Pakistan

I would also like to thank the contributors to this book: 36 authors and coauthors, from academia as well as industry from around the world, who collectively submitted 13 chapters. Without their efforts in developing quality contributions, conforming to the guidelines and meeting often the strict deadlines, this text would not have been possible.

Grateful thanks are also due to the members of my family – Rehana, Zoya, Imran, Hanya, Arif and Ozair – for their continued support and encouragement. Best wishes also to Eyaad Imran.

Department of Computing and Mathematics
University of Derby
Derby, UK

Business Management and Informatics Unit
North West University
Potchefstroom, South Africa
14 February 2016

Zaigham Mahmood

Other Springer Books by Zaigham Mahmood

Cloud Computing: Challenges, Limitations and R&D Solutions

This reference text reviews the challenging issues that present barriers to greater implementation of the Cloud Computing paradigm, together with the latest research into developing potential solutions. This book presents case studies, and analysis of the implications of the cloud paradigm, from a diverse selection of researchers and practitioners of international repute (ISBN: 978-3-319-10529-1).

Continued Rise of the Cloud: Advances and Trends in Cloud Computing

This reference volume presents latest research and trends in cloud-related technologies, infrastructure and architecture. Contributed by expert researchers and practitioners in the field, this book presents discussions on current advances and practical approaches including guidance and case studies on the provision of cloud-based services and frameworks (ISBN: 978-1-4471-6451-7).

Cloud Computing: Methods and Practical Approaches

The benefits associated with cloud computing are enormous; yet the dynamic, virtualized and multi-tenant nature of the cloud environment presents many challenges. To help tackle these, this volume provides illuminating viewpoints and case studies to present current research and best practices on approaches and technologies for the emerging cloud paradigm (ISBN: 978-1-4471-5106-7).

Software Engineering Frameworks for the Cloud Computing Paradigm

This is an authoritative reference that presents the latest research on software development approaches suitable for distributed computing environments. Contributed by researchers and practitioners of international repute, the book offers practical guidance on enterprise-wide software deployment in the cloud environment. Case studies are also presented (ISBN: 978-1-4471-5030-5).

Cloud Computing for Enterprise Architectures

This reference text, aimed at system architects and business managers, examines the cloud paradigm from the perspective of enterprise architectures. It introduces fundamental concepts, discusses principles and explores frameworks for the adoption of cloud computing. The book explores the inherent challenges and presents future directions for further research (ISBN: 978-1-4471-2235-7).

Contents

Contributors

Gopichand Agnihotram Infosys Labs, Infosys Ltd., Bangalore, India

Vyankatesh Agrawal Department of Computer Science and Engineering, Indian Institute of Technology (BHU), Varanasi, India

Gourav Arora Department of Computer Science and Engineering, Indian Institute of Technology (BHU), Varanasi, India

Zartasha Baloch IICT, Mehran University of Engineering and Technology, Jamshoro, Pakistan

Rentachintala Bhargavi School of Computing Sciences and Engineering, VIT University, Chennai, India

Derya Birant Department of Computer Engineering, Dokuz Eylul University, Izmir, Turkey

Bhaskar Biswas Department of Computer Science and Engineering, Indian Institute of Technology (BHU), Varanasi, India

Anupam Biswas Department of Computer Science and Engineering, Indian Institute of Technology (BHU), Varanasi, India

Valentin Cristea Computer Science Department, Faculty of Automatic Control and Computers, University Politehnica of Bucharest, Bucharest, Romania

Asit Kumar Das Department of Computer Science and Technology, Indian Institute of Engineering Science and Technology, Shibpur, Howrah, West Bengal, India

José Carlos Martins Delgado Department of Computer Science and Computer Engineering, Instituto Superior Técnico, Universidade de Lisboa, Lisbon, Portugal

Nirmala Dorasamy Department of Public management and Economics, Durban University of Technology, Durban, South Africa

Sarbendu Guha Infosys Labs, Infosys Ltd., Bangalore, India

Jing (Selena) He Department of Computer Science, Kennesaw State University, Marietta, GA, USA

Gabriel Iuhasz Institute e-Austria Timisoara, West University of Timisoara, Timişoara, Romania

Srijan Khare Department of Computer Science and Engineering, Indian Institute of Technology (BHU), Varanasi, India

N. Maheswari School of Computing Science and Engineering, VIT University, Chennai, Tamil Nadu, India

Mariana Mocanu Computer Science Department, Faculty of Automatic Control and Computers, University Politehnica of Bucharest, Bucharest, Romania

Catalin Negru Computer Science Department, Faculty of Automatic Control and Computers, University Politehnica of Bucharest, Bucharest, Romania

Srinivas Padmanabhuni Infosys Labs, Infosys Ltd., Bangalore, India

Anjaneyulu Pasala Infosys Labs, Infosys Ltd., Bangalore, India

Mohanavadivu Periasamy TCL Canada, Montreal, QC, Canada

Dana Petcu Institute e-Austria Timisoara, West University of Timisoara, Timişoara, Romania

Nataša Pomazalová Department of Regional Development, Mendel University, Brno, Czech Republic

Durban University of Technology, Durban, South Africa

Daniel Pop Institute e-Austria Timisoara, West University of Timisoara, Timişoara, Romania

Florin Pop Computer Science Department, Faculty of Automatic Control and Computers, University Politehnica of Bucharest, Bucharest, Romania

Satya Prateek B Infosys Labs, Infosys Ltd., Bangalore, India

Vijay V. Raghavan The Center for Advanced Computer Studies, University of Louisiana at Lafayette, Lafayette, LA, USA

Pethuru Raj IBM Global Cloud Center of Excellence, Bangalore, India

Faisal Karim Shaikh IICT, Mehran University of Engineering and Technology, Jamshoro, Pakistan

TCMCORE, STU, University of Umm Al-Qura, Mecca, Saudi Arabia

Jaya Sil Department of Computer Science and Technology, Indian Institute of Engineering Science and Technology, Shibpur, Howrah, West Bengal, India

M. Sivagami School of Computing Science and Engineering, VIT University, Chennai, Tamil Nadu, India

Gaurav Tiwari Department of Computer Science and Engineering, Indian Institute of Technology (BHU), Varanasi, India

Mukhtiar A. Unar IICT, Mehran University of Engineering and Technology, Jamshoro, Pakistan

Ying Xie Department of Computer Science, Kennesaw State University, Marietta, GA, USA

Pelin Yıldırım Department of Software Engineering, Celal Bayar University, Manisa, Turkey

About the Editor

Professor Dr. **Zaigham Mahmood** is a published author of 16 books, five of which are dedicated to Electronic Government and the other eleven focus on the subjects of Cloud Computing and Data Science, including: *Cloud Computing: Concepts, Technology & Architecture* which is also published in Korean and Chinese languages; *Cloud Computing: Methods and Practical Approaches; Software Engineering Frameworks for the Cloud Computing Paradigm; Cloud Computing for Enterprise Architectures; Cloud Computing Technologies for Connected Government; Continued Rise of the Cloud: Advances and Trends in Cloud Computing*; *Connectivity Frameworks for Smart Devices: The Internet of Things from a Distributed Computing Perspective*; and *Cloud Computing: Challenges, Limitations and R&D Solutions.* Additionally, he is developing two new books to appear later in 2017. He has also published more than 100 articles and book chapters and organised numerous conference tracks and workshops.

Professor Mahmood is the Editor-in-Chief of *Journal of E-Government Studies and Best Practices* as well as the Series Editor-in-Chief of the IGI book series on *E-Government and Digital Divide.* He is a Senior Technology Consultant at Debesis Education UK and Associate Lecturer (Research) at the University of Derby, UK. He further holds positions as Foreign Professor at NUST and IIU in Islamabad, Pakistan, and Professor Extraordinaire at the North West University, Potchefstroom, South Africa. Professor Mahmood is also a certified cloud computing instructor and a regular speaker at international conferences devoted to Cloud Computing and E-Government. His specialised areas of research include distributed computing, project management and e-government.

Professor Mahmood can be reached at *z.mahmood@debesis.co.uk*

Part I
Data Science Applications and Scenarios

Chapter 1
An Interoperability Framework and Distributed Platform for Fast Data Applications

José Carlos Martins Delgado

Abstract Big data developments have been centred mainly on the volume dimension of data, with frameworks such as Hadoop and Spark, capable of processing very large data sets in parallel. This chapter focuses on the less researched dimensions of velocity and variety, which are characteristics of fast data applications. The chapter proposes a general-purpose distributed platform to host and interconnect fast data applications, namely, those involving interacting resources in a heterogeneous environment such as the Internet of Things. The solutions depart from conventional technologies (such as XML, Web services or RESTful applications), by using a resource-based meta model that is a partial interoperability mechanism based on the compliance and conformance, service-based distributed programming language, binary message serialization format and architecture for a distributed platform. This platform is suitable for both complex (Web-level) and simple (device-level) applications. On the variety dimension, the goal is to reduce design-time requirements for interoperability by using structural data matching instead of sharing schemas or media types. In this approach, independently developed applications can still interact. On the velocity dimension, a binary serialization format and a simple message-level protocol, coupled with a cache to hold frequent type mappings, enable efficient interaction without compromising the flexibility required by unstructured data.

Keywords Internet of Things • IoT • Big data • Web services • XML • Coupling • Structural compatibility • Compliance • Conformance • Distributed programming • Variety • Velocity

J.C.M. Delgado (✉)
Department of Computer Science and Computer Engineering, Instituto Superior Técnico, Universidade de Lisboa, Lisbon, Portugal
e-mail: jose.delgado@tecnico.ulisboa.pt

Z. Mahmood (ed.), *Data Science and Big Data Computing*,
DOI 10.1007/978-3-319-31861-5_1

1.1 Introduction

One of the fundamental objectives of any distributed data system is the ability to perform the required amount of data exchange and computation in the available timeframe, which translates into a required minimum data flow and processing rates. Big data scenarios turn this into a harder endeavour due to several reasons, including the following characteristics of data [1]:

- Volume: high volume of data (more data to process)
- Velocity: high rate of incoming data (less time to process data)
- Variety: data heterogeneity (more data formats or data sources to deal with)

Big data developments have been mainly centred on the volume dimension, with dynamic frameworks such as Hadoop [2] and Spark [3], capable of processing very large data sets in parallel. This chapter focuses on the less researched dimensions of velocity and variety, which are characteristics of fast data applications [4]. Typically, these involve too many entities, interacting and exchanging too many data, at too high rates in a too heterogeneous environment. An entity can be a complex application in a server or a very simple functionality provided by a small sensor, in the context of what is usually known as the Internet of Things, abbreviated as IoT [5, 6].

The European Commission [7] estimates that by 2020, the number of globally connected devices will be in the order of 50–100 billion devices. These will generate big data, which many applications will need to process very quickly and with low latency.

Variety means supporting a diversity of data sources, formats and protocols. Not all devices are adequate to support Transmission Control Protocol/Internet Protocol (TCP/IP) and all the features required to be part of the Web. Velocity requires efficient data exchange and processing mechanisms. Together, they demand for new data-level distributed interoperability mechanisms.

Current interoperability technologies rely on text-based data description languages, such as Extensible Markup Language (XML) and JavaScript Object Notation (JSON) [57], and high-level and complex protocols such as Hypertext Transfer Protocol (HTTP) and Simple Object Access Protocol (SOAP). However, these languages have not been designed for the high throughput and low latency that fast applications require. Similarly, the big data solutions such as Hadoop emphasize the volume dimension and are not adequate for fast data [4]. In terms of interoperability, these languages and protocols constitute specific solutions, designed for the Web class of applications (many clients for each server, best effort rather than real time) and do not allow an arbitrary set of computer-based applications to interact as peers.

What is needed is a new set of solutions that support the generic interoperability of fast data applications, in the same way as web technologies have provided universal interoperability for web applications. These solutions include native support for binary data, efficient and full-duplex protocols, machine-level data and service interoperability and context awareness for dynamic and mobile environments, such as those found in smart cities [8]. Today, these features are simulated on top of Web services: applications based on representational state transfer (REST), HTTP, XML, JSON and other related technologies, rather than implemented by native solutions. The problem needs to be revisited to minimize the limitations at the source, instead of just hiding them with abstraction layers that add complexity and reduce performance.

As a contribution to satisfy these requirements, this chapter includes the following proposals:

- A layered interoperability framework, to systematize the various aspects and slants of distributed interoperability
- A language to describe not only data structures (state) but also operations (behaviour), with self-description capability to support platform-agnostic interoperability
- A data interoperability model, on which this language is based, which relies on compliance and conformance [9] instead of schema sharing (as in XML) or previously defined data types (as REST requires)
- A message-level protocol at a lower level than that of SOAP and even HTTP, with many of the features included in these protocols implemented on top of the basic interoperability model
- The architecture of a node of a distributed platform suitable for fast data applications

These features are the building blocks of a distributed interoperability platform conceived to tackle the velocity and variety dimensions of distributed applications, modelled as services. This platform is suitable not only for complex, Web-level applications but also for simple, device-level applications.

This chapter is structured as follows. Section 1.2 describes some of the existing technologies relevant to the theme of this chapter, followed by a description in Sect. 1.3 of several of the issues concerning fast data. Section 1.4 describes an interoperability framework with emphasis on fast data problems, namely, those affecting variety (interoperability and coupling) and velocity (message latency). It also presents a resource-based model to support structural compatibility, based on compliance and conformance, and a service interoperability language that implements these proposals. Section 1.5 describes the architecture of a distributed platform to support the resource-based model and the service-based language. The chapter ends by discussing the usefulness of this approach, outlining future directions of research and drawing the main conclusions of this work.

1.2 Background

Big data [1] has become a relevant topic in recent years, spurred by the ever-increasing growth rate of data produced in this globally interconnected world and by the always-pressing need of obtaining meaningful information from large data sets, in domains such as business analytics [10], healthcare [11], bioinformatics [12], scientific computing [13] and many others [14].

Big data refers to handling very large data sets, for storage, processing, analysis, visualization and control. The National Institute of Standards and Technology (NIST) have proposed a Big Data Interoperability Framework [15] to lay down a foundation for this topic.

A big push for this area came from the MapReduce programming model [16], initially used for indexing large data sets and business intelligence over data warehouses. The first motivating factor for big data was, thus, volume (data size), using immutable and previously stored data. However, agile enterprises [17] require almost real-time analysis and reaction to a large number of events and business data, stemming from many sources and involving many data formats. This means that the velocity and variety dimensions are gaining momentum [4]. A survey of systems for big data, with emphasis on real time, appears in [18].

Enterprise integration models and technologies, such as service-oriented architecture (SOA) [19], REST [20] and enterprise service bus (ESB) [21], have not been conceived for fast data processing and therefore constitute only as a best-effort approach.

Besides the dimensions already described (volume, velocity and variety), other Vs have also been deemed relevant in this context [13, 15], these being veracity (accuracy of data), validity (quality and applicability of data in a given context), value (of data to stakeholders), volatility (changes in data over time) and variability (of the data flow).

Gone are the days when the dominant distributed application scenario consisted of a Web encompassing fixed computers, both at the user (browser) and server (Web application) sides. Today, cloud computing [22] and the IoT [6] are revolutionizing the society, both at the enterprise and personal levels, in particular in urban environments [8] with new services and applications. For example, mobile cloud computing [23] is on the rise, given the pervasiveness of smartphones and tablets that created a surge in the bring your own device (BYOD) trend [24]. The increasing use of radio-frequency identification (RFID) tags [25] in supply chains raises the need to integrate enterprise applications with the physical world, including sensor networks [26] and vehicular [27] networks.

Cisco have set up a counter [28], indicating the estimated number of devices connected to the Internet. This counter started with 8.7 billion devices at the end of 2012, increased to roughly 10 and 14 billion at the end of 2013 and 2014, respectively, and at the time of writing (April 2015), it shows a figure of 15.5 billion, with a foreseen value in the order of 50 billion by 2020. The Internet World Stats site (http://www.internetworldstats.com/stats.htm), on the other hand, estimates that by mid-2014, the number of Internet human users was around three billion, almost half the worldwide population of roughly 7.2 billion people. The number of Internet-enabled devices is clearly growing faster than the number of Internet users, since the world population is estimated to be in the order of 7.7 billion by 2020 [29]. This means that the Internet is no longer dominated by human users but rather by smart devices that are small computers and require technologies suitable to them, rather than those mostly adequate to full-fledged servers.

1.3 Introducing Fast Data

Fast data has a number of inherent issues, in addition to those relating to big data. This section describes motivating scenarios and one of the fundamental issues, that of interoperability. Other issues, stemming from the variety and velocity dimensions are discussed in Sects. 1.4.2 and 1.4.3 respectively.

1.3.1 Motivating Scenarios

Figure 1.1 depicts several scenarios in which large quantities of data can be produced from a heterogeneous set of data sources, eventually with different formats and processing requirements. For simplicity, not all possible connections are depicted, but the inherent complexity of integrating all these systems and processing all the data they can produce is easy to grasp.

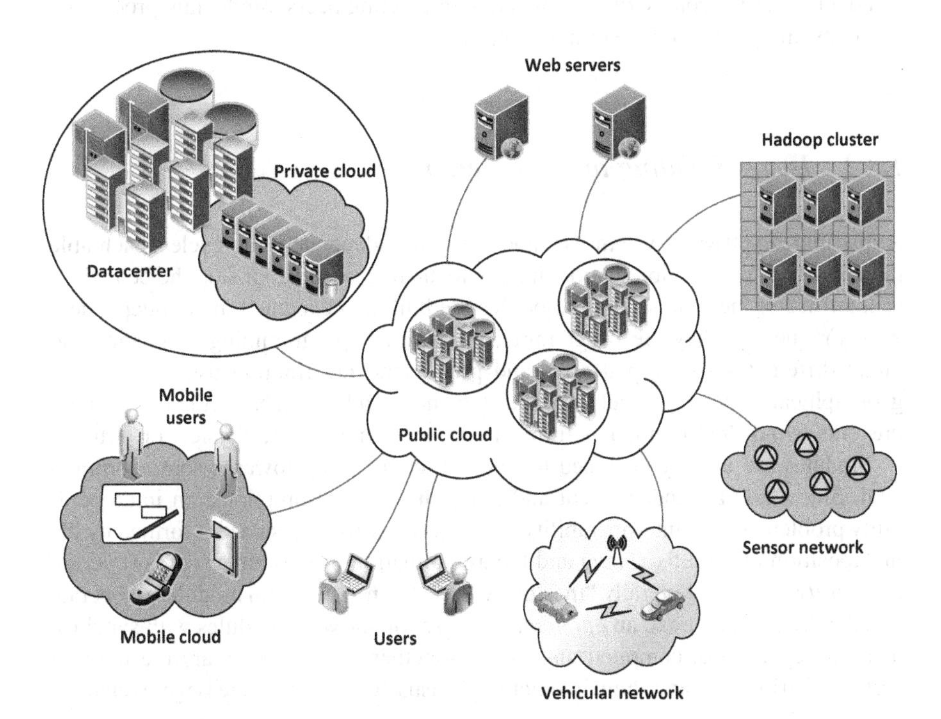

Fig. 1.1 An example of a set of heterogeneous systems with big data requirements

Most big data applications today use best-effort technologies such as Hadoop [2] and Spark [3], in which immutable data is previously loaded into the processing nodes. This is suitable for applications in areas such as business analytics [30], which attempt to mine information that can be relevant in specific contexts and essentially just deal with the volume dimension of big data. However, this is not the case for applications where many data sets are produced or a large number of events occur frequently, in a heterogeneous ecosystem of producers and consumers. In these applications, processing needs to be performed as data are produced or events occur, therefore emphasizing the variety and velocity dimensions (fast data).

No matter which dimension we consider, "big" essentially means too complex, too much, too many and too fast to apply conventional techniques, technologies and systems, since their capabilities are not enough to handle such extraordinary requirements. This raises the problem of integrating heterogeneous interacting parties to a completely new level, in which conventional integration technologies (such as HTTP, XML, JSON, Web Services and RESTful applications) expose their limitations. These are based on technologies conceived initially for human interaction, with text as the main format and sub-second time scales and not for heavy-duty, machine-level binary data exchange that characterizes computer-to-computer interactions, especially those involving big data.

New solutions are needed to deal with these integration problems, in particular in what concerns fast data requirements. Unlike processing of large passive and immutable data sets, for which frameworks such as Hadoop are a good match, fast data scenarios consist of a set of active interacting peers, producing, processing and consuming data and event notifications.

1.3.2 Issues Relating to Interoperability

A distributed software system has modules with independent life cycles, each able to evolve to a new version without having to change, suspend or stop the behaviour or interface of the others. These modules are built and executed in an independent way. Frequently, they are programmed in different programming languages and target different formats, platforms and processors. Distribution usually involves geographical dispersion, a network and static node addresses. Nevertheless, nothing prevents two different modules from sharing the same server, physical or virtual.

Modules are usually designed to interact, cooperating towards some common goal. Since they are independent and make different assumptions, an interoperability problem arises. Interoperability, as old as networking, is a word formed by the juxtaposition of a prefix (*inter*) and the agglutination of two other words (*operate* and *ability*), meaning literally "the ability of two or more system modules to operate together". In this context, an *application* is a set of software modules with synchronized lifecycles, i.e. compiled and linked together. Applications are the units of system distribution, and their interaction is usually limited to message exchange.

Applications are independent, and each can evolve in ways that the others cannot predict or control.

The interaction between modules belonging to the same application can rely on names to designate concepts in the type system (types, inheritance, variables, methods and so on). A name can have only one meaning in a given scope, which means that using a name is equivalent to using its definition. A working application usually assumes that all its modules are also working and use the same implementation language and formats, with any changes notified to all modules. The application is a coherent and cohesive whole.

The interaction of modules belonging to different applications, however, is a completely different matter. Different applications may use the same name for different meanings, be programmed in different languages, be deployed in different platforms, use different formats and without notifying other applications, migrate from one server to another, change their functionality or interface and even be down for some reason, planned or not.

This raises relevant interoperability problems, not only in terms of correctly interpreting and understanding exchanged data but also in keeping behaviours synchronized in some choreography. The typical solutions involve a common protocol (such as HTTP), self-describing data at the syntax and sometimes semantics levels and many assumptions previously agreed upon. For example, XML-based interactions, including Web services, assume a common schema. REST proponents claim decoupling between client and server (the client needs just the initial URI, Universal Resource Identifier). However, RESTful applications do require previously agreed upon media types (schemas) and implicit assumptions by the client on the behaviour of the server when executing the protocol verbs.

It is virtually impossible for one application to know how to appropriately interact with another application, if it knows nothing about that application. Not even humans are able to achieve it. Some form of coupling (based on shared and agreed knowledge, prior to interaction) needs to exist. The goal is to reduce coupling as much as possible while ensuring the minimum level of interoperability required by the problem that motivated the interaction between applications. Figure 1.2 provides an example of the kind of problems that need to be tackled in order to achieve this goal.

Figure 1.2 can be described in terms of the scenario of Fig. 1.2a which refers to the first seven steps of the process; the entire process being as follows:

1. Application *A* resorts to a directory to find a suitable application, according to some specification.
2. The directory has a reference (a link) to such an application, e.g. *B*.
3. The directory sends that reference to *A*.
4. Application *A* sends a message to *B*, which it must understand, react and respond according to the expectations of *A* (note the bidirectional arrow).
5. If message is unreachable, *A* can have predefined alternative applications, such as *B1* or *B2*. Resending the message to them can be done automatically or as a result from an exception.

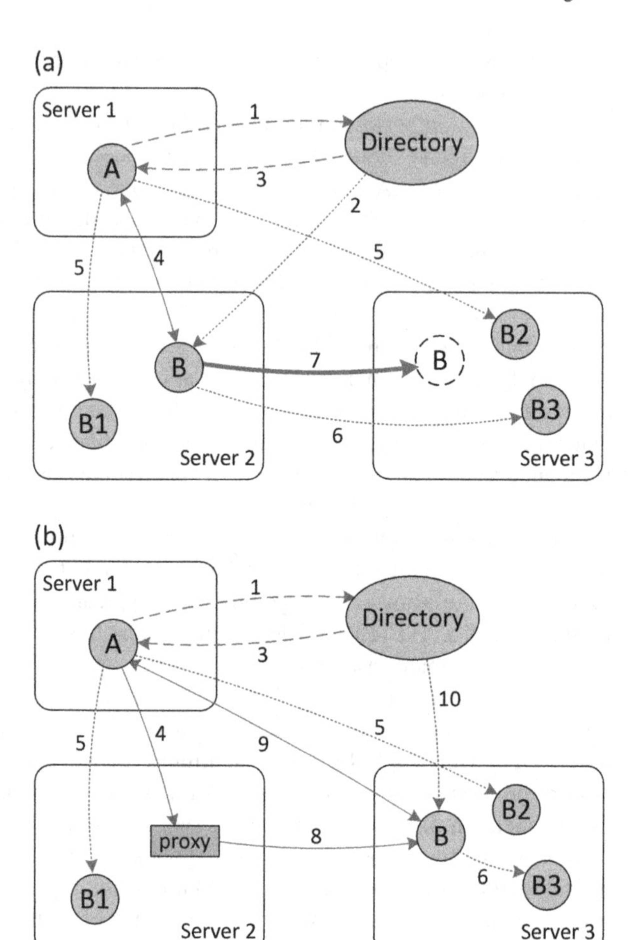

Fig. 1.2 Illustration of some interoperability aspects. (**a**) Before migration of application B. (**b**) After migration

6. If *B* is reachable but somehow not functional, *B* itself (or the cloud that implements it) can forward the message to an alternative application, such as *B3*.
7. Application *B* can be migrated dynamically to another cloud, yielding the scenario of Fig. 1.2b.
8. *B* leaves a reverse proxy as a replacement, which means that if *A* sends another message to *B* (step 4), it will be automatically forwarded to the new *B*.
9. The response, however, will be sent to the original sender, *A*, including information on the new location of *B*, which *A* will use for subsequent messages (the message protocol must support this).
10. The proxy could be garbage collected; but this is not easy to manage in a distributed system that is unreliable by nature. Therefore, the proxy can be maintained for some time under some policy and destroyed afterward. If some application still holding a reference to the old *B* sends a message to it, the protocol should respond with a suitable error stating that *B* does not exist. *A* can

then repeat steps 1 through 3, obtaining the new location of *B* from the directory.

Figure 1.2 raises some interesting interoperability issues, namely, the question of compatibility between an application and its alternatives, which must be able to behave as if they were the original application. Therefore, we need to detail what is involved in an interaction between two applications.

1.4 An Interoperability Framework for Fast Data

Having discussed the interoperability problem, a framework is needed to dissect its various slants. This will enable us to discuss what the variety and velocity dimensions really mean and to derive a metamodel and a language to deal with them. This section presents the relevant discussion.

1.4.1 Understanding Interoperability

In general, successfully sending a message from one application to another entails the following aspects (noting that requests and responses reverse the roles of the sender and receiver applications):

- *Intent.* Sending the message must have an underlying intent, inherent in the interaction to which it belongs and related to the motivation to interact and the goals to achieve. This should be aligned with the design strategy of both applications.
- *Content.* This concerns the generation and interpretation of the content of a message by the sender, expressed by some representation, in such a way that the receiver is also able to interpret it, in its own context.
- *Transfer.* The message content needs to be successfully transferred from the context of the sender to the context of the receiver.
- *Willingness.* Usually, applications are designed to interact and therefore to accept messages, but nonfunctional aspects such as security and performance limitations can impose constraints.
- *Reaction.* This concerns the reaction of the receiver upon reception of a message, which should produce effects according to the expectations of the sender.

Interoperability between two applications can be seen at a higher level, involving intentions (why interact and what reactions can be expected from whom), or at a lower level, concerning messages (what to exchange, when and how). Detailing the various levels leads to a systematization of interoperability such as the one described in Table 1.1.

Table 1.1 Levels of interoperability

Category	Level	Main artefact	Description
Symbiotic (purpose and intent)	Coordination	Governance	Motivations to have the interaction, with varying levels of mutual knowledge of governance, strategy and goals
	Alignment	Joint venture	
	Collaboration	Partnership	
	Cooperation	Outsourcing	
Pragmatic (reaction and effects)	Contract	Choreography	Management of the effects of the interaction at the levels of choreography, process and service
	Workflow	Process	
	Interface	Service	
Semantic (meaning of content)	Inference	Rule base	Interpretation of a message in context, at the levels of rule, known application components and relations and definition of concepts
	Knowledge	Knowledge base	
	Ontology	Concept	
Syntactic (notation of representation)	Structure	Schema	Representation of application components, in terms of composition, primitive components and their serialization format in messages
	Predefined type	Primitive application	
	Serialization	Message format	
Connective (transfer protocol)	Messaging	Message protocol	Lower-level formats and network protocols involved in transferring a message from the context of the sender to that of the receiver
	Routing	Gateway	
	Communication	Network protocol	
	Physics	Media protocol	
Environmental (deployment and migration)	Management	API	Environment in which each application is deployed and managed, including the portability problems raised by migrations
	Library	Utility service	
	Platform	Basic software	
	Computer	Computing hardware	

Table 1.1 can be described as follows, using the category column as the top organizing feature:

- *Symbiotic*: Expresses the purpose and intent of two interacting applications to engage in a mutually beneficial agreement. Enterprise engineering is usually the topmost level in application interaction complexity, since it goes up to the human level, with governance and strategy heavily involved. Therefore, it maps mainly onto the symbiotic category, although the same principles apply (in a more rudimentary fashion) to simpler subsystems. This can entail a tight coordination under a common governance (if the applications are controlled by the same entity), a joint venture agreement (if the two applications are substantially aligned), a collaboration involving a partnership agreement (if some goals are shared) or a mere value chain cooperation (an outsourcing contract).

- *Pragmatic*: The effect of an interaction between a consumer and a provider is the outcome of a contract, which is implemented by a choreography that coordinates processes, which in turn implement workflow behaviour by orchestrating service invocations. Languages such as Business Process Execution Language (BPEL) [31] support the implementation of processes and Web Services Choreography Description Language (WS-CDL) is an example of a language that allows choreographies to be specified.
- *Semantic*: Both interacting applications must be able to understand the meaning of the content of the messages exchanged: both requests and responses. This implies interoperability in rules, knowledge and ontologies, so that meaning is not lost when transferring a message from the context of the sender to that of the receiver. Semantic languages and specifications such as Web Ontology Language (OWL) and Resource Description Framework (RDF), map onto this category [32].
- *Syntactic*: This deals mainly with form, rather than content. Each message has a structure composed of data (primitive applications) according to some structural definition (its schema). Data need to be serialized to be sent over the network as messages using representations such as XML or JSON.
- *Connective*: The main objective is to transfer a message from the context of one application to the other regardless of its content. This usually involves enclosing that content in another message with control information and implementing a message protocol (such as SOAP or HTTP) over a communications network according to its own protocol (such as TCP/IP) and possibly resorting to routing gateways if different networks are involved.
- *Environmental*: Each application also interacts with the environment (e.g. a cloud or a server) in which it is deployed, anewed or by migration. The environment's management application programming interface (API) and the infrastructure level that the application requires will most likely have impact on the way applications interact, particularly if they are deployed in (or migrate between) different environments, from different vendors. Interoperability between an application and the environment in which it is deployed usually known as *portability*.

In most cases, not all these levels are considered explicitly. Higher levels tend to be treated *tacitly* (specified in documentation or simply assumed but not ensured), whereas lower levels are commonly dealt with *empirically* (ensured by running software but details hidden by lower level software layers).

Syntactic is the most used category, because it is the simplest and the most familiar, with interfaces that mainly deal with syntax or primitive semantics. The pragmatic category, fundamental to specify behaviour, is mainly implemented by software but without any formal specification.

Another important aspect is nonfunctional interoperability. It is not just a question of sending the right message. Adequate service levels, context awareness, security and other nonfunctional issues must also be considered when applications interact, otherwise interoperability will be less effective or not possible at all.

Finally, it should be stressed that, as asserted above, all these interoperability levels constitute an expression of application coupling. On the one hand, two uncoupled applications (with no interactions between them) can evolve freely and independently, which favours adaptability, changeability and even reliability (such that if one fails, there is no impact on the other). On the other hand, applications need to interact to cooperate towards common or complementary goals, which imply that some degree of previously agreed mutual knowledge is indispensable. The more they share with the other, the more integrated they are, and so the interoperability becomes easier, but coupling gets more complicated.

The usefulness of Table 1.1 lies in providing a classification that allows coupling details to be better understood, namely, at which interoperability levels they occur and what is involved at each level, instead of having just a blurry notion of dependency. In this respect, it constitutes a tool to analyse and to compare different coupling models and technologies.

1.4.2 The Variety Dimension

The previous section suggests that coupling is unavoidable. Without it, no interaction is possible. Our goal is to minimize it as much as possible, down to the minimum level that ensures the level of interaction required by the applications exchanging fast data. In other words, the main goal is to ensure that each application knows just enough about the other to be able to interoperate but no more than that, to avoid unnecessary dependencies and constraints. This is consistent with the *principle of least knowledge* [33].

Minimizing coupling maximizes the likelihood of finding suitable alternatives or replacements for applications, as well as the set of applications with which some application is compatible, as a consumer or as a provider of some functionality. This is precisely one of the slants of the variety problem in fast data.

Figure 1.3 depicts the scenario of an application immersed in its environment, in which it acts as a provider for a set of applications (known as *consumers*), from which it receives requests or event notifications, and as a consumer of a set of applications (called *providers*), to which it sends requests or event notifications.

Coupling between this application and the other applications expresses not only how much it depends on (or is affected by the variety in) its providers but also how much its consumers depend on (or are affected by changes in) it. Dependency on an application can be assessed by the level and amount of features that another application is constrained by. Two coupling metrics can be defined, from the point of view of a given application, as follows:

- C_F (*forward coupling*), which expresses how much a consumer application is dependent on its providers, is defined as

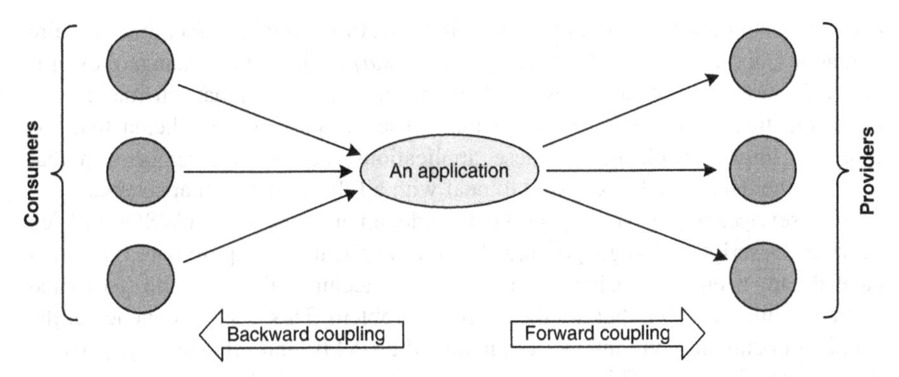

Fig. 1.3 Coupling between an application and its consumers and providers

$$C_F = \frac{\sum_{i \in P} \frac{Up_i}{Tp_i \cdot N_i}}{|P|} \tag{1.1}$$

where P is the set of providers that this application uses and $|P|$ denotes the cardinality of P, Up_i is the number of features that this application uses in provider i, Tp_i is the total number of features that provider i exposes and N_i is the number of providers with which this application is compatible as a consumer, in all uses of features of provider i by this application.

- C_B (*backward coupling*), which expresses how much impact a provider application has on its consumers, is defined as

$$C_B = \frac{\sum_{i \in C} \frac{Uc_i}{Tc \cdot M}}{|C|} \tag{1.2}$$

where C is the set of consumers that use this application as provider and $|C|$ denotes the cardinality of C, Uc_i is the number of features of this application that consumer i uses, Tc is the total number of features that this application exposes and M is the number of known applications that are compatible with this application and can replace it, as a provider.

The conclusion from metric 1.1 above is that the existence of alternative providers to an application reduces its forward coupling C_F, since more applications (with which this application is compatible, as a consumer) dilute the dependency. Similarly, the conclusion from metric 1.2 above is that the existence of alternatives to an application as a provider reduces the system dependency on it, thereby reducing the impact that application may have on its potential consumers and therefore its backward coupling C_B.

Current application integration technologies, such as Web services [34] and RESTful applications [35], do not really comply with the principle of least knowledge and constitute poor solutions in terms of coupling. In fact, both require interacting applications to share the type (*schema*) of the data exchanged, even if only a fraction of the data values is actually used. A change in that schema, even if the interacting applications do not actually use the part of the schema that has changed, implies a change in these applications, because the application that receives the data must be prepared to deal with all the values of that schema.

Web services rely on sharing a schema (a document expressed in WSDL or Web Services Description Language) and RESTful applications require data types that have already been previously agreed upon. These technologies solve the distributed interoperability problem but not the coupling problem. This is a consequence of the classical document-style interaction, heralded by XML and schema languages, as illustrated by Fig. 1.4a. This is a symmetric arrangement in which a writer produces a document according to some schema, and the reader uses the same schema to validate and to read the contents of the document. There is no notion of services, only the passive resource that the document constitutes. We are at the level of data description languages, such as XML or JSON.

Figure 1.4b introduces the notion of service, in which a message is sent over a channel and received by the receiver. It is now treated as a parameter to be passed on to some behaviour that the receiver implements, instead of just data to be read. However, the message is still a document, validated and read essentially in the same way as in Fig. 1.4a. We are at the level of Web services or RESTful applications.

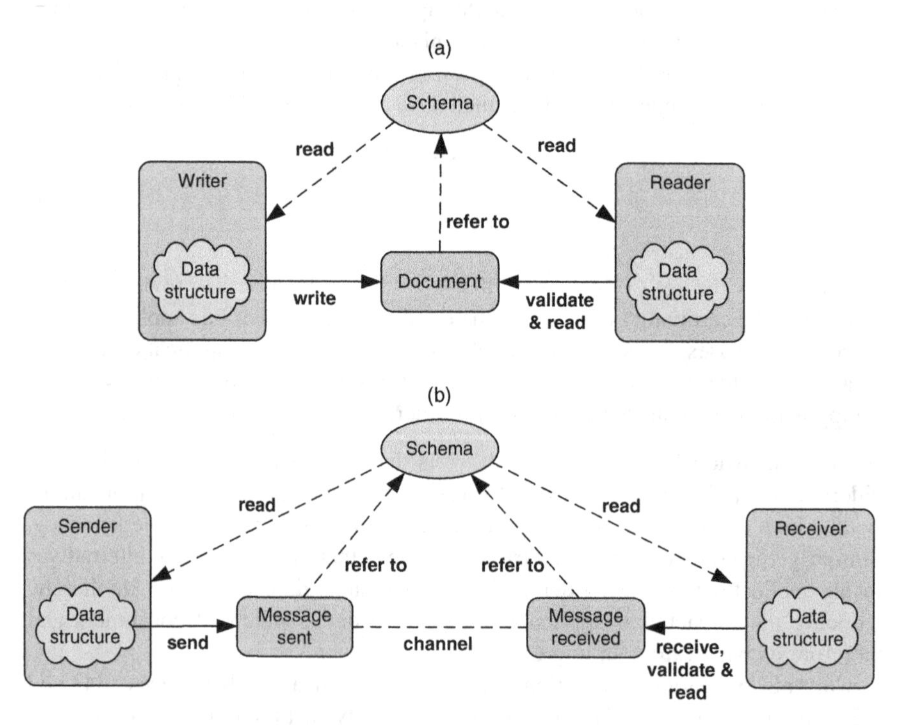

Fig. 1.4 Interaction styles: (**a**) documents; (**b**) document-based messages

The behaviour invoked can thus be exposed and implemented in various ways, but in the end, the goal is still similar.

The schemas in Fig. 1.4 refer to type specifications and need not be separate documents, as it is usual in XML schema and WSDL. In the REST world, schemas are known as media types but perform the same role. The difference is that instead of being declared in a separate document referenced by messages, they are usually previously known to the interacting applications, either by being standard or by having been previously agreed upon. In any case, the schema or media type must be the same at both sender and receiver endpoints, which imposes coupling between the applications for all its possible values of a schema, even if only a few are actually used.

Another problem concerns the variety in the networks. The protocols underlying the Web (TCP/IP and HTTP) are demanding in terms of the smaller devices. Hardware capabilities are increasingly better, and efforts exist to reduce the requirements, such as IPv6 support on low-power wireless personal area networks (6LoWPAN) [36] and the Constrained Application Protocol (CoAP) [37], to deal with constrained RESTful environments. Building on the simplicity of REST, CoAP is a specification of the Internet Engineering Task Force (IETF) working group CoRE, which deals with constrained RESTful environments. CoAP includes only a subset of the features of HTTP but adds asynchronous messages, binary headers and User Datagram Protocol (UDP) binding. It is not easy to have the Web everywhere, in a transparent manner.

1.4.3 The Velocity Dimension

Velocity in fast data is not just about receiving, or sending, many (and/or large) messages or events in a given timeframe. The fundamental problem is the reaction time (in the context of the timescale of the applications) between a request message, or an event notification, and the availability of the processed results. Real-time applications usually depend on a fast feedback loop. Messages frequently take the form of a stream. Complex event processing [38], in which a set of related events is analysed to aggregate data or detect event patterns, is a common technique to filter unneeded events or messages and thus reduce the real velocity of processing and of computing requirements.

Several factors affect velocity, e.g. processing time and message throughput. In our context, emphasizing interoperability, the main problem lies in throughput of messages exchanged. We cannot assume that data have been previously stored with a view to be processed afterwards.

Throughput depends on message latency (the time taken to serialize data and then to validate and to reconstruct it at the receiver) and on network throughput (its own latency and bandwidth). The most relevant issue for velocity is the message latency, since the characteristics of the network are usually fixed by the underlying communications hardware. Latency is affected by several factors, including the service-level platform (e.g. Web services, RESTful applications), the message protocol (e.g. HTTP, SOAP, CoAP) and the message data format (e.g. XML, JSON).

Web services, SOAP and XML are the most powerful combination and are essentially used in enterprise-class applications, which are the most demanding. But they are also the most complex and the heaviest, in terms of processing overheads. RESTful applications, HTTP and JSON are a simpler and lighter alternative, although less powerful and exhibiting a higher semantic gap in modelling real-world entities [39]. Nevertheless, application designers will gladly trade expressive power for simplicity and performance (which translates into velocity), and the fact is that many applications do not require all the features of a full-blown Web services stack. REST-based approaches are becoming increasingly popular [40].

In any case, these technologies evolved from the original Web of documents [41], made for stateless browsing and with text as the dominant media type. The Web was developed for people, not for enterprise integration. Today, the world is rather different. The Web is now a Web of services [42], in which the goal is to provide functionality, not just documents. There are now more computers connected than people, with binary data formats (computer data, images, video and so on) as the norm rather than the exception.

Yet, the evolution has been to map the abstraction of Web of services onto the Web of documents, with a major revolution brought by XML and its schemas. The document abstraction has been retained, with everything built on top of it. In the interoperability levels of Table 1.1, XML (or JSON, for that matter) covers only the syntactic category. The lack of support of XML for higher interoperability levels (viz. at the service interface level) is one of the main sources of complexity in current technologies for integration of applications. In turn, this imposes a significant overhead in message latency and, by extension, velocity.

Table 1.2 summarizes the main limitations of existing technologies that are particularly relevant for this context.

1.4.4 Modelling with Resources and Services

Any approach should start with a metamodel of the relevant entities. In this case, the organization of applications and their interactions are primordial aspects. The interoperability levels of Table 1.1 constitute a foundation for the metamodel, although this chapter concentrates on the syntactic, semantic and pragmatic categories.

The interaction between different applications cannot simply use names, because the contexts are different and only by out-of-band agreements (such as sharing a common ontology) will a given name have the same meaning in both sides of an interaction.

Table 1.2 Main limitations of relevant interoperability technologies

Technology	Main limitations
XML	Text based, verbose, complex, poor support for binary, data description only (no behaviour), syntax-level only (higher levels require other languages), high coupling (interoperability achieved by schema sharing)
JSON	Simpler but less powerful than XML and the same limitations. Coupling is also high since data types have to be agreed prior to data interaction
HTTP	Optimized for Web browsing (scalable retrieval of hypermedia information) but not for generic and distributed service-based interactions. Inefficient and specific text-based control information format. Synchronous, committed to the client-server paradigm, lack of support for the push model (server-initiated interactions) and for binary data
SOAP	Complex, significant performance overheads, XML-based (thus inheriting XML's problems), too high level, with specific solutions for nonfunctional data and routing
Web Services (SOA)	Complex (based on WSDL, SOAP and XML), many standards to cover distributed interaction aspects, high coupling (interoperability achieved by sharing the WSDL document), lack of support for structure (flat service space)
REST	Forces a create, read, update, delete (CRUD) approach to model real-world entities, causing a significant semantic gap in generic service modelling. Coupling is disguised under the structured data and the fixed syntactic interface. Structure of data returned by the server may vary freely, but the client needs to have prior knowledge of the data schemas and of the expected behaviour of the server

XML-based systems, for example, solve this problem by sharing the same schema, as illustrated by Fig. 1.4. However, this is a strong coupling constraint and contemplates data only. Behaviour (operations) needs to be simulated by data declarations, as in WSDL documents describing Web services.

We need to conceive a more dynamic and general model of applications and their interactions, which supports interoperability without requiring to share the specification of the application interface (schema). The strategy relies on structural type matching, rather than nominal type matching. This approach entails:

- A small set of primitive types, shared by all applications (universal upper ontology)
- Common structuring mechanisms, to build complex types from primitive ones
- A mechanism for structurally comparing types from interacting applications

Applications are structured, and, in the metamodel as described below, their modules are designated as *resources*. Since applications are distributed by definition, sending each other messages through an interconnecting channel is the only form of interaction. To make intra-application interactions as similar as possible to inter-application interactions, all resources interact by messages, even if they belong to the same application. Resources are the foundation artefacts of the metamodel of applications and of their interactions as depicted in Fig. 1.5.

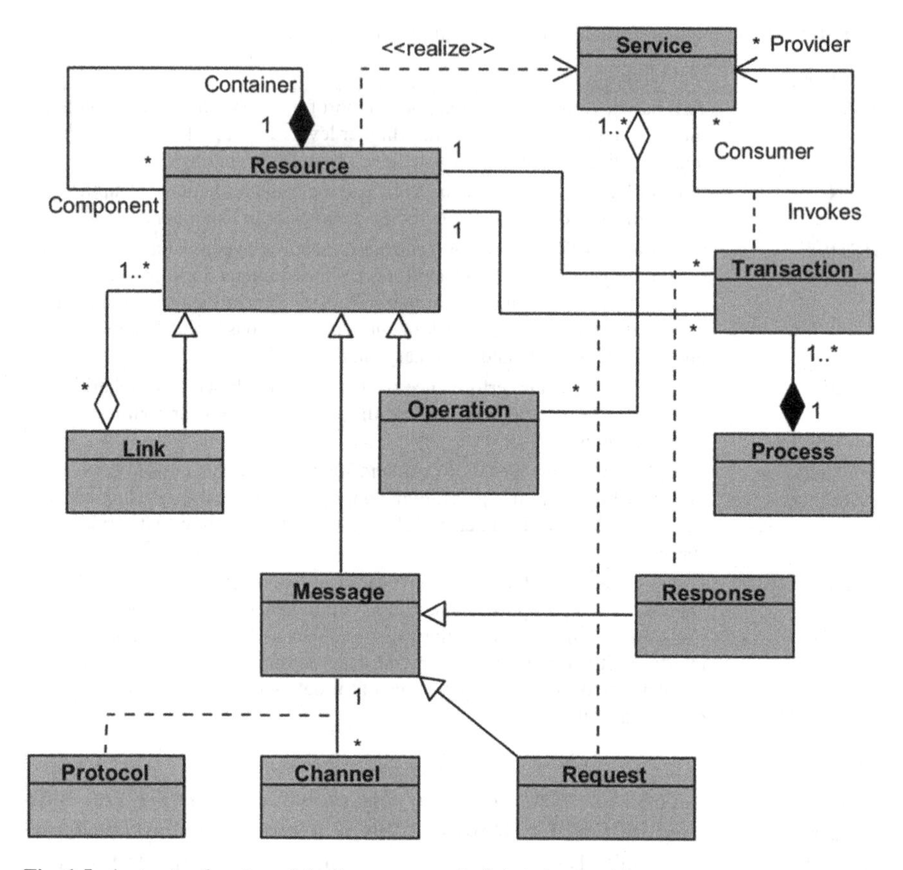

Fig. 1.5 A structural metamodel of resources and of their interactions

This metamodel can be briefly described in the following way:

- A *resource* is an entity of any nature (material, virtual, conceptual, noun, action and so on) that embodies a meaningful, complete and discrete concept, making sense by itself while being distinguishable from, and able to interact with, other entities.
- Resources that interact do so by sending each other *messages*. A message is itself a resource that migrates (through some *channel*, usually a network, using some *protocol*) from the context of the resource that sends it to the context of the resource that receives it. Messages can be either requests or responses.
- *Operations* (the equivalent of methods in objects) are first-class resources and can be sent as message components. All instantiations of an operation (one for each message received, with lazy creation) are closures [43], but distributed messages (between applications) must be self-contained (no free variables). The advantages of using operations as full-fledged resources instead of methods or functions are uniformity of treatment (everything is a resource), support for higher-order programming (operations as arguments to other operations) and

transparent event processing (the occurrence of an event sends a message to a resource, which can be an operation or any other kind of resource).

- A *link* can target more than one resource, constituting a list of remote references, for consumer-side redundancy. If a resource targeted by the link is unavailable, the support system can automatically try to resend the message to the next in the list.
- A *service* is a set of logically related reactions of a resource to the messages it can receive and react to (depends on the operations that are components of that resource). A service can be seen as pure behaviour, although the implementation of concrete reactions may depend on state, which needs a resource as an implementation platform, or have side effects, changing that state.
- Sending a request message from a resource *A* to a resource *B* (which eventually responds) constitutes a *service transaction* between these two resources, in which the resources *A* and *B* perform the roles of service *consumer* and service *provider*, respectively. A service transaction can entail other service transactions, as part of the chain reaction to the message by the service provider. A service is defined in terms of reactions to messages (external view) and not in terms of state transitions or activity flows (internal view).
- A *process* is a graph of all service transactions that can occur, starting with a service transaction initiated at some resource *A* and ending with a final service transaction, which neither reacts back nor initiates new service transactions. A process corresponds to a use case of resource *A* and usually involves other resources as service transactions flow.

The metamodel depicted in Fig. 1.5 has been conceived to cover the main levels of interoperability described in Table 1.1 and as a contribution to solve the limitations of existing technologies, summarized in Table 1.2.

Interoperability, in terms of the current technological developments, is based on data description languages such as XML and JSON, which cover just the syntactic category of Table 1.1. Semantics has been an afterthought, with languages such as RDF and OWL [44]. Pragmatics (behaviour) has been dealt with through generic programming languages or, under the process paradigm, with BPEL [31]. Services have been modelled essentially with WSDL documents, which are no more than an interface specification simulated on top of a data description language (XML). The REST architectural style uses a fixed service specification at the syntactic level and restricts variability to resources and their structure. In essence, no solution today covers Table 1.1 adequately.

Resources in Fig. 1.5 model real-world entities, with their interfaces modelled as services. Dynamically, services interact by messages, requests and responses, over some channel with some protocol. These constitute transactions that build up processes. Structured resources refer to others by component names, or roles, in the sense of Description Logic [45], and allow semantics to be incorporated in the model from scratch (details not described here due to space limitations). Operations are modelled as resources, in the same manner as entities typically modelled by a class. This makes the metamodel simpler and more canonical and allows a finer grain of application modelling.

1.4.5 Handling the Variety Dimension

Section 1.4.2 has shown that reducing coupling is a good approach to deal with the variety problem, by increasing the potential base of consumers and providers. Most resources are made interoperable by design, i.e. conceived and implemented to work together, in a consumer-provider relationship. Therefore, coupling is high, and it is not easy to accommodate changes and deal with variants of these resources. For instance, Web services are a typical interoperability solution but work by sharing schemas (Fig. 1.4b) and establish coupling for all the possible documents satisfying each schema, even if they are not actually used. Searching for an interoperable service is done by schema matching with *similarity* algorithms [46] and ontology matching and mapping [47]. This does not ensure interoperability and manual adaptations are usually unavoidable.

This chapter introduces a different perspective on interoperability, stronger than similarity but weaker than sharing schemas and ontologies. The trick is to allow partial (instead of full) interoperability, by considering only the intersection between what the consumer needs and what the provider offers. If the latter subsumes the former, the degree of interoperability required by the consumer is feasible, regardless of whether the provider supports additional features or not. When this is true, the consumer is said to be *compatible* with the provider or, more precisely, that a resource *X* is compatible with a resource *Y* regarding a consumer-provider relationship. In this respect, two resource relationships are of relevance:

- *Consumer-provider*: Compatibility between a consumer and a provider is known as *compliance* [48]. The consumer must satisfy (*comply with*) the requirements established by the provider to accept requests sent to it, without which these cannot be validated, understood and executed. It is important to note that any consumer that complies with a given provider can use it, independently of having been designed for interaction with it or not. The consumer and provider need not share the same schema. The consumer's schema needs only to be compliant with the provider's schema in the features that it actually uses (partial compliance). Since distributed resources have independent lifecycles, they cannot freely share names and types, and schema compliance must be tested structurally, feature by feature, between messages sent by the consumer and the interface offered by the provider;
- *Provider-provider*: The issue is to ascertain whether a provider *Y*, serving a consumer *X*, can be replaced by another provider *Z* in a way that the consumer-provider relationship enjoyed by *X* is not impaired. In other words, the issue is whether *Z* is replacement compatible with *Y*. Replacement compatibility between two providers is known as *conformance* [49, 50]. The provider must

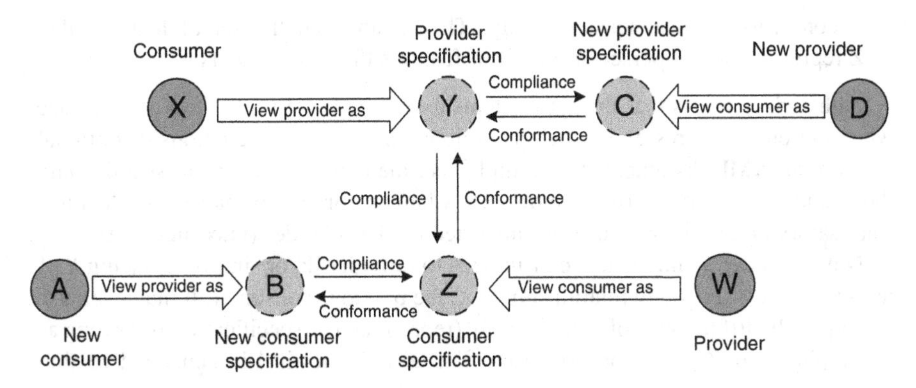

Fig. 1.6 Resource compatibility, by use and replacement

fulfil the expectations of the consumer regarding the effects of a request (including eventual responses), therefore being able to take the form of (*to conform to*) whatever the consumer expects it to be. Note that a provider may be conformant to another with respect to one consumer but not with respect to another consumer. It all depends on the set of features used by the consumer. The reasons for replacing a provider with another may be varied, such as switching to an alternative in case of failure or lack of capacity of *Y*, evolution of *Y* (in which case *Z* would be the new version of *Y*), or simply a management decision. The important aspect to note is that *Z* does not need to support all the features of *Y*, but just those that *X* actually uses (partial conformance).

Compliance and conformance are not symmetric relationships (e.g. if *X* complies with *Y*, *Y* does not necessarily comply with *X*). Figure 1.6 illustrates compliance and conformance, as well as its effects on reducing coupling, either by limiting dependency to the required features or by increasing the set of possible alternative providers or consumers.

A resource *X*, in the role of consumer, has been designed to interact with a resource, in the role of provider, with a specification *Y*, which represents the set of features that *X* requires as its provider. Nothing prevents the actual provider from having additional features, as long as it includes this set.

Now consider a resource *W*, in the role of provider, which has been designed to expect a consumer *Z*, which represents the set of features that *W* expects that the actual consumer will access. Nothing prevents the consumer from accessing less features, as long as they are included in this set.

X and *W* were not designed to interoperate, since what one expects is not what the other is. They have been developed independently, or at least do not constitute a full match of features (in other words, do not share the same schema). However, interoperability is still possible, as long as the following holds:

- *Y* complies with *Z*, at least partially. This means that the set of features that *Y* represents is a subset of the set of features that *Z* represents. In turn, *X* cannot access a feature that *W* does not expect to be accessed.

- Z conforms to Y, at least partially. This means that the set of features that Z represents is a superset of the set of features that Y represents.

Although it seems that these two items assert the same thing, it is not the case when optional features exist, such as optional arguments to functions or optional fields in an XML document. For compliance, the consumer needs to specify only the mandatory features (minimum set), whereas for conformance, all features (mandatory and optional) must be supported by the provider (maximum set).

Now consider adding a new consumer A to provider W, or simply replacing X. A does not have to bear any relationship to X, the previous consumer. It may even use a completely different set of features of W (represented as specification B) as long as B complies with Z and Z conforms to B. This also applies if A is a changed version of X.

A similar reasoning can be made if X now uses a new provider, D, with the set of requirements on consumers represented by specification C. Again, interoperability is still possible as long as Y complies with C and C conforms to Y. Apart from this, D and W need not bear any relationship. This also applies if D is a changed version of W.

Reducing dependencies to the set of actually used features, instead of having to share the full set of features (the schema) is an improvement over current technologies, typically based on XML or JSON documents. It also ensures interoperability, whereas similarity-based approaches [46] do not.

We can now extend Fig. 1.4 with another interaction style, compliance-based messages, as illustrated by Fig. 1.7.

Messages do not obey some external schema. Each message has one specific value (most likely structured, but it is not a type, only one specific value) and its own exclusive schema, which is nothing more than a self-description, without the value variability that a type exhibits. This value and its description can be validated against an infinite number of schemas, those that have this particular value included in the set of their instances.

The receiver in Fig. 1.7 exposes a schema that defines the values it is willing to accept. When a message is received, its schema is checked against the receiver's own schema. If it *complies* (satisfies all the requirements of the receiver's schema), the message can be accepted and processed. The advantage of this is that a resource can send a message to all the resources with schemas that the message complies

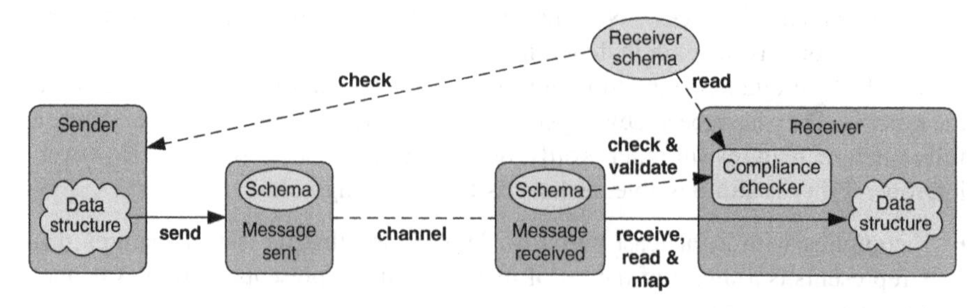

Fig. 1.7 Extending the interaction styles of Fig. 1.4 with compliance-based messages

with and, conversely, a resource can receive messages from any resource that sends messages compliant with its receiving schema.

In other words, coupling occurs only in the characteristics actually used by messages and not in all the characteristics of the schemas used to generate the message or to describe the service of the receiving resource. Since the schemas of the message and of the receiver are not agreed upon beforehand, they need to be checked *structurally*. Resources of primitive types have predefined compliance rules. Structured resources are compared by the names of components (regardless of order of declaration or appearance) and (recursively) by the compliance between structured resources with matching names. Since the order of appearance of named component resources may different in the message and in the receiving schema, there is the need to map one onto the other (bottom of Fig. 1.7).

This is a form of polymorphism that increases the range of applicability of both sender and receiver, constituting a means to reduce coupling to only what is actually used. Sender and receiver no longer need to be designed for each other. Compliance ensures *usability*, whereas conformance ensures *replaceability*. When a resource is able to interact with another, although not entirely interoperable with it, we say that they have *partial interoperability*.

We also need to handle the velocity dimension, but, since it involves performance, this is included in the description of the platform, in Sect. 1.5.1.

1.4.6 A Service Interoperability Language with Support for Variety

Data description languages, such as XML and JSON [57], merely describe data and their structure. If we want to describe resources and their services, we can use WSDL, but the resulting verbosity and complexity has progressively turned away developers in favour of something much simpler, REST [20]. If we want a programming language suitable for distributed environments, we can use BPEL [31] but again with an unwieldy XML-based syntax that forces programmers to use visual programming tools that generate BPEL and increase the complexity stack.

JSON is much simpler than XML, and its popularity has been constantly increasing, but the evolution of the dynamic nature of the Web, as shown by JavaScript and HyperText Markup Language, version 5 (HTML5) [51], hints that data description is not enough anymore and distributed programming is a basic requirement. In the IoT, with machine-to-machine interactions now much more frequent than human-to-server interactions, this is even of greater importance.

Current Web-level interoperability technologies are greatly constrained by the initial decision of basing Web interoperability on data (not services) and text markup as the main description and representation format. Using them to implement the metamodel of Fig. 1.5 leads to a complex stack of specifications and to syntactic and semantic adaptations and compromises, to bridge the gap between the metamodel and the chosen language and/or platform. A language conceived from scratch to implement this metamodel, providing native support for its characteristics, will be a better match. We have designed Service Interoperability Language (SIL) [56] to provide this support.

Each resource in SIL is represented by a SIL Public Interface Descriptor (SPID), which corresponds to a Web service's WSDL but much more compact and able to describe both structure and operations. It is obtained automatically from the resource description itself by including only the public parts (public component resources, including operation headings). Unlike XML or even JSON, there is no separate schema document to describe a resource.

Program 1.1 shows a simple example of resources described in SIL, involving sensors in the context of the IoT [6]. It includes a temperature controller (tempController), with a list of links to temperature sensors with history (tempSensorStats). Each of these has a link to a remote temperature sensor (tempSensor). The lines at the bottom illustrate how these resources can be used and should be included in some resource that uses them.

1.1 Describing and Using Resources Using SIL

```
tempSensor: spid {// descriptor of a temperature sensor
 getTemp: operation (-> [-50.0 .. +60.0]);
};

tempSensorStats: {// temperature sensor with statistics
 sensor: @tempSensor; // link to sensor (can be remote)
 temp: list float; // temperature history
 startStats <||; // spawn temperature measurements
 getTemp: operation (-> float) {
 reply sensor@getTemp<--; // forward request to sensor
 };
 getAverageTemp: operation ([1 .. 24] -> float) {
 for (j: [temp.size .. temp.size-(in-1)])
 out += temp[j];
 reply out/in; // in = number of hours
 };
 private startStats: operation () {// private operation
 while (true) {
 temp.add<-- (getTemp<--); // register temperature
 wait 3600; // wait 1 hour and measure again
 }
 }
};

tempController: {// controller of several temperature
sensors
 sensors: list @tempSensorStats; //list of links to sensors
 addSensor: operation (@tempSensor) {
 t: tempSensorStats; // creates a tempSensorStats resource
```

```
t.sensor = in; // register link to tempSensor
sensors.add<-- @t; // add sensor to list
};
getStats: operation (-> {min: float; max: float;
average: float})
{
out.min = sensors[0]@getTemp<--;
out.max = out.min;
total: float := out.min; // initial value
for (i: [1 .. sensors.length-1) { // sensor 0 is done
t: sensors[i]@getTemp<--; // dereference sensor i
if (t <out.min) out.min = t;
if (t >out.max) out.max = t;
total += t;
};
out.average = total/sensors.length;
reply; // nothing specified, returns out
}
};

// How to use the resources
// tc contains a link to tempController
tc@addSensor<-- ts1; // link to a tempSensor resource
tc@addSensor<-- ts2; // link to a tempSensor resource
x: tc@sensors[0]@getAverageTemp<-- 10; // average last
10 hours
```

This program can be described in the following way:

- The temperature sensor (tempSensor) is remote, and all we have is its SPID, the equivalent to a Web service's WSDL. For example, the SPID of tempSensorStats can be expressed by the following lines:

```
tempSensorStats: spid {// temperature sensor with
statistics
 sensor: @tempSensor; // link to sensor (can be remote)
 temp: list float; // temperature history
 getTemp: operation (-> float);
 getAverageTemp: operation ([1 .. 24] -> float);
}
```

- Resources and their components are declared by a name, a colon and a resource type, which can be primitive, such as integer, a range (e.g. [1 .. 24]), float or user defined (enclosed in braces, i.e. "{. . .}"). There are some resemblances to JSON, but component names are not strings, and operations are supported as first-class resources.
- The definition of a resource is similar to a constructor. It is executed only once, when the resource is created, and can include statements. This is illustrated by

the statement "startStats<||" in tempSensorStats. Actually, this is an asynchronous invocation ("<||") of private operation startStats, which is an infinite loop registering temperature measurements every hour. Asynchronous invocations return a *future* [52], which will be later replaced by the returned value (in this example, the returned future is ignored). Synchronous invocation of operations is done with "<– –", followed by the argument, if any.

- Operations have at most one argument, which can be structured (with "{. . .}"). The same happens with the operation's reply value, as illustrated by operation getStats. Inside operations, the default names of the argument and the value to return are in and out, respectively. The heading of operations specifies the type of the argument and of the reply value (inside parentheses, separated by "->");
- Links to resources (indicated by the symbol "@") are not necessarily URIs, not even strings. They can also be structured and include several addresses, so that a resource in a network (e.g. the Internet) can reference another in a different network, with a different protocol (e.g. a sensor network). It is up to the nodes and gateways of the network, supporting these protocols, to interpret these addresses so that transparent routing can be achieved, if needed
- Resource paths, to access resource components, use dot notation, except if the path traverses a link, in which case a "@" is used. For example, the path used in the last line of Program 1.1 computes the average temperature, along the last 10 h, in sensor 0 of the controller.

Program 1.2 illustrates the compliance and conformance concepts, by providing two additional temperature sensors (weatherSensor and airSensor) in relation to Program 1.1. Only the additional and relevant parts are included here, with the rest as in Program 1.1.

1.2 Example of Partial Interoperability with Structural Compliance and Conformance

```
tempSensor: spid {
 getTemp: operation (->[-50.0 .. +60.0]);
};

weatherSensor: spid {
 getTemp: operation (->[-40.0 .. +50.0]);
 getPrecipitation: operation (-> integer);
};

airSensor: spid {
 getTemp: operation (->[-40.0 .. +45.0]);
 getHumidity: operation (-> [10 .. 90]);
};

// tc contains a link to tempController
tc@addSensor<-- ts1; // link to a tempSensor resource
tc@addSensor<-- ts2; // link to a tempSensor resource
tc@addSensor<-- as1; // link to an airSensor resource
tc@addSensor<-- ws1; // link to an weatherSensor resource
```

```
tc@addSensor<-- ws2; // link to an weatherSensor resource
tc@addSensor<-- as2; // link to an airSensor resource

x: tc@sensors[0]@getAverageTemp<-- 10; // average last
10 hours
s: {max: float; average: float};
s = tc@getStats<--; // only max and average are assigned to s

temp: [-50.0 .. +50.0];
temp = ws1@getTemp <--; // complies. Variability ok
temp = ts1@getTemp <--; // does not comply. Variability
mismatch
```

In Program 1.2:

- weatherSensor and airSensor conform to tempSensor, since they offer the same operation and the result is within the expected variability. This means that they can be used wherever a tempSensor is expected, which is illustrated by adding all these types of sensors to tempController (through tc, a link to it) as if they were of type tempSensor. Nonrelevant operations are ignored.
- The result of invoking the operation getStats on tempController is a resource with three components (as indicated in Program 1.1), whereas s has only two. However, the assignment is still possible. Structural interoperability ignores the extra component.
- The last statement triggers a compliance check by the compiler, which issues an error. The variability of the value returned by operation getTemp in tempSensor (referenced by ts1) is outside the variability range declared for component temp.

Other features of SIL, not illustrated here for simplicity, include:

- *Delegation*: An operation can execute some statements and then delegate the rest of the execution of the request to another resource, by forwarding either the received message or another one. The delegate will respond directly to the first sender, the one that originated the request.
- *Context-awareness*: Messages can be sent in two parts, data and context (separated by a with keyword). The latter is typically used to pass nonfunctional information. Both parts are normal resources and obey the same compliance and conformance rules as other resources. This applies to both request and response messages, which means that SIL supports forward and backward context passing. The latter is useful both in normal responses and exceptions. Context awareness is another way in which variety can be supported.
- *Reliability*: The links in Programs 1.1 and 1.2 can target a list of resources. If for some reason the protocol fails to communicate with the first resource in the link, it will try automatically the following one. An exception is generated if all alternatives fail. There is opportunity for recovery, at the level of the exception handler (covering a set of statements) or at the application level.
- *Interface to other languages*: Resources in SIL can invoke methods in object-oriented languages and have its operations invoked. The solution is similar to XML-based systems, including Web services, and involves annotations. The

SPID of a resource can be used to generate at compile time an interface in Java, for instance, or reflection can be used to support access to resources and objects only at runtime.

1.5 A Distributed Interoperability Platform

This section proposes a platform for distributed interoperability that revisits the problem and constitutes a native solution for computer-based applications in a fast data context, instead of being based on the classical technologies conceived originally for human browsing. Backward compatibility is lost, but this exploratory approach has the potential to compensate with flexibility and adequacy.

1.5.1 Handling the Velocity Dimension

We used several techniques to reduce message processing latency, including:

- Unlike XML or JSON, SIL is compiled. This corresponds to having two resource representations, text (source) and compiled (binary), with coherence between the two maintained by the compiler.
- A binary format based on the tag, length and value (TLV) scheme used by ASN.1 [53], which allows a very efficient traversal of a serialized data structure.
- Messages can be sent with an adjustable level of metadata, from source (including SPID) to almost no metadata at all (just structures composed of primitive resources). Metadata means overhead and does not need to be sent in full in all messages.
- Compliance checks (mappings between a message and the data structure expected by the receiver) can be cached, so that subsequent messages of the same type can be optimized.
- Using efficient full-duplex transport level protocols with support for binary data, such as Web sockets [51], with automatic session management.
- Using a generic message-level protocol as the foundation, instead of an application-level protocol such as HTTP, even in its HTTP/2 incarnation. The protocol just deals with basic messages and not resource-level operations.

We have developed a compiler for SIL based on ANother Tool for Language Recognition (ANTLR) [54], which is able to produce the SPID automatically and to convert source to instructions and data in a binary format (TLV), or *silcodes* (similar to Java's bytecodes). An interpreter, similar to a Java virtual machine (JVM), executes these silcodes. The current implementation, in pure Java, is not optimized and has a performance roughly 50 times slower than a JVM. Much of that time is spent just on virtual method dispatch, the mechanism used to execute the various silcodes. A C-based interpreter, for example, would be much faster, although harder to develop. To maintain flexibility and control of implementation, we did not use a JVM.

The format of a compiled resource is exactly the same as a serialized resource, to be sent as a message or stored in a persistent medium. It supports three levels of metadata (self-description) information, in addition to the data itself: source code, ontology (names and relationships) and no metadata. The latter is used when a previous data-type checking was cached by the message receiver and a token returned to the sender, to be used in subsequent messages. The protocol returns an error if metadata is absent and there is no valid token in the message. The sender then repeats the message, this time with metadata. The ontology-level metadata is the usual case without cache optimization. The source code can be included, if desired, to provide user-level readability.

Table 1.3 gives an idea of the size of resource representations in various situations. The example refers to a data-only resource, so that XML and JSON can be used for comparison. The two lines refer to the same resource, but with longer and shorter component names, which have an impact on metadata size. That is why the sizes in the last column are the same, and the size reduction with less metadata is greater when names are longer. The sizes presented vary with the concrete example, but this gives an idea of the values involved.

An end-to-end application performance comparison with SOA- and REST-based solutions has not been done yet. Trivial examples tend to assess essentially the application servers and network latency, which are not the distinguishing aspects. A meaningful comparison needs to separate concerns, such as protocols, data parsing, platform-agnostic execution (SIL) versus local programming language (e.g. Java), professional versus research implementation and so on. This will be done in the near future. In the meantime, it has been shown that parsing binary data markup (TLV) can be at least an order of magnitude faster than parsing XML text markup [55].

Unlike HTTP, which goes down from the transport level up to the service level, with verbs such as GET or PUT, the SIL protocol deals only with messages, addressed to resources. Its goal is to support message-level interaction and nothing else. What to do with the message is the receiver's responsibility. Unlike SOAP, which defines headers for additional information such as security, the SIL protocol delegates to higher interoperability levels the responsibility to add extra information. This is done by wrapping simple messages within other messages, with the extra information, in what constitutes an open extension to the basic protocol. Both interacting applications need to support the protocol extension mechanism, which may be standardized or be specific to a pair of applications. The basic protocol remains simple.

Table 1.3 Comparison of the sizes (in bytes) of several resource representations, from source to binary only

			Silcodes		
Names	XML	JSON	With source	With ontology	Without metadata
Long	2595	1631	1432	953	382
Short	1972	1193	1014	621	382

The SIL protocol requires only an underlying addressable message-level transport protocol, which does not have to be reliable. Message level reliability, if not provided by the transport protocol, can be implemented at the resource or application level. This is intended as a means to support heterogeneity, namely, in non-TCP/IP protocols, such as those used in sensor networks [26], or in the IoT, at the UDP level, with protocols such as CoAP [37]. In the opposite direction, security can be provided by the underlying protocol itself, if it is complex enough to support secure transport.

Table 1.4 illustrates some of the message types in this protocol, which implements basic transactions in a universal way. Any resource can be both sender and receiver of messages.

There is no notion of session. Messages are independent of each other, with coherence maintained by higher interoperability levels, namely, the pragmatic and symbiotic categories in Table 1.1. However, the underlying transport protocol can open a channel, such as a socket in the Web socket protocol [51], and maintain it open for some predefined time (TTL or time to live) or until some consecutive inactivity period occurs. These timings can be defined by any of interacting applications (either of them can close the channel). If subsequently a message needs to be sent between the interacting applications, the channel is open again, automatically. Timings can be adaptive, depending on message traffic.

The correlation between request and response messages is made by a token value generated at the sender's server (by a pseudo-random number generator) for each message sent. It is meaningful in the context of that server only. The receiver of the request must copy this token to the response and forward it in delegation messages, so that the delegate can also include it in the response. The implementation of futures also uses this mechanism.

Table 1.4 Examples of message types

Message category/type	Description
Request	*Initial request in each transaction*
React	React to message, no answer expected
React & respond	React to message and answer/notify
Asynchronous react & respond	Asynchronous data message. Return future immediately
Amendment	*Further information on an already sent request*
Cancel	Cancel the execution of the request
Response	*Response to the request*
Answer	Data returned as a response
Resource fault	Data returned as the result of an exception
Protocol fault	Error data resulting from protocol or partner failure
Notification	*Information of completion status*
Denial	Confirm rejection of request
Done	Request completed but has no value to reply
Cancelled	Confirm cancellation of request

1.5.2 Architecture of the Platform

The platform is essentially a distributed collection of SIL nodes, each confined to one computing server. Each SIL node can connect to more than one network, as shown in Fig. 1.8, which details the architecture of one node. This requires the provision of some server or handler to receive and send messages through each network's protocol. The protocol at message level (Table 1.1) is identical in all cases.

The architecture of a SIL node can be briefly described as follows:

- The application server (Jetty, in our implementation, but any other would do) is the interface to the Internet. It can support several protocols, although we have only catered for two, HTTP (to handle Web requests) and Web sockets (to support the message protocol of SIL). There is a list of protocol handlers and each in turn checks whether it recognizes the message format. The first one to do so gets the message for further processing. All SIL messages begin with "SIL", encoded in UTF-8.
- Other types of network protocols, such as those of sensor networks, can be supported as long as there are handlers to use those protocols at the transport level.

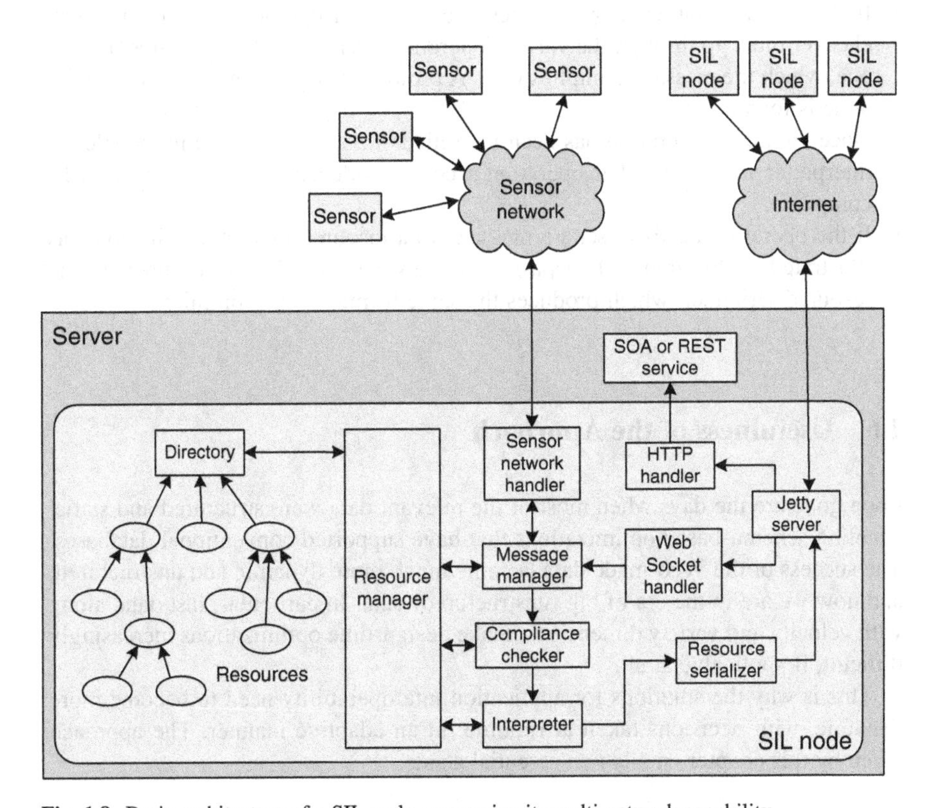

Fig. 1.8 Basic architecture of a SIL node, expressing its multi-network capability

- A message received (by any of the protocols) is handled by the message manager, which determines the recipient of the message, the type of the message (according to the SIL message protocol, described in Table 1.4), and whether a compliance token is present (described in Sect. 1.5.1);
- The resource manager implements the access to the structured resources registered in the directory, obtaining internal links (indices to a resource table) to resources targeted by messages or by distributed links (the equivalent of URIs).
- The directory is the root resource in each SIL node and implements several operations, such as searching for a resource with a service that conforms to a given SPID. The resource tree depicted in Fig. 1.8 shows only containment relationships. Any resource can have a distributed link to another but only if it is registered as globally accessible in a directory. This means that resources can be locally reached from others during execution of a SIL program, but only registered resources, directly in the directory, can be addressed by a global, distributed link.
- The compliance checker performs type compliance between the message and the addressed operation or resource. It can do so in text or binary formats, since each has all the information needed, as long as the metadata is also present. Naturally, this is faster when done in binary. Messages that include a compliance token, obtained in a previous checking, can skip this step.
- If the resource targeted by the message is not an operation, the compliance checker must go through the various operations defined in that resource, to find one, which the message complies with. A protocol fault (Table 1.4) is returned if none is found.
- Once the target operation has been identified, a thread is created in the silcode interpreter to execute that operation's code, produced previously by the SIL compiler.
- If the operation needs to send a message to a resource in another SIL node, as illustrated by Program 1.1, it passes the message (which is a resource) to the resource serializer, which produces the same format as the compiler.

1.6 Usefulness of the Approach

Long gone are the days when most of the relevant data were structured and static, enabling schema-based optimizations that have supported conventional databases. The success of the Web-made data become much more dynamic and unstructured, and now we are in the era of big (unstructured) data. In particular, fast data, along with velocity and variety dimensions, make design-time optimizations increasingly difficult, if applicable at all.

This is why the solutions for application interoperability need to become more dynamic, with decisions taken at runtime, in an adaptive manner. The approach taken by this chapter entails two essential slants:

- To reduce the interoperability requirements at the schema level, through partial compliance and conformance. Interacting applications need not be designed to work together, either by sharing schemas or by agreeing on specific data types at design time. Any application that happen to comply with another can use it as a consumer, and any application that conforms to another can replace it as a provider. Although not a magical or universal solution, it can definitely contribute to better support variants of some basic design. This is a contribution to the variety dimension.
- To optimize statistically while maintaining the flexibility. Checking for compliance and conformance is just too slow to be done in all messages. However, messages arriving at a high rate tend to repeat a data pattern, which means that the mapping between the data types of the messages and the receiver's interface can be cached and managed automatically. A simple token such as a 64-bit integer, returned by the receiver in the first message's response and used subsequently by the sender in future messages of the same type, can optimize performance without losing the benefits of dynamic type checking. When a mapping is cached, metadata need not be checked and only binary data is relevant (Table 1.3). The more repetitive message patterns are, the more effective this technique is. This is a contribution to the velocity dimension.

1.7 Future Research

The proposals contained in this chapter constitute exploratory work. There is a prototype of the language compiler, interpreter and node platform, but much work is still needed before the approach can be validated in a sufficiently large scale. The following activities are currently in development:

- Assessing the fault-tolerant capabilities of the platform. SIL supports fault-tolerant resource links (with alternative resources), but state checkpointing and recovery are not implemented yet.
- Assessing the scalability capabilities of the platform and assuming a large number of small-grained applications, characteristic of the IoT. This will be carried out by simulation.
- Tackling the volume dimension with a general-purpose mindset. Programming models such as MapReduce force a specific organization of data processing, from input data until results are produced. Our approach is to use any type of service (a task, in MapReduce parlance) or combination thereof and start with the last service (the one producing the final result) and proceed backward, invoking other services asynchronously. The final result, a collection of data (resource) initially containing just futures, is progressively filled in with the partial results returned by these services. Services are automatically replicated by each request. The nuance is to allow automatic replication and spreading of resources and their services, throughout a pool of SIL nodes. SIL already

supports distributed resources, since components of a resource can be remote, accessible by a link. The idea is to allow automatic replication of immutable components (a feature of SIL not described in the example of this paper), as well as code. This allows not only the code-to-data migration approach of Hadoop but also data-to-code (in which new data, such as streaming messages, are send to remote nodes to be processed there) or even resource-to-node (in which complete resources, code and data, are migrated to new nodes just to spread the computing load);
- Carrying out a comparative study, with qualitative and quantitative assessment, between the distributed platform proposed in this chapter and other dynamic platforms for real-time data processing [18].

1.8 Conclusions

Currently available platforms for big data are mainly centred on the volume dimension and usually assume that data already exist, are immutable and have been loaded into the system. This may be adequate for batch applications but not for processing dynamic data, namely, data streams. Although there are platforms trying to process large sets of data or events in real time or near it [18], these are still very focused on data and events and do not yet constitute a general-purpose platform to interconnect a large number of heterogeneous services. This is the scenario when we consider the Web and the IoT, in which the number of interacting entities and messages is very large but do not constitute a coherent, or even related, large set of data or events. This chapter intends to constitute a step towards such a general-purpose platform.

Current integration technologies evolved from basic browsing to web services by privileging backward compatibility over adequacy to computer-to-computer interactions. As a result, they exhibit many limitations for systems in which the number of entities is large and data granularity (messages) is small. New integration approaches, with native solutions to the problems of today, need to be devised.

This chapter contends that new solutions to fast data should adopt the following approaches:

- Languages should describe not only passive data (XML, JSON) but also operations, describing structured resources that offer services, so that most of the interoperability levels of Table 1.1 are covered natively, instead of having to build languages over languages, which increases the complexity and reduces the performance.
- A resource-based metamodel should be used as the foundation, to minimize the semantic gap in modelling real-world entities, in structure, state and behaviour.
- Interoperability should be allowed to be partial, based on compliance and conformance. This reduces service coupling with regard to sharing of schemas

(XML, Web services) or of types (JSON, REST), providing a better support for heterogeneity of services and data types.

- The distributed platform should be based on nodes as universal and as independent from the network protocol as possible. This way, it can be overlaid on different networks in a more transparent way.
- The resource serialization format should be natively binary and adjust automatically the level of metadata included, to minimize message size.
- The message protocol should be simple but efficient. Features that are more complex should be built on top of the basic mechanisms, according to the levels as shown in Table 1.1.

References

1. Zikopoulos P et al (2012) Understanding big data. McGraw-Hill, New York
2. White T (2012) Hadoop: the definitive guide. O'Reilly Media, Inc., Sebastopol
3. Zaharia M, Chowdhury M, Franklin M, Shenker S, Stoica I (2010) Spark: cluster computing with working sets. In: Proceedings of the 2nd USENIX conference on Hot topics in cloud computing, Boston, MA, pp 10
4. Mishne G, Dalton J, Li Z, Sharma A, Lin J (2013) Fast data in the era of big data: twitter's real-time related query suggestion architecture. In: Proceedings of the international conference on management of data, Ahmedabad, India, pp 1147–1158
5. Luigi A, Iera A, Morabito G (2010) The internet of things: a survey. Comput Netw 54:2787–2805
6. Gubbi J, Buyya R, Marusic S, Palaniswami M (2013) Internet of things (IoT): a vision, architectural elements, and future directions. Futur Gener Comp Syst 29(7):1645–1660
7. Sundmaeker H, Guillemin P, Friess P, Woelffle S (2010) Vision and challenges for realising the internet of things. European Commission Information Society and Media. http://bookshop.europa.eu/en/vision-and-challenges-for-realising-the-internet-of-things-pbKK31110323/. Accessed 20 Apr 2015
8. Schaffers H et al (2011) Smart cities and the future internet: towards cooperation frameworks for open innovation. In: Domingue J et al (eds) The future internet. Springer, Berlin/Heidelberg, pp 431–446
9. Delgado J (2014) The role of compliance and conformance in software engineering. In: Ghani I, Kadir W, Ahmad M (eds) Handbook of research on emerging advancements and technologies in software engineering. IGI Global, Hershey, pp 392–420
10. Ohlhorst F (2012) Big data analytics: turning big data into big money. Wiley, Hoboken
11. Murdoch T, Detsky A (2013) The inevitable application of big data to health care. J Am Med Assoc 309(13):1351–1352
12. Holzinger A (2014) Biomedical informatics: discovering knowledge in big data. Springer International Publishing, Cham
13. Demchenko Y, Zhao Z, Grosso P, Wibisono A, de Laat C (2012) Addressing Big Data challenges for scientific data infrastructure. In: Proceedings of the IEEE 4th international conference on cloud computing technology and science, Taipei, Taiwan, pp 614–617
14. Mayer-Schönberger V, Cukier K (2013) Big data: a revolution that will transform how we live, work, and think. Houghton Mifflin Harcourt, New York
15. NIIST (2014) DRAFT NIST Big Data interoperability framework: volume 6, reference architecture. http://bigdatawg.nist.gov/_uploadfiles/BD_Vol6-RefArchitecture_V1Draft_Pre-release.pdf. Accessed 20 Apr 2015

16. Dean J, Ghemawat S (2008) MapReduce: simplified data processing on large clusters. Commun ACM 51(1):107–113
17. Heisterberg R, Verma A (2014) Creating business agility. Wiley, Hoboken
18. Liu X, Iftikhar N, Xie X (2014) Survey of real-time processing systems for big data. In: Proceedings of the 18th international database engineering & applications symposium, Porto, Portugal, pp 356–361
19. Erl T (2007) SOA: principles of service design. Prentice Hall PTR, Upper Saddle River
20. Fielding R (2000) Architectural styles and the design of network-based software architectures. Doctoral dissertation, University of California at Irvine. http://www.ics.uci.edu/~fielding/pubs/dissertation/fielding_dissertation_2up.pdf. Accessed 20 Apr 2015
21. Chappell D (2004) Enterprise service bus. O'Reilly Media, Inc., Sebastopol
22. Armbrust M et al (2010) A view of cloud computing. Commun ACM 53(4):50–58
23. Fernando N, Loke S, Rahayu W (2013) Mobile cloud computing: a survey. Futur Gener Comp Syst 29(1):84–106
24. Keyes J (2013) Bring your own devices (BYOD) survival guide. CRC Press, Boca Raton
25. Aggarwal C, Han J (2013) A survey of RFID data processing. In: Aggarwal C (ed) Managing and mining sensor data. Springer, New York, pp 349–382
26. Potdar V, Sharif A, Chang E (2009) Wireless sensor networks: a survey. In: Proceedings of the international conference on advanced information networking and applications workshops, Bradford, UK, pp 636–641
27. Hartenstein H, Laberteaux K (eds) (2010) VANET: vehicular applications and inter-networking technologies. Wiley, Chichester
28. Karen T (2013) How many internet connections are in the world? Right. Now. http://blogs.cisco.com/news/cisco-connections-counter/. Accessed 20 Apr 2015
29. United Nations (2013) World population prospects: the 2012 revision, key findings and advance tables. Working Paper No. ESA/P/WP.227. United Nations, Department of Economic and Social Affairs, Population Division. http://esa.un.org/unpd/wpp/Documentation/pdf/WPP2012_%20KEY%20FINDINGS.pdf. Accessed 20 Apr 2015
30. Liebowitz J (ed) (2013) Big data and business analytics. CRC Press, Boca Raton
31. Juric M, Pant K (2008) Business process driven SOA using BPMN and BPEL: from business process modeling to orchestration and service oriented architecture. Packt Publishing, Birmingham
32. Shadbolt N, Hall W, Berners-Lee T (2006) The semantic web revisited. IEEE Intell Syst 21 (3):96–101
33. Palm J, Anderson K, Lieberherr K (2003) Investigating the relationship between violations of the law of demeter and software maintainability. In: Proceedings of the workshop on software-engineering properties of languages for aspect technologies. http://www.daimi.au.dk/~eernst/splat03/papers/Jeffrey_Palm.pdf. Accessed 20 Apr 2015
34. Papazoglou M (2008) Web services: principles and technology. Pearson Education Limited, Harlow
35. Webber J, Parastatidis S, Robinson I (2010) REST in practice: hypermedia and systems architecture. O'Reilly Media, Sebastopol
36. Jacobsen R, Toftegaard T, Kjærgaard J (2012) IP connected low power wireless personal area networks in the future internet. In: Vidyarthi D (ed) Technologies and protocols for the future of internet design: reinventing the web. IGI Global, Hershey, pp 191–213
37. Castellani A, Gheda M, Bui N, Rossi M, Zorzi M (2011) Web services for the internet of things through CoAP and EXI. In: Proceedings of the international conference communications workshops, pp 1–6. doi:10.1109/iccw.2011.5963563
38. Cugola G, Margara A (2012) Processing flows of information: from data stream to complex event processing. ACM Comp Surv 44(3):article 15
39. Delgado J (2012) Bridging the SOA and REST architectural styles. In: Ionita A, Litoiu M, Lewis G (eds) Migrating legacy applications: challenges in service oriented architecture and cloud computing environments. IGI Global, Hershey, pp 276–302

40. Adamczyk P, Smith P, Johnson R, Hafiz M (2011) REST and web services: in theory and in practice. In: Wilde E, Pautasso C (eds) REST: from research to practice. Springer, New York, pp 35–57
41. Berners-Lee T (1999) Weaving the web: the original design and ultimate destiny of the world wide web by its inventor. HarperCollins Publishers, New York
42. Tolk A (2006) What comes after the semantic web – PADS implications for the dynamic web. In: Proceedings of the 20th workshop on principles of advanced and distributed simulation, Beach Road, Singapore, pp 55–62
43. Järvi J, Freeman J (2010) C++ lambda expressions and closures. Sci Comp Program 75 (9):762–772
44. Grau B et al (2008) OWL 2: the next step for OWL. Web Semant Sci Serv Agents World Wide Web 6(4):309–322
45. Baader F et al (eds) (2010) The description logic handbook: theory, implementation, and applications, 2nd edn. Cambridge University Press, Cambridge
46. Jeong B, Lee D, Cho H, Lee J (2008) A novel method for measuring semantic similarity for XML schema matching. Expert Syst Appl 34:1651–1658
47. Euzenat J, Shvaiko P (2007) Ontology matching. Springer, Berlin
48. Kokash N, Arbab F (2009) Formal behavioral modeling and compliance analysis for service-oriented systems. In: de Boer F, Bonsangue M, Hallerstede S, Leuschel M (eds) Formal methods for components and objects. Springer, Berlin/Heidelberg, pp 21–41
49. Kim D, Shen W (2007) An approach to evaluating structural pattern conformance of UML models. In: Proceedings of the ACM symposium on applied computing, Seoul, Korea, pp 1404–1408
50. Adriansyah A, van Dongen B, van der Aalst W (2010) Towards robust conformance checking. In: Proceedings of the business process management workshops, Hoboken, NJ, pp 122–133
51. Lubbers P, Albers B, Salim F (2010) Pro HTML5 programming: powerful APIs for richer internet application development. Apress, New York
52. Schippers H (2009) Towards an actor-based concurrent machine model. In: Rogers I (ed) Proceedings of the 4th workshop on the implementation, compilation, optimization of object-oriented languages and programming systems, Genova, Italy, pp 4–9
53. Dubuisson O (2000) ASN.1 communication between heterogeneous systems. Academic, San Diego
54. Parr T (2013) The definitive ANTLR 4 reference. The Pragmatic Bookshelf, Raleigh
55. Sumaray A, Makki S (2012) A comparison of data serialization formats for optimal efficiency on a mobile platform. In: Proceedings 6th international conference on ubiquitous information management and communication, article 48. doi:10.1145/2184751.2184810
56. Delgado J (2014) Structural services: a new approach to enterprise integration. In: Wadhwa M, Harper A (eds) Technology, innovation, and enterprise transformation. IGI Global, Hershey, pp 50–91
57. Galiegue F, Zyp K (eds) (2013) JSON schema: core definitions and terminology. Internet engineering task force. https://tools.ietf.org/html/draft-zyp-json-schema-04. Accessed 20 Apr 2015

Chapter 2
Complex Event Processing Framework for Big Data Applications

Rentachintala Bhargavi

Abstract The fundamental requirement for modern IT systems is the ability to detect and produce timely reaction to the occurrence of real-world situations in the system environment. This applies to any of the Internet of Things (IoT) applications where number of sensors and other smart devices are deployed. These sensors and smart devices embedded in IoT networks continually produce huge amounts of data. These data streams from heterogeneous sources arrive at high rates and need to be processed in real time in order to detect more complex situations from the low-level information embedded in the data. Complex event processing (CEP) has emerged as an appropriate approach to tackle such scenarios. Complex event processing is the technology used to process one or more streams of data/events and identify patterns of interest from multiple streams of events to derive a meaningful conclusion. This chapter proposes CEP-based solution to continuously collect and analyze the data generated from multiple sources in real time. Two case studies on intrusion detection in a heterogeneous sensor network and automated healthcare monitoring of geriatric patient are also considered for experimenting and validating the proposed solutions.

Keywords Complex event processing • IoT • Big data • Data stream • Event • Intrusion detection • Geriatric health monitoring

2.1 Introduction

A wide range of IoT applications have been developed and deployed in recent years. IoT has provided promising solution to several real-time applications by leveraging the growing ubiquity of radio frequency identification (RFID) and wireless mobile, sensors, and other smart devices [7]. Big data is a term encompassing the use of techniques to capture, process, analyze, and visualize potentially large datasets in a reasonable time frame not accessible to standard IT technologies [11]. The platforms, tools, and software used for this purpose are

---

R. Bhargavi (✉)
School of Computing Sciences and Engineering, VIT University, Chennai, India
e-mail: bhargavi.r@vit.ac.in

Z. Mahmood (ed.), *Data Science and Big Data Computing*,
DOI 10.1007/978-3-319-31861-5_2

collectively called “big data technologies.” These technologies refer to and present the ability to crunch vast collections of information, analyze it instantly, and draw conclusions. Big data technologies deal with petabytes of records, files, and transactional data either arriving as streams or in batches. The RFIDs, sensors, smart devices, etc., embedded in IoT networks produce huge amounts of data/events continuously [19]. This data has the characteristics like volume, variety, velocity, variability, veracity, and complexity. Some of the characteristics of the data/event streams and their implications are as follows:

- Data streams are continuous and sequential and are ordered by a time stamp or any other attribute value of the data item. Therefore, data items which belong to the same stream are processed in the order they arrive.
- Data streams are generated by external sources and are sent to a processing system. Hence, the data stream processing system does not have any direct control over the data sources.
- The input characteristics and the rate of a data stream are unpredictable. The input rate can be very irregular and, at times, bursty in nature. Also, the nature of the input does not allow one to make multiple passes over the data while processing.
- The amount of data is very large and unbounded. Therefore, processing requirements may not permit persistence followed by processing. However, data or summary of data can be stored for archival or other purposes.
- The data types of the data items can be structured, semi-structured, or unstructured.
- Data items in a data stream are not error free because the data sources are external. Some data may be corrupted or discarded due to network problems.

These data streams need to be processed and analyzed to identify some interesting patterns and take actions if necessary. Processing these continuous data/event streams in real time to identify the patterns among them is a herculean task and has raised new research challenges over the last few years [3, 5]. A large number of solutions exist in terms of systems, middleware, applications, techniques, and models proposed by researchers to solve different challenges [13]. Data generated from multiple sources have logical and spatiotemporal relations among them. There is a need for data fusion as the data from a single source may not be enough for taking accurate decision. Conventional process-oriented control flow software architectures do not explicitly target the efficient processing of continuous event streams. Complex event processing is the technology used to process and analyze one or more streams of data/events and identify patterns of interest from multiple streams of events to derive a meaningful conclusion and respond by taking appropriate action [2].

The remaining parts of this chapter are organized as follows. Section 2.2 presents CEP preliminaries and event modeling. Semantic intrusion detection using complex event processing is elaborated in Sect. 2.3. Section 2.4 discusses the CEP-enabled geriatric health monitoring, and Sect. 2.5 concludes the chapter.

2.2 Complex Event Processing

Complex event processing (CEP) is a relatively new concept, but it has received wider acceptance due to its systematic and multilevel architecture-driven concept approach [9]. An event is an object that is a record of an activity; it signifies the activity. A key stroke or the output reading produced by a sensor, etc., are examples of an event. CEP allows one to set a request for an analysis or some query and then have it executed continuously over a period of time against one or many streams of events in a highly efficient manner. CEP is all about the processing of events that combines data from many sources to infer events or patterns that represent more complicated circumstances. In contrast to Hadoop's two-stage disk-based MapReduce paradigm, CEP's push-based paradigm supports faster processing of data streams.

2.2.1 CEP Architectural Layers

The architectural layers of a CEP system are shown in Fig. 2.1. A typical system consists of four layers, as briefly described below.

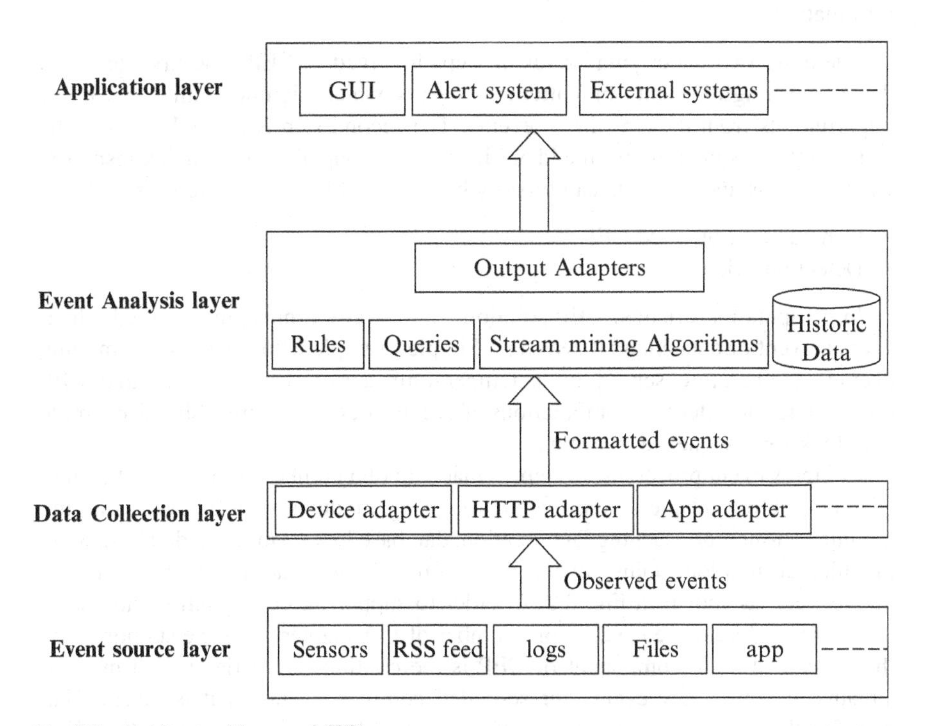

Fig. 2.1 Architectural layers of CEP

Event Source Layer This layer consists of event sources. These sources can be sensors, RFID readers, RSS feed, feed from network monitoring systems, web logs, etc.

Data Collection Layer Data collection layer is responsible for collecting the data coming from various sensors and filtering the data. Data collection could be distributed or centralized based on the requirement of the application. Basic capabilities in WSN involve mechanisms like routing, tunneling, data aggregation, and clustering to collect information from nodes and forward them to a sink node.

Event Analysis Layer The event analysis layer is responsible for data mining on streams of data. An important feature of this layer is pattern matching and correlation across multiple event streams. In this layer, few fixed continuous queries act on the incoming data streams to identify patterns of interest. Comparison of real-time data with historical or static data is also often required for event analysis. In many instances, historical or static data is also often required along with the real-time data for analysis. Thus, the event analysis layer should contain facilities for connecting to data base.

Application Layer This layer consists of event listeners, i.e., modules/systems that receive the processed events. Examples of event listeners are event storage systems, mobile phones and pagers, or other systems that can take actions based on the results of event processing (e.g., GUI receives the processed events and displays the information).

There are two underlying stages or steps involved in CEP. The first step is to detect meaningful events or pattern of events which signifies either threats or opportunities from the streams of events. The second step is to send alerts to the responsible person/entity for the identified threat or opportunity for quick response. CEP solutions and concepts can broadly be classified into two categories:

- Computation-oriented CEP
- Detection-oriented CEP

In computation-oriented CEP solutions, online algorithms are executed whenever an event or data enters the system. Simple example is to calculate the moving average temperature sensed by a temperature sensor. Detection-oriented CEP concentrates on detecting combinations of events or event patterns. Simple example is to look for a sequence of events.

Complex event processing system is made up of a number of modules like input adapters, output adapters, and event processing modules such as event filtering modules, in-memory caching, aggregation, database lookup module, database write module, correlation, joins, event pattern matching, state machines, dynamic queries, etc, as shown in Fig. 2.2. In order to support more flexibility and adaptability for different use cases, more number of I/O adapters must be supported by the CEP. The main component of CEP is the continuous queries which monitor streams of simple/raw events for so-called complex events, that is, events that manifest themselves in certain temporal, spatial, or logical combinations. Querying

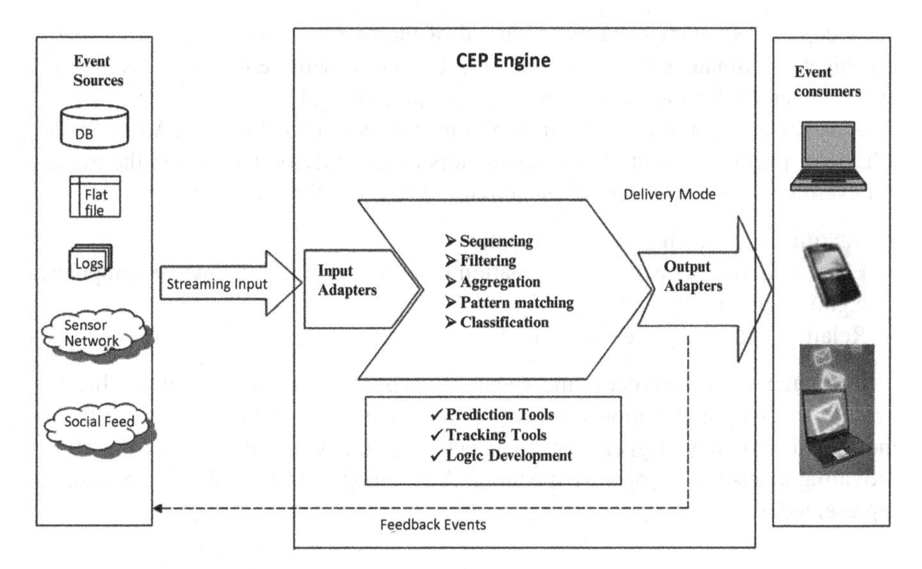

Fig. 2.2 Logical view of CEP

events over data streams is different from traditional querying with database, in the way that traditional database querying is pull based, whereas continuous querying of events is push based.

Many of the real-time distributed applications require continuous monitoring and processing and analysis of information or data in a timely fashion as it flows from periphery to the system. Intrusion detection in surveillance and healthcare monitoring are a couple of such applications. Traditional pull-based approaches can hardly address the requirements of timeliness, response generation, etc. Hence, this chapter proposes CEP-based solution to continuously collect and analyze the data generated from multiple sources in real time. Two case studies: Intrusion detection in a heterogeneous sensor network and automated healthcare monitoring of geriatric patient are considered for experimenting and validating the proposed solutions.

2.2.2 *Event Modeling*

An event is "anything that happens, or is contemplated as happening" [10] in the real world, and normally it is of interest to some groups of people. A key stroke, a sensor outputs a reading, etc., are couple of examples of an event. Sometimes these events, in turn, may produce secondary events internally. Real-world occurrences can be defined as events that happen over space and time. Events are of two types: basic or primitive events and complex events. Events have event attributes. An event attribute is a property of the event. For example, the entry of an identified person in a restricted area could be treated as an activity. Then the form of the event instance could be composed by unique id of the person, time, and location (geographical coordinates). An event is an object that represents, encodes, or records an event, generally for the purpose of computer processing [10].

A basic event is atomic and indivisible and occurs at a point in time. Attributes of a basic or primitive event are the parameters of the activity that caused the event. An event instance is a record of an activity, which has three features such as:

- Significance, which gives its semantic
- Form, which gives activity information that will be processed by the computer, e.g., unique id, time, etc.
- Relation with other event instances

Computer systems process the events by representing them as event objects. From a software application perspective, an event is something that needs to be monitored and may trigger a specific action. Specifying an event is therefore providing a description of the happening. A common model of an event is a tuple represented as

$$E = E(\mathrm{id}, a, t)$$

where

id is the unique ID of an event.
$a = \{a_1, a_2, \ldots\ldots, a_m\}$, $m > 0$, is a set of attributes.
t is the time of occurrence of the event.

For example, an RFID event can be described with some set of dimensions which includes source of event, location of event, time at which the event occurred, and a possible set of operations for combining events. An RFID event is denoted as $E = e$ (o, r, t) where o is the tag EPC, r is the reader ID, and t is the time stamp of the event.

Complex events are composed of basic events. Complex events are defined by connecting basic events using temporal, spatial, or logical relations. A common model for a complex event is as follows:

$$E = E(\mathrm{id}, a, c, t_b, t_e), t_b <= t_e$$

where

$c = \{e_1, e_2, \ldots\ldots, e_n\}$, $n > 0$ is the vector that contains basic events and complex events that cause this event to happen.
t_b, t_e are starting and ending times of the complex event.

Attributes of complex events are derived from the attributes of the constituent primitive events. Event constructors and event operators are used to express the relationship among events and correlate events to form complex events.

Any basic event or a complex event is specified by an event expression. An event expression is a mapping from histories (domain) to histories (range) [6], i.e.:

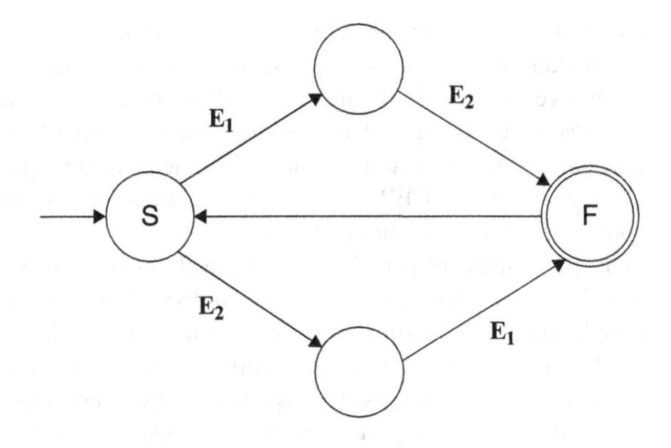

Fig. 2.3 Finite automata representation of composite event E1 ^ E2

$$E : \text{histories} \rightarrow \text{histories}.$$

Since event expressions are equivalent to regular expressions, it is possible to implement event expressions using finite automata. For example, composite/complex event $= E_1 \wedge E_2$ can be represented using the finite automata as shown in Fig. 2.3.

Event expression is formed by combining events with the event constructors. Many of the event processing engines support different types of logical and temporal constructs.

2.3 Semantic Intrusion Detection Using CEP

There is an increasing demand for security solutions in the society. This results in a growing need for surveillance activities in many environments. Recent events, like terrorist attacks, have resulted in an increased demand for security in society. There is a growing interest in surveillance applications, because of the availability of cheap sensors and processors at a reasonable cost. Intelligent remote monitoring systems allow users to survey sites from remote location. Sensor networks bridge the gap between the physical world and the virtual world of processing and communication. It is envisioned that sensor networks can reduce or eliminate the need for human involvement in information gathering and processing in surveillance applications if the ability of the sensor networks can be suitably harnessed. Distributed sensors and smart devices in the surveillance application produce huge data [14]. Processing, analyzing, and detecting abnormal patterns from this data are very complex in nature. There is a need for multi-sensor data fusion in sensor network applications as the data from homogeneous sensors may not be enough for taking accurate decision [17]. Also the data generated from multiple sensors have logical and spatiotemporal relations among them. Special processing and mining algorithms are proposed in the literature for distributed and in-network processing.

Existing solutions for surveillance and intrusion detection are based on pull-based architecture where the data from sensors are stored, and later, the user runs queries to retrieve the data. There are also solutions based on machine learning techniques, but these solutions again work on the static data. Traditional approaches which are pull based can hardly address the requirements of timeliness, response generation, etc. The proposed CEP-based semantic intrusion detection system (CSIDS) [1] addresses the abovementioned problems.

In the context of surveillance, the evaluation of available technologies shows that the security bottleneck is not the hardware but rather the real-time analysis and correlation of data provided by various sensors. The objective of proposed CEP-based semantic intrusion detection system is early detection and prevention of security compromises/risks by identifying abnormal situations and quickly responding with appropriate action. To achieve this goal, sensors (such as cameras, RFID readers, etc.) are installed in the places to be monitored, and a control center receives the information transmitted by sensors. This information is processed and analyzed by an event processing engine to detect the intrusive patterns. There are a few factors that make this task rather difficult. First, the raw data to deal with is very huge, and second, the information sources are heterogeneous. Therefore, there is a need to fuse or aggregate the data coming from such sensors to construct a global view of the situation and even more to go one step beyond to the intent assessment.

The proposed CSIDS allows fusion of information generated by heterogeneous sensors that support the goal of providing a global situational view for intrusion detection. The goal of this fusion is to transform lower-level data into higher-level quality information and to improve the certainty of situation recognition, i.e., reduce the false alarm rate. The fusion is realized by taking into consideration the semantics of the information. This means that there should be a representation or model of the situation that needs to be detected (such as an intrusion in a restricted area). In this context, a situation is viewed as a combination of several activity elements (or smaller situations), each of them appearing at some place and time. As a consequence, the finer the granularity of these elements, the more difficult the situation modeling task will become. This task is further complicated if any of the elements appear at different time instants and do not relate to the same situation. Furthermore, the set of heterogeneous sensors capturing these sub-situations have an asynchronous behavior, meaning each of the sensors computes the situation independently, unconnected of each other.

CSIDS supports online detection of event patterns which represent anomalies. The association of an event pattern, a constraint, and an action is referred to as a rule. The approach consists of time, content, and context-based selection of a subset of events described by their pattern. Finally, after the event pattern is matched, one or more appropriate actions are executed such as creation and sending of a new event or sending an alert message to the concern personnel, etc.

The architecture of the complex event processing system for semantic intrusion detection system is shown in Fig. 2.4. CSIDS has multiple event receivers. Each event receiver receives the data/events coming from a different data/event source. The event receiver on receiving the data from the source converts them into event

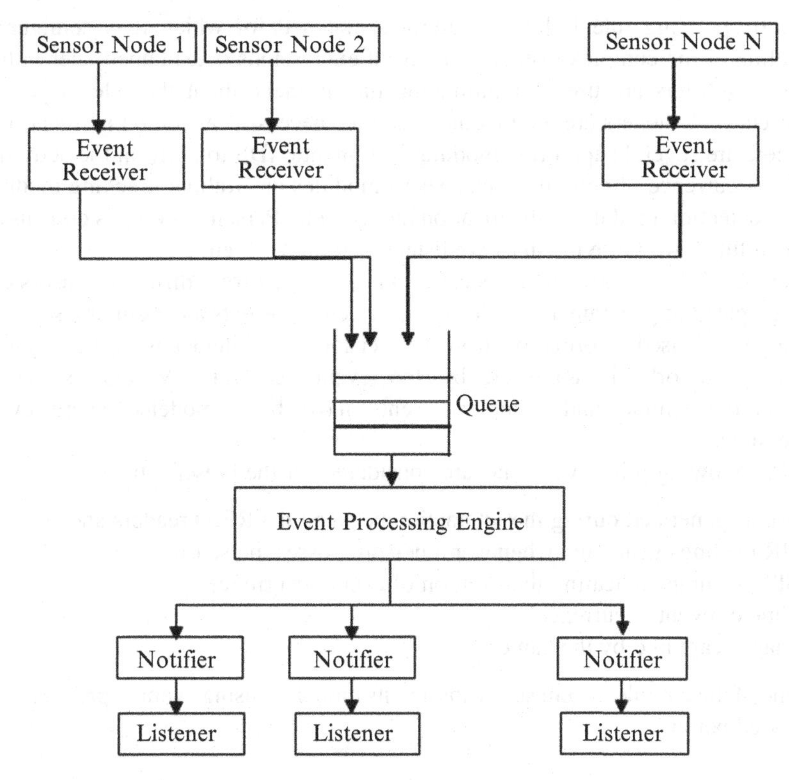

Fig. 2.4 CEP architecture for semantic intrusion detection system

streams. Data generated by different sources follows a different format; hence, event receivers also convert the data from different sources to the specific format suitable for processing further by the event processing engine. The events generated by the event receivers are inserted in to a FIFO or a queue. Events are organized in the queue in the order of their detection time. To avoid the out of order arrival of the events which is caused by the network delays, the events are stored internally in the queue for some time T. Hence, the events are dequeued after a time T for processing. Any event getting generated at time t will be processed after t+T time by the CEP engine. Any event arriving at the queue with a time stamp smaller than the time stamp of the already dequeued events is ignored. This leads to missing of events. Missing events due to network delays can be avoided by having larger value of T, i.e., by storing the events internally for longer duration. But this will delay the event processing time. Hence, there is a trade-off between missing events due to network delay and latency.

Event processing engine processes the event streams. The events generated from heterogeneous sensors are collected and aggregated using logical and spatiotemporal relations to form complex events which model the intrusion patterns. All the rules and patterns are to be registered initially. The listeners are intimated whenever the corresponding rule is hit or pattern is matched for which it is configured. Modeling of the complex events using event expressions is explained later in this section. Notifiers are used for intimating the listeners about the rule or pattern occurrences. Listeners are the modules that take necessary action on notifications.

There are several supporting modules in semantic IDS to perform the activities on the occurrence of certain events. These modules are Kalman tracking module, person detection module, authentication module, etc. Person tracking is done using Kalman filter. All these modules are listeners to the CEP engine.

Figure 2.5 shows how the physical events generated from different sensors can be aggregated to generate a complex event which represents a pattern or a scenario of interest. Sensed information from the sensors that interact with the physical environment/world is collected by the event receivers. Various scenarios representing simple and complex events have been modeled using event expressions.

The following primitive events are considered for the present study:

- Events generated during the interaction between the RFID readers and tags
- PIR readings generated whenever a person crosses the sensor
- GPS readings indicating the location of event occurrence
- Time of event occurrence
- Images captured by the camera

One of the complex event scenarios and its modeling using event expressions are discussed below.

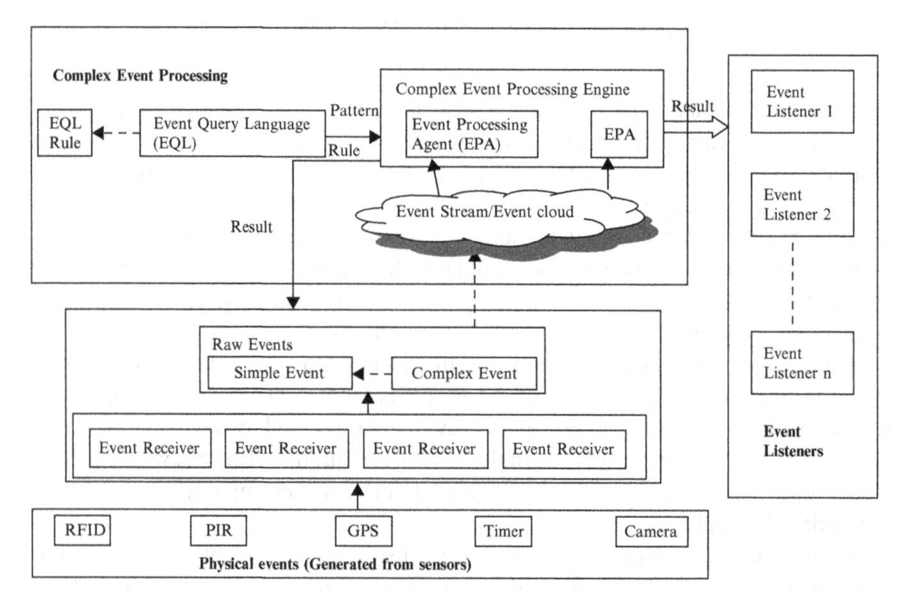

Fig. 2.5 CSIDS

Example: Unauthorized Entry
When a person who does not possess a valid RFID tag tries to enter a location just behind an authorized person, it is called as tailgating. This scenario can be captured as explained below.

CEP queries the wireless data for an event where PIR is present but RFID is zero. Once such an event is identified, CEP gets the image from the database corresponding to that event's time stamp and gives the image to Haar face detection module. The Haar face detection algorithm counts the number of persons present and gives the count back to CEP. This scenario can be modeled as follows:

E1: Office entry
E1 = (s1, o1, t1) type (s1) = RFID event
E2: Office entry
E2 = (s2, o1, t2) type (s2) = PIR event
E3: Multiple people sensed by image sensor

The complex pattern can now be formulated as

$$\text{every}((E2 : \neg(E1)) \wedge E3)$$

Now, the rule can be developed to raise an alert on the above complex event using IF – THEN condition:

IF (true)
Get the image with the same time stamp and node ID from the data base.
Send the location of the image in the data base to the image processing module for checking number of people.
Receive the response from the image processing module.
Generate alert if needed.
End

2.3.1 Implementation and Validation

The proposed CSIDS is validated using a sensor network. A wireless sensor node is devolved which consists of a passive infrared sensor, radio frequency identifier reader, GPS receiver, and timer. Ten such sensor nodes are used for deploying the wireless sensor network. Some of the nodes are connected with cameras. The camera is triggered by the PIR sensor and is connected to a personal computer. The images captured are stored in PC and transmitted to the server via wired network. A unique ID is assigned to each sensor node. Data from the RFID and PIR sensors are sent to the sink using wireless communication (using ZigBee module). Data from the cameras is sent to the server using Ethernet. This is done to overcome the bandwidth, memory, processing, and power limitations of sensor node. The complete CSIDS is developed using Java. Esper event processing engine is used to implement and execute the rules. Patterns representing intrusion are modeled as complex events which in turn are aggregated from base events and other complex events using logical and spatiotemporal relations. Rules have been developed to identify the unauthorized entry of the people in to the security zones, tailgating issue, and few other scenarios. It is observed that the proposed CSIDS identifies the intrusion patterns efficiently in near real time. It is also observed that the proposed system out performs the pull-based solutions in terms of detection accuracy and detection time. The quality of results depends on the quality of sensors and processes of their data. For example, if the sensors take more time to sense the data, process, and communicate, then this introduces a delay between the occurrence of real situation and its recognition by the application which can become considerable in sensitive situations.

2.4 CEP-Enabled Geriatric Health Monitoring

Healthcare monitoring is another application which can best exploit the advantages of CEP. This section explains the proposed CEP-based geriatric health monitoring system (CGHMS). Advancements in wireless body area networks have led researchers to exploit the usage of these technologies in healthcare applications [4, 15, 18]. Availability of wearable and cost-effective physiological and motion sensors allow automated health monitoring of elderly people. There are a number of healthcare applications using sensor networks. Within the hospital, there are three applications: (1) to track people and objects around the hospital [16], (2) to safeguard use of equipment, and (3) to assist medical personnel with their jobs. Another important application is elder care in home. Global increase in the ratio of elderly people [8] in the world requires alternatives to the traditional homecare technologies that are available today. This in turn indicates that there will be challenges related to giving proper care to the elderly people since care giving requires enough people and resources. Hence, there is a need for alternatives that automate the home care application domain.

The important requirement of healthcare monitoring is automatic anomaly detection of vital parameters. Another important requirement is that when something dangerous or critical occurs, it is important that the system detects this immediately and that the delay is minimal. There are several existing solutions for monitoring the activity of a person or abnormality of a particular health parameter. But there is no solution for identifying the abnormal situations like fall of a person by combining the vital parameters, activities of the person, and the context information. CEP-enabled geriatric healthcare monitoring system (CGHMS) collects data generated from physiological and environmental sensors and detects the abnormalities in vital parameters and fall of a person.

The main objective of the CEP-enabled geriatric healthcare monitoring is to collect health parameters from the patient, detect the presence of an abnormality in vital parameters, and provide feedback to the rules in decision-making about the health condition and falls of a geriatric patient. In the geriatric healthcare monitoring domain, complex event processing involves analyzing raw sensor data and recognizing next level of events like low BP, high BP, and high temperature to complex events like person has fallen, etc. The data/event streams are filtered correlated and aggregated to identify any abnormal patterns. Rules/patterns of interest corresponding to various scenarios are stored in the knowledge base. Rules are executed on the incoming data to detect the anomalies. Every rule has associated action to be executed like sending SMS to caregiver or doctor or patient. Figure 2.6 shows the architecture of the CEP-enabled geriatric health monitoring system. The physical and semantic data flows are also shown. The proposed CGHMS uses wearable BP sensor, pulse oximeter, and BioHarness 3 device, a Zephyr product which contains both biosensors and triaxial accelerometer for measuring the vital parameters like respiration/breathing rate, heart rate, ECG, and movement along X-, Y-, and Z-axis, respectively. CGHMS also uses RFID reader which can be connected and used with IPAQ PDA device. RFID tags and environmental sensors are placed throughout the area to be monitored to get the context information. Data generated by these sensors are collected and sent to a central server where CEP engine. The knowledge base consists of CEP rules which model the various abnormal health condition events. The CEP engine executes all the rules whenever it receives the sensor data/facts to detect abnormality situations.

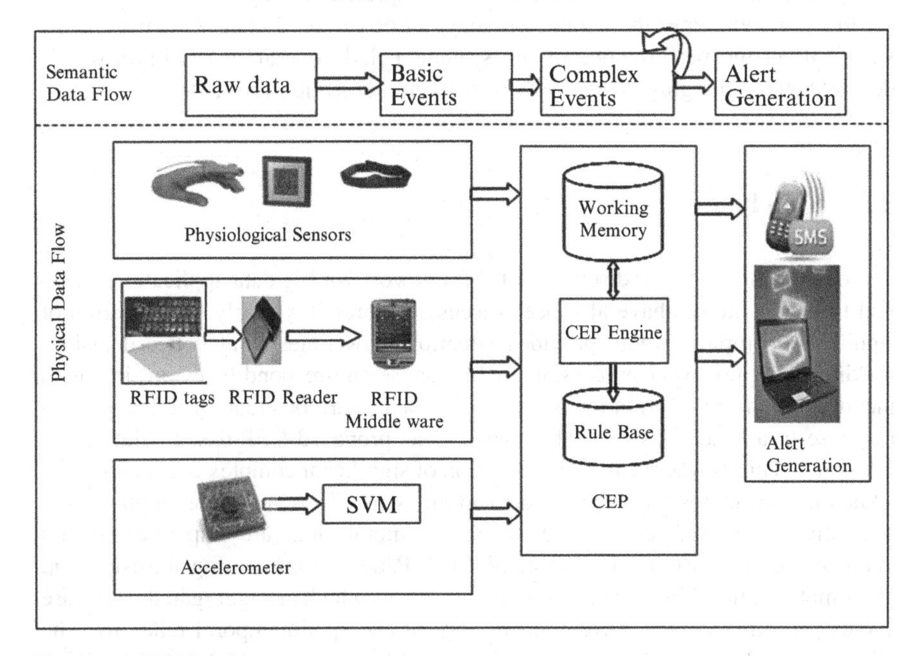

Fig. 2.6 CGHMS architecture

Every rule has associated action to be executed like sending SMS to the caregiver or doctor or patient.

2.4.1 Implementation and Validation

The complete CGHMS is implemented in Java. Java-based Drools Expert engine is used for rule development. Drools APIs are extended to support CGHMS. To reduce the false alarms in fall detection, the proposed CGHMS uses vital parameter data, activity of the person, RFID information which gives the contextual information, and camera. For activity classification using triaxial accelerometer and gyroscope, SVM classifier is used [12]. The proposed CGHMS provides continuous accessibility to vital parameters and other relevant data and analysis of the data. The CGHMS improves the quality of life of the elderly people by monitoring the well-being and alerting in case of emergency. The proposed CGHMS is tested and validated in real-time environment using Zephyr BioHarness device; pulse oximeter; BP sensor, for measuring the vital parameters; RFID tags and reader to get the context information; activity of a person using accelerometer and gyroscope; and camera. It is observed that CGHMS detects the abnormalities in the vital parameters and fall of a person more precisely compared to using the individual sensors. As discussed earlier, the performance of the proposed CGHMS depends on the quality of the sensors used. Also, the wearable sensor device must be worn by the person correctly in the right position; otherwise, the captured data by the sensors may be incorrect or sometimes the parameters may not be captured at all. Other problems which affect the performance of the system include operating condition of the device, battery charging, availability of the communication links, etc.

2.5 Conclusion

To conclude, this chapter discusses CEP framework for big data applications. Two real-time case studies have also been discussed thoroughly. Early identification of significant complex events provides situational awareness and better decision-making. Complex event processing enables sense-and-respond behavior, in which incoming events or information is used to assess the current situation and generate a response in a timely fashion. This chapter has proposed CEP-based solutions to Internet of Things where early identification of significant complex events provides situational awareness and better decision-making. Two real-time applications, semantic intrusion detection and healthcare monitoring, are implemented and validated to demonstrate the power of CEP. Patterns representing intrusion and abnormal situations are modeled as complex events which are aggregated from base events and other complex events using logical and spatiotemporal relations. The CEP-based solutions provide the capabilities of heterogeneous information fusion coming from several kinds of sensors to get a global situation view and take necessary action in near real time. The proposed semantic intrusion detection system is found to outperform the existing solutions in terms of detection latency (i.e., time delay between the event occurrence and event detection) and detection accuracy, i.e., event identification with less false positives and false negatives. The proposed CEP-based solutions require domain knowledge of the application. All the situations/scenarios of interest should be well understood and modeled as rules using the events and available constructors. Failing to which, the pattern representing the complex event (event of interest) may not be captured as expected. Missing events is another common problem in case of wireless sensors. Missing basic events will lead to the failure of getting the high-level complex situations captured. The quality of results depends on the quality of sensors. For example, if the sensors take more time to sense the data, process, and communicate, then this introduces a delay between the occurrence of real situation and its recognition by the application which can become considerable in sensitive situations.

In any of the complex event processing-based applications, the system is initially configured with all the rules (defined by the domain experts) which define the complex events or patterns to be identified and alerted. But there are several applications in which new rules need to be added or deleted at run time. Hence, the proposed solutions can be extended to support dynamic rule addition and rule deletion. CEP is a rule-based technology for detecting known patterns of events and reacting to the identified situations in real time. Incremental or dynamic learning is used to learn the dynamic environments effectively. Hence, the proposed solutions can be extended to incorporate dynamic learning, and the newly learned knowledge/concepts can be given as feed back to the CEP engine to add new rules and identify new patterns or situations of interest. This makes the complete system more adaptive and intelligent.

References

1. Bhargavi R, Vaidehi V (2013) Semantic intrusion detection with multisensor data fusion using complex event processing. Sadhana Acad Proc Eng Sci 38(2):169–185, ISSN: 0256–2499
2. Bastian Hoßbach and Bernhard Seeger (2013) Anomaly management using complex event processing: extending data base technology paper. In: Proceedings of the 16th international conference on extending database technology (EDBT '13). ACM, New York, pp 149–154
3. Cantoni V, Lombardi L, Lombardi P (2006) Challenges for data mining in distributed sensor networks. ICPR 1:1000–1007
4. Poon CYC, Liu Q, Gao H, Lin W-H, Zhang Y-T (2011) Wearable intelligent systems for E-health. J Comput Sci and Eng 5(3):246–256
5. Elnahrawy E (2003) Research directions in sensor data streams: solutions and challenges. DCIS, Technical Report DCIS-TR-527, Rutgers University
6. Gehani NH, Jagadish HV, Shmueli O (1992) Composite event specification in active databases: model and implementation. In: VLDB '92: Proceedings of the 18th international conference on very large data bases. Morgan Kaufmann Publishers Inc., San Francisco, pp 327–338

7. Jin J, Gubbi J, Marusic S, Palaniswami M (2014) An information framework for creating a smart city through internet of things. Internet Things J IEEE 1(2):112–121. doi:10.1109/JIOT.2013.2296516
8. Kinsella K, He W (2009) An aging world: 2008. International population reports, U.S. Department of Health and Human Services
9. Luckham DC (2010) The power of events: an introduction to complex event processing in distributed enterprise systems. Addison Wesley Longman Publishing Co., Inc., Boston
10. Luckham DC, Schulte R (2008) Event processing glossary – version 1.1.. Event processing technical society. URL: http://www.ep-ts.com/component/option.com_docman/task.doc_download/gid.66/Itemid.84/
11. NESSI White Paper, December 2012.
12. Palaniappan A, Bhargavi R, Vaidehi V (2012) Abnormal human activity recognition using SVM based approach. In: Proceedings of IEEE international conference on recent trends in information technology (ICRTIT 2012), Chennai, India, pp 97–102, April 19–21
13. Perera C, Zaslavsky A, Christen P, Georgakopoulos D (2014) Context aware computing for the internet of things: a survey. Commun Surv Tutorials IEEE 16(1):414–454
14. Tian He, Krishnamurthy S, Liqian Luo, Ting Yan, Lin Gu, Stoleru R, Gang Zhou, Qing Cao, Vicaire P, John AS, Tarek FA, Jonathan Hui, Krogh B (2006) VigilNet: an integrated sensor network system for energy-efficient surveillance. ACM Trans Sen Netw 2(1):1–381
15. Vaidehi V, Bhargavi R, Ganapathy K, Sweetlin Hemalatha C (2012) Multi-sensor based in-home health monitoring using complex event processing. In: Proceedings of IEEE international conference on recent trends in information technology (ICRTIT 2012), Chennai, India, pp 570–575, April 19–21
16. Yao W, Chu C-H, Zang Li Yao W, Chu C, Li Z (2011) Leveraging complex event processing for smart hospitals using RFID. J Netw and Comput Appl 34(3):799–810
17. White Jr FE (1987) Data fusion lexicon. Data fusion subpanel of the joint directors of laboratories, Technical Panel for C3, Naval Ocean Systems Centre, San Diego
18. Wood A, Virone G, Doan T, Cao Q, Selavo L, Wu Y, Fang L, He Z, Lin S, Stankovic S (2006) ALARM-NET: wireless sensor networks for assisted-living and residential monitoring. Technical report, Department of Computer Science, University of Virginia, Wireless Sensor Network Research Group
19. Zaslavsky A, Perera C, Georgakopoulos D (2012) Sensing as a service and big data. In: International conference on advances in cloud computing (ACC-2012), Bangalore, India, July 2012, pp 21–29

Chapter 3
Agglomerative Approaches for Partitioning of Networks in Big Data Scenarios

Anupam Biswas, Gourav Arora, Gaurav Tiwari, Srijan Khare, Vyankatesh Agrawal, and Bhaskar Biswas

Abstract Big Data systems are often confronted with storage and processing-related issues. Nowadays, data in various domains is growing so enormously and so quickly that storage and processing are becoming the two key concerns in such large systems of data. In addition to the size, complex relationship within the data is making the system highly sophisticated. Such complex relationships are often represented as network of data objects. Parallel processing, external memory algorithms, and data partitioning are at the forefront of techniques to deal with the Big Data issues. This chapter discusses these techniques in relation to storage and processing of Big Data. The Big Data partitioning techniques, such as agglomerative approaches in particular, have been studied and reported. Network data partitioning or clustering is common to most of the network-related applications where the objective is to group similar objects based on the connectivity among them. Application areas include social network analysis, World Wide Web, image processing, biological networks, supply chain networks, and many others. In this chapter, we discuss the relevant agglomerative approaches. Relative advantages with respect to Big Data scenarios are also presented. The discussion also covers the impact on Big Data scenarios with respect to strategic changes in the presented agglomerative approaches. Tuning of various parameters of agglomerative approaches is also addressed in this chapter.

Keywords Big Data • Agglomerative approaches • Clustering • Partitioning • MapReduce • Network partitioning

A. Biswas (✉) • G. Arora • G. Tiwari • S. Khare • V. Agrawal • B. Biswas
Department of Computer Science and Engineering, Indian Institute of Technology (BHU), Varanasi, India
e-mail: anupam.rs.cse13@iitbhu.ac.in; gourav.arora.cse11@iitbhu.ac.in; gaurav.tiwari.cse11@iitbhu.ac.in; srijan.khare.cse11@iitbhu.ac.in; v.agrawal.cse11@iitbhu.ac.in; bhaskar.cse@iitbhu.ac.in

Z. Mahmood (ed.), *Data Science and Big Data Computing*,
DOI 10.1007/978-3-319-31861-5_3

3.1 Introduction

Big Data maintenance and processing are the two major issues in modern-day data mining. Complex relationships present within the data have resulted in data systems becoming even more complicated. Often such relationships are represented as networks in different domains [1–4]. In the network data representation, various objects within the data are represented with nodes and relationships among those objects that are represented with connections. The network representation of complex relationships present within the data is relatively easily interpretable. However, the main issues relating to memory and processing are retained due to huge size of these networks. In this context, various parallel processing techniques become highly useful for efficient processing. GPU techniques such as Pregel [5], Giraph [6] and Seraph [7] are very popular for parallel processing of network data. These techniques are based on the computation model called the bulk-synchronous parallel (BSP) model [8]. On the other hand, to deal with memory-related issues such as maintenance of larger network data than the memory, external memory algorithms are often frequently utilized.

Both parallel processing and external memory techniques require partitioning of networks [5–9]. In a parallel processing scheme, smaller parts of a network are processed simultaneously in multiple processing units. Results of the multiple processing units are accumulated to generate overall output. On the contrary, external memory schemes partition the network to fit into the memory. Thus, efficient partitioning of networks is very important from both perspectives. Meaningful partitions of networks can reduce computation cost. Moreover, such partitions can be easily handled with low-memory requirement. The objective of meaningful partitioning is to divide the network into subnetworks such that nodes within the subnetwork are close or similar. It means that nodes belonging to subnetworks have more connectivity than the connectivity among subnetworks. Such subnetworks are often referred to as clusters or communities.

Numerous network data partitioning or clustering approaches have been developed, which can be broadly categorized as node centric [3, 10], group centric [11–14], network centric [15], and hierarchical [11, 16–20]. Node-centric approaches specify certain properties that needed to be satisfied by all the nodes of any cluster. Group-centric approaches relax such restrictions from all nodes of clusters to some members of the cluster. Instead of bounding individual nodes, group-centric approaches consider properties for groups during partitioning of networks. Network-centric approaches consider properties that are defined by covering the entire network. Hierarchical approaches can be further subcategorized as divisive and agglomerative. Divisive approaches follow top-down method, i.e., these approaches keep partitioning the network until it maximizes prespecified criterion. In contrast, agglomerative approaches follow natural way of group formation. Starting from nodes, gradually groups grow by merging similar groups. In order to ensure a meaningful clustering, several metrics [21–23] are used to evaluate clustering algorithms. In this context, agglomerative approaches are found better in generating accurate clustering than other approaches [11, 16–20]. In this chapter, we have presented agglomerative approaches in detail with their advantages and limitations in relation to certain Big Data scenarios.

Remainder of the chapter is organized as follows. Section 3.2 explains various Big Data scenarios and their inherent issues. Section 3.3 illustrates parallel processing techniques of Big Data for the resolution of related issues. Section 3.4 explains various external memory operations to tackle memory-related issues. Section 3.5 discusses how agglomeration can be helpful for Big Data scenarios in parallel processing and external memory operation techniques. Section 3.6 describes various agglomerative approaches with a generic model. Section 3.7 discusses relative impact in handling large graphs with various strategic changes in the agglomerative approaches. Section 3.8 lists the specific parameters of various agglomerative algorithms and discusses beneficiary treatments of those parameter in Big Data scenarios. In Sect. 3.9, we discuss various advantages, disadvantages, and issues related to incorporation of agglomerative approaches in Big Data scenarios. Finally, the conclusion is presented in Sect. 3.10.

3.2 Big Data Scenarios and Issues

Big Data involves three Vs (volume, variety, and velocity) [24] in its definition, which resembles large, heterogeneous, and rapidly and continuously growing data sets. In this section, we study applications of various domains that deal with large data and associated issues. For example, social networks such as Facebook and Twitter are growing enormously and are therefore generation huge volumes of user data. Graph-based analysis of links and text analysis-related application to such social networks are very common. Facebook has more than 802 million active users and has to process over 1.26 billion queries every day [25]. Twitter also has almost 500 million active users who generate more than 58 million tweets and 2.1 billion search queries every day [26]. Another instance is the searching of the World Wide Web. Popular search engine Google processes 3.5 billion searches per day [27]. Dealing with such large data is often confronted with the major issues mainly related to handling of processing and storage of data.

Telecommunication traffic management system [28] receives billions of common channel signaling (CCS) messages and calls from multiple subnetworks that require gigabytes of memory. Analysis of characteristics of both the telephone call arrival process and the signaling message arrival process is key requisite for the traffic system. Major concern is to analyze call arrival, call holding, message arrival, message routing, etc. Such analysis requires processing of gigabytes of data. The size and complexity of traffic data require sophisticated methodology for data management, data translation and manipulation, and data analysis. Today's Internet traffic management system [29] requires even more sophisticated and efficient processing of such enormous amounts of data collected from the network.

In the image processing domain, content-based image retrieval (CBIR) systems [30], duplicate image detection (DID) systems [31], and face recognition systems [32] involve massive image data processing. Billions of images are available on Facebook and Picasa. These image processing systems require processing each of the images, which implies necessity of memory as well as efficient processing.

From the perspective of the data analysis specialist, the Big Data scenario infers various specialized difficulties. The most evident of them is the computational issue of needing to manage large data in memory (memory management) and to do as such in a sensible amount of time (efficient processing). Efficient processing of algorithms depends on super-linearity on size of the data and dimensions of data. Increasing any one of these two variables requires memory optimization, parallelism, and new creative approaches. Other fundamental issues with aforementioned scenarios are to deal with dimensionality, variety, and heterogeneity of data.

In practical applications of real-world problems, dealing with large data with billions of instances and high dimensions focuses mainly on efficient processing and creative memory management. In recent years, researchers have developed brilliant models that optimize the processing part by executing in parallel. Prime focus of this chapter is on large graph processing or processing of large-scale linked data. MapReduce [33] and bulk-synchronous parallel (BSP) [8] are the widely accepted models for processing of large graphs in parallel with multiple machines. Similarly, for the storage issues, numerous solutions are proposed which have proven to be very efficient in handling large data. External memory operations are very popular for efficient memory management. In coming sections, various BSP models for parallel processing and external memory operations are elaborated.

3.3 Parallel Processing

This section explains the bulk-synchronous parallel (BSP) model [8] for the processing of large graphs in parallel. In addition to this, techniques such as Pregel [5] and Seraph [7] that incorporate BSP model for graph processing are also illustrated.

3.3.1 Bulk-Synchronous Parallel (BSP)

Main objective of both MapReduce and BSP models is the same, i.e., processing of graphs in parallel. The difference between the two models lies in the flow of data while processing takes place. A typical BSP model adopted from [7] is presented in Fig. 3.1. BSP partitions the input data once and for all during initialization. Nodes process the part of data that are assigned to them independently by consuming

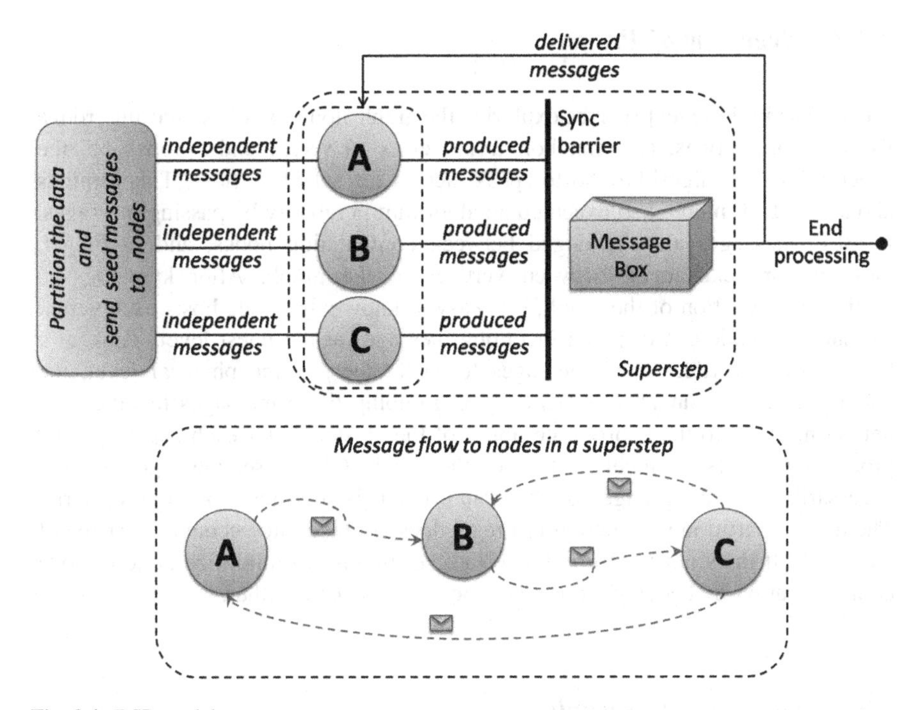

Fig. 3.1 BSP model

messages from a private box. Data does not flow between the nodes; rather instruction messages are exchanged between nodes that are smaller and efficiently transmitted. This is an incarnation of message passing interface (MPI) pattern. There is no restriction on the outcome of the processing phase. It can be the modification of the local partition's data or output of a new data set or in particular production of new messages, which will be shipped to the appropriate nodes and consumed on the next iteration and so on.

Processing moves ahead with the completion of smaller units called *supersteps*. It indicates the completion of processing in all nodes for the current iteration. Thus, after each superstep, all nodes join a common synchronization barrier. The role of such barriers is to decouple the production and consumption of messages. All nodes pause their activity, while the BSP plumbing enqueues messages into each node's private mailbox and signals ready for consumption when processing resumes at the next superstep. In this way, a message produced by node A for node B will not interrupt the latter during its processing activity. Resources are mostly dedicated to computing. In case of nonuniform processing of nodes, faster nodes will sit idle while waiting for the other ones to complete a superstep. The superstep that produces no more messages that will be shipped to any other nodes is the terminating step of the algorithm.

3.3.2 Overview of Pregel

The idea behind Pregel [5] is the exploring the graph along its edges. Starting from a fixed set of vertices, one can hop from vertex to vertex and "propagate" the execution of the algorithm, so to speak, across the set of vertices. This reminds about how BSP model had executed an algorithm iteratively by passing messages. In BSP, messages are exchanged between computation nodes, and in Pregel, messages are exchanged between vertices of the graph. After knowing the partitioning function of the graph, it is easy to know what node hosts each vertex so that each node can dispatch incoming messages at the finest level. Thus, at a logical level, vertices send messages to each other; at the physical level, the underlying BSP foundation takes care of grouping those messages together and delivering them to the appropriate nodes of the cluster. After each superstep, the processing covers a bigger portion of the graph. Of course, messages are not necessarily sent along edges of the graph, but it is the most common scenario. The default partitioning function in Pregel does not take into account the shape of the graph. It does not try to optimize the partitioning to minimize node-to-node communication by keeping strongly connected vertices together.

3.3.3 Overview of Seraph

Seraph [7] is an efficient, low-cost system for concurrent graph processing. Existing systems such as Pregel can process single graph job efficiently, but they incur high cost while dealing with multiple concurrent jobs. These systems limit the sharing of graphs in memory for multiple jobs. In other words, these systems strongly associate graph structure with job-specific vertex values. Therefore, each individual job needs to maintain a separate graph data in memory that imply inefficient use of memory. Seraph solves the above issues as follows. Concurrent jobs can share graph structure, which reduces memory requirement even if the fact that graph structure and job-specific data are strongly associated. Small amount of data is required in Seraph; with that, the entire computation states can be recovered. Seraph enables multiple jobs to share the same graph structure data in memory [7] as illustrated in Fig. 3.2. To maximize job parallelism, Seraph incorporates three new features as follows:

- In order to isolate mutations in graph from concurrent jobs, Seraph adopts "copy-on-write" semantic. When the graph structure needs to be modified by a job, Seraph copies the local region that corresponds to that job, and mutations are applied without affecting remaining jobs.
- A "lazy snapshot" protocol is used. It maintains a continuous graph updates. Further, consistent snapshot is generated for each new submitted job.
- Delta graph check pointing and state regeneration are used to implement fault tolerance. This is because it deals with both, job-level and worker-level failures.

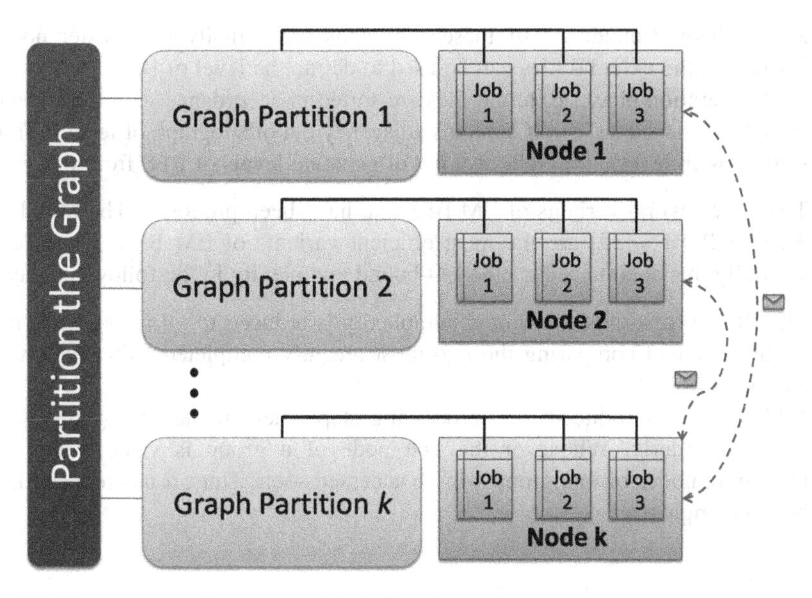

Fig. 3.2 The execution of concurrent jobs in Seraph

A job scheduler controls the execution progress and message flow of each concurrent job. Therefore, avoiding aggressive resource competition requires a job scheduler to be a part of Seraph.

3.4 External Memory Operations

In the case of large graph networks that cannot be stored completely in the main memory, external disk is used. However, this incurs I/O interaction overheads. External memory breadth-first search (EM BFS) algorithm is one of the competent algorithms in efficiently handling such overheads. This section briefly discusses different variants of the EM BFS algorithm.

3.4.1 External Memory BFS

EM BFS consists mainly of two phases: partitioning of graphs into disjoint sub-graphs and breadth-first search (BFS) generation phase.

- Partitioning phase: Partitioning of graph is done generally in parallel to speed up the process. Adjacency list representation of graph is restructured in this phase. With randomly selected nodes called master nodes, multiple connected sub-graphs are prepared and stored in external file and spatial indexing is used to access them. Boundaries of these subgraphs are actually the master nodes. Therefore, the external files can be used to define the level of BFS.

- BFS generation phase: External files are sorted in accordance with their identifiers. External file at level i contains adjacency list of subgraph of level i. Thus, starting with level 0 through level n will generate levels of BFS from 0 to n.

There are several variants of EM BFS that have been proposed. The MM BFS [34] and MR BFS [35] are the most efficient variants of EM BFS. These have successfully managed to reduce the I/O-bound complexity in the following ways:

- MR BFS: Worst-case I/O-bound complexity is reduced to $\mathrm{O(n + sort(n + m))}$ by I/O efficient comparing the two most recently completed BFS levels with current BFS.
- MM BFS: It introduced a preprocessing step where nodes are grouped and accessed orderly. Whenever the first node of a group is visited, then the remaining nodes of that group will be accessed soon. Thus, it has reduced I/O-bound complexity.

3.5 Agglomeration in Big Data Scenarios

Discussions in Sects. 3.3 and 3.4 indicate that the handling of big data issues with both the parallel processing and external memory operation requires partitioning of graphs or networks. This section illustrates how the agglomerative approaches are becoming beneficial for the partitioning of large networks. In agglomerative approaches, partitioning is started from arbitrary nodes and gradually enlarged to obtain final clusters present in the entire network. Clusters are identified locally with agglomeration approaches. Most of these approaches do not require access to the entire network at the same time, which helps during processing large networks. Large networks are processed in parallel by partitioning them into smaller sub-networks. Thus, clustering is very important for parallelization of processing. However, most of the clustering algorithms other than agglomerative approaches require processing of the entire network at the same time, which takes us on the processing and memory-related issues. On the contrary, agglomerative approaches partition the networks through local processing of parts of the networks.

Local nature of the agglomerative approaches gives a place for implication of these approaches to execute and start clustering from multiple parts of the network in parallel. For example, local nature of merging clusters based on incremental modularity in fast unfolding algorithm [11] can be noticed easily. Processing of two clusters for merging requires only information about those clusters. If it indicates that the modularity increases with merging, then simply merge those clusters without bothering about other parts of the network. Hence, merging operation in fast unfolding algorithm can be treated as local operation and can be executed in parallel. Similarly, visiting of nodes in random walk algorithm [20] during similarity matrix preparation requires processing only those nodes, which come under the path with prespecified length. Visiting of nodes in multiple random paths can be performed in parallel. Thus, clustering process itself can be parallelized easily with agglomerative approaches.

The local nature of agglomerative approaches can be parallelized if local operations are defined explicitly. However, some of the algorithms are a little bit complex, and the local nature of these algorithms is implicitly defined within the process. Local operations cannot be extracted from the main process of the algorithm. These approaches limit the implication of the local operations in parallel. For example, the evaluation of membership of nodes in LICOD algorithm [10] requires the identification of the shortest path. The entire network has to be processed to find the shortest path. Therefore, the algorithm cannot be parallelized. Similarly, leader-follower algorithm [16] also requires the shortest path. Moreover, the merging of clustering in LICOD is also not local, since it incorporates of Borda voting scheme. Therefore, these algorithms cannot be parallelized. On the contrary, top leader algorithm [18] can be parallelized though the local nature is implicit. Identification of top k nodes requires access to the entire network. However, after identification of top k nodes, the portion that finds associated members to these nodes can be easily executed in parallel.

Partitioning of large networks and processing those parts in parallel are the primary aspects to deal with most of the Big Data scenarios where linked information are presented in the form of a network. Therefore, efficient and proper partitioning of network is prerequisite for processing the networks in parallel. If partitioning scheme itself can be parallelized, then it avails additional advantage for dealing with Big Data. Agglomerative approaches have privilege over other clustering or partitioning schemes, since these approaches can be parallelized. However, due to implicit definition of local operation or inseparable local operations, parallelization of the algorithms cannot be performed for some cases. Nevertheless, these approaches reduce significantly the processing time. Specially, the approaches such as fast unfolding and SCAN [19] are very fast and overwhelmingly accepted approaches for dealing with Big Data scenarios.

3.6 Agglomerative Approaches

This section elaborates on various agglomerative approaches with a generic model for partitioning graphs. Agglomerative approaches follow bottom-up approach where each observation starts in its own cluster and pairs of clusters are merged as they move up in the hierarchy. Unlike divisive approaches, here clusters merge instead of splitting. The generic agglomerative approach is briefly described, first, as follows.

3.6.1 Generic Model

A generic model of agglomerative approach is shown in Fig. 3.3. Initially, each node is considered as a cluster. At each level, these clusters are merged with other clusters. Generally, merging of nodes is done in a greedy manner. Each level of merging carries cost or benefit of merging. If merging is associated with any cost function, then definitely that has to be minimized; otherwise, it has to be maximized. Generally, functions related to benefit of merging are considered to merge clusters. The function associated with each level can be presented in the dendrogram. These functions are also used to determine when to stop merging clusters. For example, in Fig. 3.3, the values of benefit of merging associated with each level are $\{S_1, S_2, S_3, \ldots, S_n\}$, here $n = 4$. Merging of clusters will stop at level 2 if $S_2 > S_3$, i.e., the benefit of merging decreases with further merging of clusters. Hence, to gain maximum benefit, algorithm will stop at level 2.

Apart from the cost and benefits, distance among clusters and number of clusters are also used as stopping criteria. Each agglomeration occurs at a greater distance between clusters than the previous agglomeration. One can decide to stop clustering either when the clusters are too far apart to be merged (distance criterion) or when there is a sufficiently small number of clusters (number criterion). Generally, merging of clusters takes place in two ways depending on formation of clustering. The model described in Fig. 3.3 is the global approach where clusters are formed on processing of the entire graph. On the contrary, Fig. 3.4 presents generic model for the local approach. Clusters are formed locally without processing the entire graph. Clusters are formed sequentially one after another. However, each cluster is formed by merging smaller clusters in successive levels until a maximum benefitted cluster is obtained. Such merging is centric to a particular node that advocates benefit

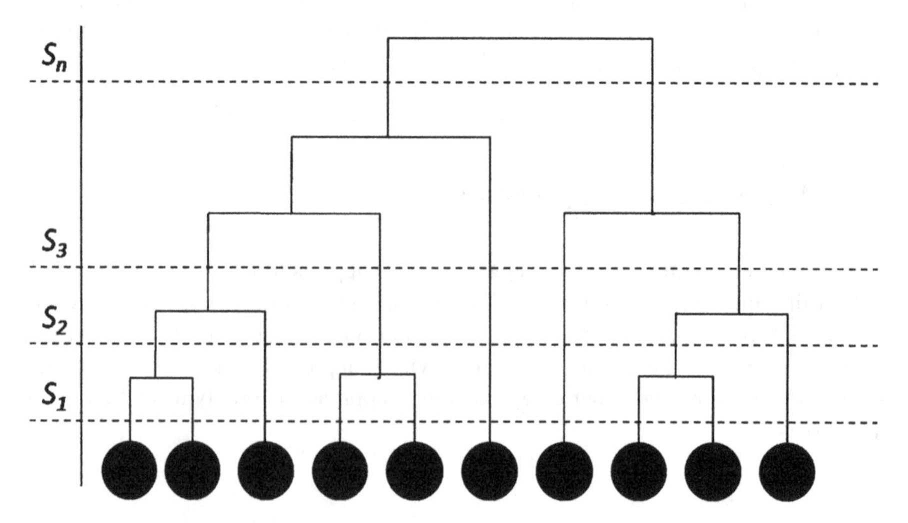

Fig. 3.3 Generic model of agglomerative approach

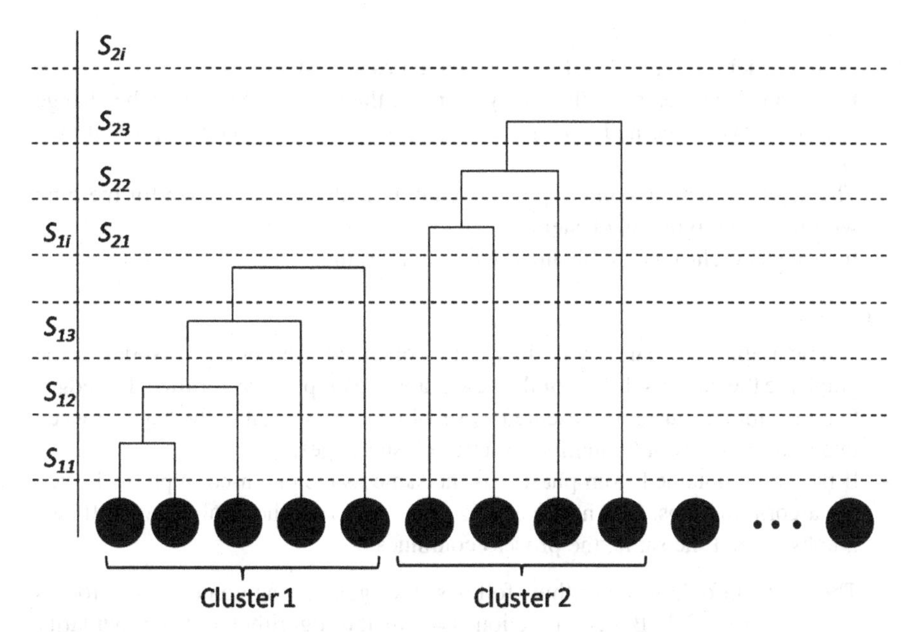

Fig. 3.4 Generic local model of agglomerative approach

associated with the cluster. Once the cluster associated with the advocating node achieves maximum possible benefit, it starts new cluster associated with other advocating node. In Fig. 3.4, S_{1i} is the maximum benefit of the cluster 1. After achieving the value S_{1i}, processing for the cluster 1 ends and processing for cluster 2 begins. Most of the agglomerative algorithms follow local agglomeration approach. In the next subsections, we illustrate strategies incorporated in different agglomerative approaches in perspective of the generic model.

3.6.2 Fast Unfolding

Fast unfolding algorithm [11] uses the modularity function as benefit function to obtain perfect clusters. The algorithm keeps on merging the clusters until maximum modularity is reached. During intermediate states of the algorithm, the actual size of graph reduces to the number of clusters formed at the current level, which are the new nodes of the graph. These nodes are again sent back to the algorithm to reduce it further to obtain maximum modularity. The algorithm has two phases. In the first phase, a particular node is assigned to a neighboring node based on the change in modularity, while in the second phase, the reduced graph is obtained from the first phase. This process continues until there is no change in the graph obtained from the first phase and the second phase.

Phase 1:

- Initially each individual node is represented as a cluster.
- Each node is processed with its neighbors for the value of ΔQ that is the change in modularity of the node in case neighbor is merged with node representing a cluster.
- The neighbor with maximum positive gain in modularity is chosen for merging with the node to obtain clusters.
- Intermediate clusters are obtained with the merging.

Phase 2:

- A new graph is prepared with the clusters obtained from phase 1. Nodes of the graph are the clusters. Edges of the new graph are prepared as follows. The edges that lie within a cluster represented by self-loop. The weight of the edge between two clusters is given by number of inter-cluster edges.
- If the graph obtained from phase 2 is the same as the one used in phase 1, then the algorithm stops. The nodes at this stage represent the final clusters. If two graphs are not the same, the process continues.

The fast unfolding algorithm follows the generic agglomeration process discussed in Fig. 3.3. Benefit function used in the algorithm is the modularity function. Stopping criterion is the difference between two graphs that are processed in two successive levels.

3.6.3 SCAN

SCAN [19] is another popular agglomerative clustering approach. In addition to clusters, it also detects hubs and outliers. It follows local agglomerative approach as presented in Fig. 3.4. At the initial stage, the core nodes are identified. All those nodes that are structurally reachable from the core nodes are assigned to respective core nodes that form a cluster. The process continues until all the nodes are either classified or marked as nonmember. The nodes that are identified as nonmembers are further classified as hubs and outliers. The algorithm uses the following special terminologies.

- Structural similarity: Represented by $\sigma(\mathrm{u}, \mathrm{v})$ where u and v are nodes and is evaluated as number of common neighbors of u and v divided by geometric mean of number of neighbors of u and v.
- e-neighborhood: The number of neighbors of v whose structural similarity with v is greater than e belongs to its e-neighborhood.
- Core: A vertex v is defined as core if its e-neighborhood contains at least μ vertices.
- Direct structure reachability: The nodes that are neighbors of the core are directly structure reachable from that core.

- Structure reachability: A node v is structure reachable from some node w, if there exist a path such that all the intermediate nodes are cores.
- Hub: A node is defined as a hub if some of its neighbors are in one cluster while some in other.
- Outliers: An isolated node is treated as an outlier if all its neighbors lie in one cluster or does not belong to any as such.

Unlike the fast unfolding algorithm, SCAN does not have clear levels of agglomeration. It merges nodes based on structural reachability and structural similarity as follows.

- Nodes are arbitrarily selected and find their neighbors to decide if the nodes represent core or not.
- If a node is not a core node, then it is marked as a nonmember. Otherwise, a new cluster is formulated corresponding to the core node.
- The neighbors of the core node that belong to its e-neighborhood are inserted into a queue Q.
- Nodes that are in Q are dequeued one by one until it becomes empty.
- For each dequeued element from Q, direct structure reachable nodes are identified. If any of these nodes is a nonmember node, then assign that node to current cluster and enqueue it into the queue.
- After completing the processing of the entire graph, nonmembers that do not belong to any cluster are marked as hubs or outliers.

One of the major advantages of SCAN algorithm is that it not only forms the clusters but also identifies the nodes that are not formally in any cluster as hubs and outliers. The algorithm is very fast. The time complexity of the algorithm is $O(m)$ for a graph with n vertices and m edges.

3.6.4 Leader-Follower

The leader-follower algorithm [16] is designed based on the idea of *leaders* and their *followers*. Nodes are divided into either leader or follower by measuring *distance centrality*. It measures the closeness of a node to all other nodes. Followers are again subdivided as loyal or not loyal. A loyal follower must have its distance centrality value at least equal to each of its neighbors.

Leaders are arranged in increasing order of their distance centrality. Initially, all the leaders represent potential clusters. The representative cluster for a leader expands as follows. The expansion process follows local agglomerative approach as presented in Fig. 3.4. The neighbors that are loyal followers of the leader are assigned to the representative cluster of that leader. Many of the leaders get no followers and constitute cluster with single node. Such leaders are identified as fake leaders. These fake leaders are removed by assigning them to other clusters. The neighbors of the leader are considered for such assignment. The leader that has the majority of its neighbors are loyal follower is selected as the leader for assigning fake leaders. A *special case* occurs when the neighbors of a fake leader are themselves leaders. This is handled by the same procedure of finding majority leader and breaking ties randomly that results in random leader assignment. Major steps of the algorithm are summarized as follows:

- Initialization is done by calculating the distance centrality of each node. This method utilizes Floyd-Warshall algorithm for computing the shortest path among all pairs of nodes.
- Identify leaders and followers by measuring distance centrality.
- Each representative cluster for a leader is expanded by assigning loyal followers.
- Single-node clusters, i.e., fake leaders are assigned to appropriate cluster as explained in the above discussion.

Leader-follower algorithm can identify smaller clusters. The algorithm does not need to specify any parameter as it naturally discovers the underlying structure. Most importantly, it does not generate singleton clusters because all such clusters are assigned to other clusters. The algorithm results unexpected clusters, as most of the data sets do not follow the definition of cluster utilized in the algorithm. There are numerous other leader-based algorithms in existence. These are covered in the following sections in terms of their efficiency with respect to Big Data scenarios.

3.6.5 HC-PIN

HC-PIN [17] is another agglomerative algorithm that was designed to identify clustering in protein-protein interaction networks. The algorithm follows local approach as in Fig. 3.4 for expanding clusters. However, at the initial stage, it processes the entire network for calculating the edge clustering value (ECV). In the essence of agglomeration, the algorithm initially considers each individual node as cluster. For each of the edges, ECV is calculated, and those are arranged in a list in decreasing order. The ECV of an edge is the square of the product of a number of common neighbors of the nodes divided by the product of a number of neighbors of each node. Higher ECV for an edge indicates greater the tendency that both the associated nodes of that edge to be included in the same cluster.

The edges are removed one by one from the ECV list. The weighted degree for the clusters of each of the representative nodes is computed. Weighted degree is defined as the sum of weights of all the incoming edges to a particular node. For a cluster, it is the summation of weighted degrees for all the nodes belong that cluster. A parameter λ is used for decision-making. The ratio of weighted in-degree to out-degree for each of the clusters is compared with λ parameter. Another parameter s, is used at the end to determine the minimum size of each cluster. Only clusters that have minimum size as s are included in the result. The algorithm is summarized as follows:

- All the vertices are initialized as a singleton cluster.
- ECV value is calculated for each of the edges of the network. All the ECV values along with associated edges are arranged in a list in decreasing order.
- Edges are removed one by one from the list, and associated nodes of that edge are compared. The comparison-based decision is made to merge both the nodes either in the cluster or in the different clusters.
- Weighted in-degree and out-degree of all clusters are computed. These weighted values are compared with parameter λ in order to determine if it satisfies necessary structural requirement of the cluster or not. If a cluster does not fulfill this objective, then the nodes of that cluster are assigned to other cluster.
- In addition, clusters are also refined with parameter s to ensure the sizes of clusters that are identified.

HC-PIN is a successful algorithm in terms of accuracy of clusters it generates.

3.6.6 Other Approaches

Apart from the algorithms discussed in the above subsections, there are numerous agglomerative approaches that have been developed which fall within either of the two models. For instance, random walk algorithm [20] has been developed utilizing very similar mechanism as SCAN and HC-PIN algorithms. The leader-based algorithms such as top leader [18] and LICOD [10] have been developed along the same lines as leader-follower algorithm. Recently, another agglomerative algorithm, ENBC [36], has been proposed. The ENBC algorithm incorporates both models discussed above. The algorithm is also very efficient in generating accurate clusters. Details of the strategic changes of these algorithms and their impact on Big Data scenarios are discussed in the next section.

3.7 Agglomerative Strategic Changes for Big Data Scenarios

Strategies incorporated in handling Big Data problems have a vital role for making processing efficient and effective. As part of large graph processing with parallel processing and external memory schemes requires partitioning. Section 3.5 has discussed several advantages of incorporating agglomerative clustering in partitioning of large graphs. Prominent agglomerative approaches are studied in Sect. 3.6 with generic models. This section is devoted to discussing the agglomerative approaches by accounting their strategic changes and relative impact in handling large graphs with such strategic changes.

Numerous strategic changes in agglomerative approach have been proposed in the quest for more effective and efficient clustering. Most promising strategic change in agglomerative approach is the local processing of graphs. Several algorithms have been proposed to make the local processing more efficient. One common approach is to derive a mechanism which defines similarity among the nodes of the entire graph and identifies clusters locally. Normally, clusters start from any arbitrary node and gradually expand by utilizing similarities among nodes. Most of such algorithms can be roughly viewed as two phases, i.e., *similarity matrix creation* and *cluster formation*. For example, in random walk algorithm [20], a similarity matrix is prepared that keeps track of frequencies of pairs of nodes that appear in the random path. Initially, the algorithm traverses randomly in fixed length path started from any random node. The strategy here is the nodes that are in densely connected region will appear frequently in the random path. Thus, those nodes will get higher similarity values. Similar kind of strategy is followed in HC-PIN [17], where edge clustering value (ECV) is computed for all pairs of edges and maintained in a list. Cluster formation part of such strategies is local, but the similarity matrix preparation requires processing of the entire network. Therefore, it requires special mechanism to generate similarities locally for large graphs.

An alternative of similarity computation is identification of leader nodes locally among the representative groups of nodes within the graph. Leader-based algorithms are designed based on this principle. These approaches can also be viewed in two phases, i.e., leader identification and cluster formation. In the LICOD algorithm [10], leader nodes are identified utilizing the degree centrality. Since leader node represents potential cluster, the group of identified leader nodes is optimized to get minimized set of nodes. Followers of these leader nodes are identified and based on their membership values assigned to the representative cluster of a leader node. Leader-follower algorithm [16] divides followers further into two groups: loyal followers and non-loyal followers. Distinction of follower results in more refined clusters. In contrast to both LICOD and leader-follower algorithms, top leader algorithm [18] identifies top k potential leaders instead of optimized set of leader nodes. With this strategy, the optimization cost for obtaining leader nodes is diminished, and as a result, top leader algorithm [18] requires comparatively less time. On the other hand, SCAN algorithm [19] uses same mechanism as leaders with new terminology called core nodes. It utilizes notion of structural reachability instead of followers to define clusters represented for core nodes.

Both the similarity matrix-based approaches and leader-based approaches require twofold processing of graphs. For large graphs, such processing costs double. In that case, instead of two phases of clustering, merging both phases into one will be more beneficial for large graphs. Fast unfolding algorithm [11] utilizes such strategy with multilevel clustering. Clusters obtained in each level agglomeration can be treated as potential clusters with specific properties. The algorithm is very efficient for large-scale graphs with billions of nodes and edges. On the other hand, ENBC algorithm [36] utilizes single-level agglomeration as in the similarity matrix-based and leader-based approaches. However, ENBC identifies clusters by merging similarity computation phase and cluster formation phase into single phase. It starts from any arbitrary node that represents potential cluster and that expands based on structural similarity of neighbors. The process of cluster formation followed in ENBC is local and more efficient than both similarity matrix-based and leader-based approaches. Therefore, it is suitable for large graph processing both in perspective of parallel processing and external memory operations.

3.8 Parameter Tuning for Big Data Scenarios

Most of the agglomerative algorithms have parameters to control various aspects of clusters such as size and quality. From the perspective of Big Data, these aspects are very important. As mentioned above, graph clustering is a prominent intermediary step for handling Big Data with both parallel processing and external memory operations. Sizes of clusters have to be balanced in a sense that neither the large nor the small clusters are suitable for parallel processing and external memory operations. Cluster sizes have to be as homogeneous as possible because too much variance in clustering is not suitable or ineffective for both parallel processing and external memory operations. Agglomerative algorithms used different parameters for controlling cluster sizes.

Controlling of cluster sizes is done in two ways as follows. First approach is to define the limit to the total number of nodes that can belong in a cluster. In HC-PIN algorithm, this approach is utilized very effectively with parameter s for controlling cluster size. Second approach on the other hand sets limit on the total number of clusters. Top leader algorithm has this approach by utilizing parameter k. It indirectly sets the size of cluster as follows. If the value of k is very large, then the sizes of cluster will become very small. Again, if the value of k is very small, then the sizes of cluster will become huge. Therefore, depending on the size of the network, the parameter k can be set with a value that may result in reasonable cluster sizes.

Quality of clustering is very important for structurally feasible clustering. It indicates how a cluster is confident to claim that it is a cluster. Generally, it is defined as the ratio of number of edges within the cluster to the number of edges outside the cluster. Structurally feasible clusters are easy to handle and require less effort specially for parallel processing. Agglomerative approaches are very rich in controlling quality of clustering with different parameters to ensure structurally feasible clusters. The ways of controlling quality of clustering through parameters vary with the strategy of the algorithm. Different algorithms ensure qualities through different parameters as follows:

- **SCAN:** The algorithm has two parameters e and μ for ensuring quality of clustering. The parameter e is a threshold that represents the minimum structure similarity for a node u to be in e-neighborhood of node v. Structure similarity defines the closeness of two nodes. The parameter μ shows whether a node v is the more central one by setting this threshold of μ representing minimum μ elements in e-neighborhood of node v, thereby declaring v as core.

- Top leader: It also has two parameters γ and δ for assuring quality of clustering. With the parameter γ, the algorithm identifies outliers that do not fit in any cluster. Higher γ value implies more connections, i.e., higher edge density for a node to be not an outlier. The parameter δ defines the depth up to which the common neighbors between a node and a leader can have. It represents the required strength of connection between the node and a leader for the node to get included in the cluster of leader.
- HC-PIN: The parameter λ is used in comparison with the weighted degree of clusters. It indicates the kind of cluster quality that is desired by the algorithm. Varying the λ parameter, different hierarchical structures can be obtained with different qualities. In addition, another parameter s determines the minimum size of clusters. With parameter s, small and unrelated clusters easily distinguished from large, significant, and dominant clusters and process further. Such processing of insignificant clusters enhances overall qualities of clusters. Some algorithms often report small group of nodes as a separate cluster that might have been a part of larger cluster. HC-PIN has undermined such difficulty with proper settings of values of parameter s.
- LICOD: The algorithm has three parameters σ, δ, and ε. The parameter σ determines of leaders, which actually represents the node that is surrounded by densely connected region. The parameter δ is used for merging of the identified leaders. It determines the closeness between two leaders. If two leaders are closed, those are kept in same cluster rather than creating different clusters for each. Therefore, the parameter δ has very important role in gaining better quality clusters. The parameter ε is used to control the extent to which a node might belong in more than one cluster. By varying ε parameter, the degree of overlapping of the clusters can be controlled. For nonoverlapping clustering, the parameter ε has no role in determining quality of clusters.

Due to a variety of parameters, most agglomerative algorithms provide large degree of control to obtain variety of clusters with specific properties as desired. The different characteristics of the clusters can be highlighted *by* changing these parameters. Therefore, with agglomerative approaches, graphs can be partitioned easily as per the requirements of both parallel processing and external memory operations for dealing with large graphs.

3.9 Discussion

Big Data scenarios need to be handled in the context of both processing time and memory requirements. Data clustering has a vital role in dealing with both these issues. Representation of linked data in the form of network is another aspect of Big Data scenarios. Parallel processing and external memory operations are prominent solutions of these Big Data problems. Partitioning of such network is an intermediary step for parallel processing as well as memory management. For a meaningful and efficient partitioning of networks, graph clustering techniques are used. However, as regards implications of clustering in dealing with Big Data scenarios, the following major issues must be tackled:

- Efficient and proper utilization of partitions or clusters is very important in parallel processing.
- Dealing with size of clusters is a major issue. Smaller clusters may reduce memory requirement but increases processing overhead. Too large clusters may not fit into memory and increase time complexity.
- Meaningful clusters are foremost important for efficient parallel processing.
- Evaluation of clusters is also very important for measuring clustering and algorithms.
- Selection of clustering algorithm is another trivial problem especially with the large numbers of algorithm available for clustering. Each of these algorithms possesses specific advantages, which makes the selection process more difficult.

Agglomerative approaches are helpful in dealing with Big Data scenarios. These approaches are local in nature and hence can be easily parallelized. The local nature of these approaches can be utilized in efficient processing and memory management. Beside this, agglomerative approaches have numerous other advantages that are summarized as follows:

- Most of these approaches are very fast. Time complexities as well as space complexity are low.
- These approaches generate clusters that are more accurate than other approaches.
- Merging of clusters in level by level allows generation of structurally feasible and meaningful clusters.
- Most of the operations incorporated in agglomerative approaches are local, which do not require accessing of the entire networks. Therefore, it reduces average memory requirement during intermediate steps of algorithms.
- Explicit local operations of agglomerative approaches can be executed in parallel that can reduce clustering time significantly.
- Control parameters used in agglomerative approaches allow generating customized clusters. The algorithms are flexible to generate clusters with specific properties.

Clearly, agglomerative approaches have several advantages in the perspective of Big Data scenarios. However, incorporation of some of these approaches results in the following disadvantages:

- Too many parameters used in algorithms may cause difficulty in customization of clustering.
- Algorithm cannot be parallelized if operations of algorithms are not local or local operations are implicit and inseparable from the main algorithm.
- Local operations in some algorithms caused large numbers of smaller sized clusters.

3.10 Conclusion

Big Data handling systems are often confronted with two major issues: efficient processing and memory management. Parallel processing and external memory operation techniques are the most favored processes to solve these issues where partitioning of data is the vital prerequisite for these techniques. Both these techniques have undertaken processing and memory-related issues to an extent. However, the rapid growth of data and complex relationships within the data have unleashed more difficulty in handling. For such Big Data scenarios, incorporation of network representation and clustering has been addressed and discussed in this chapter. The utilization of clustering is advantageous to get a meaningful and efficient partitioning for parallel processing and memory-related issues. Nevertheless, this kind of utilization requires bearing extra overheads related to clustering. Agglomerative approaches can be very helpful due to their local nature for clustering network data. Besides, several other advantages are also discussed while dealing with Big Data scenarios. Despite numerous limitations and concerns, agglomerative approaches have emerged as a core catalyst for dealing with processing and memory-related issues, especially for Big Data scenarios of modern-day data systems.

References

1. Wei T, Lu Y, Chang H, Zhou Q, Bao X (2015) A semantic approach for text clustering using wordnet and lexical chains. Expert Sys with Appl 42:2264–2275
2. Li S, Wu D (2015) Modularity-based image segmentation. IEEE Trans Circuit Syst Video Technol 25:570–581
3. Nikolaev AG, Razib R, Kucheriya A (2015) On efficient use of entropy centrality for social network analysis and community detection. Soc Networks 40:154–162
4. Li S, Daie P (2014) Configuration of assembly supply chain using hierarchical cluster analysis. Procedia fCIRPg 17:622–627
5. Malewicz G, Austern MH, Bik AJ, Dehnert JC, Horn I, Leiser N, Czajkowski G (2010) Pregel: a system for large-scale graph processing. In: SIGMOD'10, 2010
6. The Apache Software Foundation (2014) http://giraph.apache.org/. Accessed 28 Apr 2015
7. Xue J, Yang Z, Qu Z, Hou S, Dai Y (2014) Seraph: an efficient, low-cost system for concurrent graph processing. In: Proceedings of ACM HPDC'2014, Vancouver, Canada, 23–26 June
8. Vial T (2012) http://blog.octo.com/en/introduction-to-large-scale-graph-processing/. Accessed 28 Apr 2015.
9. Ajwani D, Dementiev R, Meyer U (2006) A computational study of external-memory BFS algorithms, SODA 2006 ACM-SIAM Symposium on Discrete Algorithms, Miami, Florida, USA, January 2006
10. Kanawati R (2011) Licod: leaders identification for community detection in complex networks. In: SocialCom/PASSAT, pp 577–582
11. Blondel VD, Guillaume JL, Lambiotte R, Lefebvre E (2008) Fast unfolding of communities in large networks. J Stat Mech: Theory Exp 10:P10008

12. Fan W, Yeung K (2015) Similarity between community structures of different online social networks and its impact on underlying community detection. Commun Nonlinear Sci Numer Simul 20:1015–1025
13. Zhang T, Ramakrishnan R, Livny M (1996) BIRCH: an efficient data clustering method for very large databases. In: ACM SIGMOD Record, vol. 25, No. 2, pp. 103–114, ACM.
14. Wang W, Yang J, Muntz R (1997) STING: a statistical information grid approach to spatial data mining. In: VLDB, vol 97, pp 186–195
15. Shang R, Luo S, Li Y, Jiao L, Stolkin R (2015) Large-scale community detection based on node membership grade and sub-communities integration. Physica A Stat Mech Appl 428:279–294
16. Shah D, Zaman T (2010) Community detection in networks: the leaderfollower algorithm. In: Workshop on networks across disciplines in theory and applications, NIPS, November 2010
17. Wang J, Li M, Chen J, Pan Y (2011) A fast hierarchical clustering algorithm for functional modules discovery in protein interaction networks. IEEE/ACM Trans Comput Biol Bioinform 8(3):607–620
18. Khorasgani RR, Chen J, Zaïane OR (2010) Top leaders community detection approach in information networks. In: Proceedings of the 4th workshop on social network mining and analysis, ACM.
19. Xu X, Yuruk N, Feng Z, Schweiger TAJ (2007) SCAN: a structural clustering algorithm for networks. In: Proceedings of the 13th ACM SIGKDD international conference on knowledge discovery and data mining, San Jose, CA, USA, August 12–15, 2007
20. Steinhaeuser K, Chawla NV (2010) Identifying and evaluating community structure in complex networks. Pattern Recogn Lett 31(5):413–421
21. Hubert L, Arabie P (1985) Comparing partitions. J Classif 2(1):193–218
22. Fred A, Jain A (2003) Robust data clustering. In: Proceedings of the CVPR, 2003
23. Wikipedia (2015) Precision and recall http://en.wikipedia.org/wiki/Precision_and_recall. Accessed 28 Apr 2015
24. Laney D (2001) 3d data management: controlling data volume, velocity, and variety, application delivery strategies. META Group Inc, Stamford
25. Minas M, Subrahmanyam K, Dennis J (2015) Facebook use and academic performance among college students: a mixed-methods study with a multi-ethnic sample. Comput Hum Behav 45:265–272
26. Debra AG, Kullar R, Newland JG (2015) Review of Twitter for infectious diseases clinicians: useful or a waste of time? Clin Infect Dis 60(10):1533–1540
27. Google Search Statistics (2015) http://www.internetlivestats.com/google-search-statistics/. Accessed 28 Apr 2015
28. Duffy DE, McIntosh AA, Rosenstein M, Willinger W (1993) Analyzing telecommunications traffic data from working common channel signaling subnetworks. Comput Sci Stat 1993:156–156
29. Joo H, Hong B, Choi H (2015) A study on the monitoring model development for quality measurement of internet traffic. Inf Syst 48:236–240
30. Joan BE (2015) Content-based image retrieval methods and professional image users. J Assoc Inform Sci Technol 67(2):2330–1643
31. Sebastiano B, Farinella GM, Puglisi G, Ravì D (2014) Aligning codebooks for near duplicate image detection. Multimedia Tools Appl 72(2):1483–1506
32. Karmakar D, Murthy CA,(2015) Face recognition using face-autocropping and facial feature points extraction. In: Proceedings of the 2nd international conference on perception and machine Intelligence, Kolkata, West Bengal, India, pp 116–122
33. Sabeur A, Lacomme P, Ren L, Vincent B (2015) A MapReduce-based approach for shortest path problem in large-scale networks. Eng Appl Artif Intell 41:151–165
34. Mehlhorn K, Meyer U (2002) External-memory breadth-first search with sublinear I/O. In: Proceedings 10th annual European Symposium on Algorithms (ESA), vol 2461 of LNCS, pp 723–735. Springer

35. Munagala K, Ranade A (1999) I/O-complexity of graph algorithms. In: Proceedings 10th symposium on discrete algorithms, ACM-SIAM, pp 687–694
36. Biswas A, Biswas B (2015) Investigating community structure in perspective of ego network. Expert Sys Appl 42(20):6913–6934

Chapter 4
Identifying Minimum-Sized Influential Vertices on Large-Scale Weighted Graphs: A Big Data Perspective

Ying Xie, Jing (Selena) He, and Vijay V. Raghavan

Abstract Weighted graphs can be used to model any data sets composed of entities and relationships. Social networks, concept networks, and document networks are among the types of data that can be abstracted as weighted graphs. Identifying minimum-sized influential vertices (MIV) in a weighted graph is an important task in graph mining that gains valuable commercial applications. Although different algorithms for this task have been proposed, it remains challenging for processing web-scale weighted graph. In this chapter, we propose a highly scalable algorithm for identifying MIV on large-scale weighted graph using the MapReduce framework. The proposed algorithm starts with identifying an individual zone for every vertex in the graph using an α-cut fuzzy set. This approximation allows to divide the whole graph into multiple subgraphs that can be processed independently. Then, for each subgraph, a MapReduce-based greedy algorithm can be designed to identify the minimum-sized influential vertices for the whole graph.

Keywords MapReduce framework • Minimum-sized influential vertices • Social influences • Data mining • Large-scale weighed graph, big data • Big data analytics

4.1 Introduction

Weighted graphs can be used to model any data sets composed of entities and relationships. Social networks, concept networks, and document networks are among the types of data that can be abstracted as weighted graphs. Identifying minimum-sized influential vertices (MIV) in a weighted graph is an important task in graph mining that gains valuable commercial applications. Consider the

Y. Xie • J. He (✉)
Department of Computer Science, Kennesaw State University, Marietta, GA, USA
e-mail: yxie2@kennesaw.edu; jhe4@kennesaw.edu

V.V. Raghavan
The Center for Advanced Computer Studies, University of Louisiana at Lafayette, Lafayette, LA, USA
e-mail: raghavan@louisuana.edu

Z. Mahmood (ed.), *Data Science and Big Data Computing*,
DOI 10.1007/978-3-319-31861-5_4

following hypothetical scenario as a motivating example. A small company develops a new online application and would like to market it through online social networks. It is worth to mention that word-of-mouth or viral marketing through online social networks differentiates itself from other marketing strategies because it is based on trust among individuals' close social circle of families, friends, and co-workers. Research shows that people trust the information obtained from their close social circle far more than the information obtained from general advertisement channels such as TV, newspaper, and online advertisements [1]. The company has a limited budget such that it can only select a small number of initial users to use it (by giving the initial users gifts or payments). The company wishes that these initial users would like the application and start influencing their friends on the social networks to use it. And their friends would influence their friends' friends and so on, and thus through the word-of-mouth effect a large population in the social network would adopt this application. Using the social network shown in Fig. 4.1 as the example, the links between each individual represent the interactions between the pair of the individuals, while the numbers on the links mean the social influence between the pair of the individual. It is obvious that the company may first choose Andy as the initial user, since Andy can influence Tony, John, and Ricky with some influences. In sum, the MIV problem is who to be selected as the initial users so that they eventually influence the largest number of people in the network. The objective of the MIV problem is to minimize the size and the set of the initial users because of budget limitations.

The problem is first introduced for social networks by Domingos and Richardson in [2] and [3]. Subsequently, Kempe et al. [4] proved this problem to be NP-hard and propose a basic greedy algorithm that provides good approximation to the optimal solution. However, the greedy algorithm is seriously limited in efficiency because it needs to run Monte-Carlo simulation for considerably long time period to guarantee an accurate estimate. Although a number of successive efforts have been

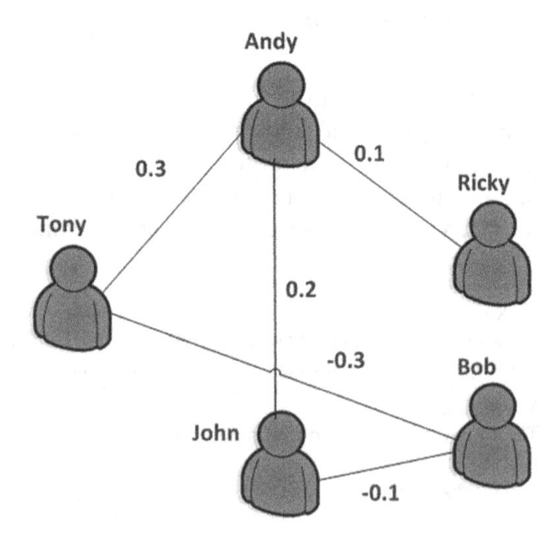

Fig. 4.1 A sample of a social network

made to improve the efficiency, the state-of-the-art approaches still suffer from excessively long execution time due to the high-computational complexity for large-scale weighted graph. Furthermore, the graph structures of real-world social networks are highly irregular, making MapReduce acceleration a nontrivial task. For example, Barack Obama, the US president, has more than 11 million followers in Twitter, while for more than 90 % of Twitter users, the follower number is under 100 [5]. Such irregularities may lead to severe performance degradation.

On the other hand, MapReduce framework has recently been widely used as a popular general-purpose computing framework and has also been shown promising potential in accelerating computation of graph problems such as breadth-first searching and minimum spanning tree [6-9] due to its parallel processing capacity and ample memory bandwidth. Therefore, in this chapter, we explore the use of MapReduce framework to accelerate the computation of MIV in large-scale weighted graphs.

The proposed framework starts with identifying an individual zone for every vertex in the graph. The individual zone of a given vertex is the set of vertices that the given vertex can influence. To design a scalable algorithm to address this, we approximate individual zone by using the concept of α-cut fuzzy set. This approximation allows us to reduce the complexity of multi-hop influence propagation to the level of single-hop propagation. Subsequently, we aim to find a minimum-sized set of vertices whose *influence* (the formal definition will be presented in Sect.4.3) reaches a predefined threshold. To reach this goal, a MapReduce-based greedy algorithm is designed by processing *individual zones* (the formal definition will be given in Sect. 4.3) for all vertices.

The contribution of this work can be summarized as follows:

- A fuzzy propagation model was proposed to describe multi-hop influence propagation along social links in weighted social networks.
- An α-cut fuzzy set called individual zone was defined to approximate multi-hop influence propagation from each vertex.
- MapReduce algorithms were designed to locate each individual zone and then identify minimum-sized influential vertices (MIV) using a greedy strategy.

This remainder of this chapter is organized as follows: Sect. 4.2 reviews related literatures of the MIV problem. Network model and the formal definition of the MIV problem are given in Sect. 4.3. The MapReduce-accelerated framework is presented in Sects. 4.4 and 4.5. Finally, the work is concluded and future working direction is given in Sect. 4.6.

4.2 Related Works

In this section, we first summarize the current research status of influence maximization problem in social network and then summarize some works; try to use GPU framework to accelerate the process to find the solution of the influence maximization problem. Finally, we remark on the differences of the work described in the chapter and emphasize the contribution of this work.

4.2.1 Influence Maximization Problem in Social Network

Finding the influential vertices and then eventually influencing most of the population in the network are first proposed by Domingos et al. in [2, 3]. They model the interaction of users as a Markov random field and provide heuristics to choose users who have large influence in the network. Kempe et al. [4] formulate the problem as a discrete optimization problem and propose a greedy algorithm. However, the greedy algorithm is time-consuming. Hence, recently huge amount of researchers try to improve the greedy algorithm in two ways. One is reduce the number of individual searched in the graph. The other is improving the efficiency of calculating the influence of each individual. Leskovec et al. [10] propose an improved approach which is called CELF to reduce the number of individual searched in the graph. Later, Goyal et al. [11] propose an extension to CELF called CELF++, which can further reduce the number. Kimura et al. [12] utilize the Strong Connected Component (SCC) to improve the efficiency of the greedy algorithm.

Although many algorithms are proposed to improve the greedy algorithm, they are not efficient enough for the large scale of current social networks. Hence, some works are proposed to fit for large-scale networks. Chen et al. proposed a method called MixGreedy [13] that reduces the computational complexity by computing the marginal influence spread for each node and then selects the nodes that offer the maximum influence spread. Subsequently, Chen et al. [14] use local arborescence of the most probable influence path between two individuals to further improve the efficiency of the algorithm. However, both of the algorithms provide no accuracy guarantee. In [15], Liu et al. propose ESMCE, a power-law exponent supervised Monte-Carlo method that efficiently estimates the influence spread by randomly sampling only a portion of the nodes. There have been also many other algorithm and heuristics proposed for improving the efficiency issues for large-scale social networks, such as [16–19]. However, all of the aforementioned improvements are not effective enough to reduce execution time to an acceptable range especially for large-scale networks.

4.2.2 GPU Framework

Completely different from the previously mentioned work, Liu et al. [20] present a GPU framework to accelerate influence maximization in large-scale social networks called IMGPU, which leveraging the parallel processing capability of graphics processing unit (GPU). The authors first design a bottom-up traversal algorithm with GPU implementation to improve the existing greedy algorithm. To best fit the bottom-up algorithm with the GPU architecture, the authors further develop an adaptive K-level combination method to maximize the parallelism and reorganize the influence graph to minimize the potential divergence. Comprehensive experiments with both real-world and synthetic social network traces demonstrate that the proposed IMGPU framework outperforms the state-of-the-art influence maximization algorithm up to a factor of 60.

4.2.3 *Remarks*

In this chapter, we focus on addressing the MIV problem on large-scale weighted graphs using the MapReduce framework. The proposed method first improves the algorithm efficiency by dividing the whole graph into some subgraphs using fuzzy propagation model. Subsequently, a MapReduce greedy algorithm is presented to search the best candidates in each subgraph to achieve high parallelism. The proposed framework shows potential to scale up to extraordinarily large-scale graphs especially from big data perspective.

4.3 Graph Model and Problem Definition

In this section, we first introduce how to model a weighted graph and then formally define the minimum-sized influential vertices (MIV) problem.

4.3.1 *Graph Model*

We model a weighted graph by an undirected graph $G(V,E,W(E))$, where V is the set of N vertices, denoted by v_i, and $0<i<N$. i is called the vertex ID of v_i. An undirected edge $e_{ij}=(v_i, v_j)$ e E represents weights between the pair of vertices. $W(E)=\{p_{ij}$, if (v_i, v_j) e E, then $0<p_{ij}<1$, else $p_{ij}=0\}$, where p_{ij} indicates the weights between vertices v_i and v_j. For simplicity, we assume the links are undirected (bidirectional), which means two linked vertices have the same weight (*i.e.*, p_{ij} value) on each other. Figure 4.2 shows an example of a weighted graph. There are nine vertices in the graph. The weights over edges represent the social influences between the pair of vertices.

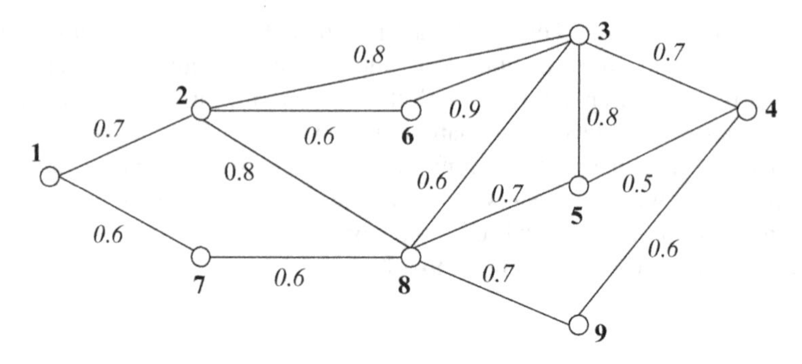

Fig. 4.2 A sample of a weighted graph

4.3.2 Problem Definition

The objective of the MIV problem is to identify a minimum-sized subset of influential vertices in the weighted graph. Such that, eventually large number of vertices in the graph can be influenced by these initially selected vertices. As we mentioned in Sect. 4.1, we first partition the whole graph into subgraphs by applying fuzzy propagation modes. The subgraph is called *individual zone* for each vertex. Subsequently, we formally define individual zone as follows:

Definition 1. Individual zone ($Zone_v$). For weighted graph $G(V;E;W(E))$, the individual zone is a fuzzy set (U_v, M_v), where U_v is the set of vertices and M_v is a function $U_v \rightarrow [0;1]$, such that for any vertex $v \in U_v$, we have:

$$M_v(x) = \begin{cases} 1, & \text{if} x = v; \\ \prod_{e_{ij} \in P_{xv}} W(e_{ij}), & \text{otherwise}, \end{cases}$$

where P_{xv} is the path from vertex x to vertex v and e_{ij} is an edge in the path. We further define $W(e_{ii}) = 1$.

Another important terminology *influence* must be formally defined before the problem definition.

Definition 2. Influence (ξ_v). For weighted graph $G(V;E;W(E))$, the influence of vertex v is denoted by ξ_v, which is the sum of the membership value of all vertices in $Zone_v$.

Now, we are ready to define the minimum-sized influential vertices (MIV) problem as follows:

Definition 3. Minimum-sized influential vertices (MIV). For weighted graph $G(V;E;W(E))$, the MIV problem is to find minimum-sized influential vertices $\chi \subseteq V$, such that $\forall v \in \chi, \xi_v > N \times x\%$, where $x\%$ is a predefined threshold.

4.4 MapReduce Algorithm for Identifying Individual Zones

To scale the MIV problem to a large-scale weighted graph, we approximate individual zones by using α-cut fuzzy sets. That is, given a vertex v, the α-cut individual zone of v contains and only contains all vertices whose membership value toward v is greater than or equal to the given parameter α. For simplicity, in the description of the MapReduce algorithms shown in Sects. 4.4 and 4.5, individual zone actually means α-cut individual zone.

A given weighted graph will be represented by using adjacency lists, which are similar representations used in MapReduce-based algorithms for breath-first searching and minimum spanning tree [6-8]. For instance, the adjacency lists for the weighted graph shown in Fig. 4.2 are described as follows:

1. 2(0.7), 7(0.6)
2. 1(0.7), 3(0.8), 6(0.6), 8(0.8)
3. 2(0.8), 4(0.7), 5(0.8), 8(0.6), 6(0.9)
4. 3(0.7), 5(0.5), 9(0.6)
5. 3(0.8), 4(0.7), 8(0.7)
6. 2(0.6), 3(0.9)
7. 1(0.6), 8(0.6)
8. 2(0.8), 3(0.6), 5(0.7), 7(0.6), 9(0.7)
9. 8(0.7), 4(0.6)

As already shown, each line in the adjacency lists represents the one-hop neighbors for each vertex. The values in the parenthesis are the weights over each link.

For a big weighted graph, we divide its adjacency lists into k equal-sized files, where k = *total size/block size* (*total size* is the total size of the adjacency lists of the graph data, and *block size* is the block size of the HDFS of the Hadoop cluster). Assume there are n vertices for each of the k files. Then we select m vertices from each file, respectively, to form a set of target vertices for which we identify individual zones. In other words, the execution of the following MapReduce algorithm is able to identify individual zones for $k*m$ target vertices. Therefore, we just need to run n/m times of this algorithm in order to identify individual zones for all vertices. Fortunately, the n/m times of executing this algorithm are totally independent to each other and, thus, can run in a completely parallel manner.

As an illustration, we assume $k=2$ and $n=4$ for the weighted graph given in Fig. 4.2. Then, let $m=2$, *i.e.*, for each of the 2 files, we select 2 vertices as the target vertices. Assume for the first run of the algorithm, we select vertices 1 and 2 from file 1 and vertices 5 and 6 from file 2. Then we will have the following two input files to identify individual zones for vertices 1, 2, 5, and 6.

File 1:

1. 2(0.7), 7(0.6) | 1, 0, 0, 0 | G, W, W, W| 0, −1, −1, −1
2. 1(0.7), 3(0.8), 6(0.6), 8(0.8) |0, 1, 0, 0 | W, G, W, W| −1, 0, −1, −1

3. 2(0.8), 4(0.7), 5(0.8), 8(0.6), 6(0.9) |0, 0, 0, 0| W, W, W, W| −1, −1, −1, −1
4. 3(0.7), 5(0.5), 9(0.6) |0, 0, 0, 0| W, W, W, W| −1, −1, −1, −1

File 2:

5. 3(0.8), 4(0.7), 8(0.7)|0,0,1,0|W,W,G,W| −1, −1, 0, −1
6. 2(0.6), 3(0.9)|0,0,0,1|W,W,W,G| −1, −1, −1, 0
7. 1(0.6), 8(0.6)| 0,0,0,0|W,W,W,W| −1, −1, −1, −1
8. 7(0.6), 2(0.8), 3(0.6), 5(0.7), 9(0.7)| 0,0,0,0|W,W,W,W| −1, −1, −1, −1
9. 8(0.7), 4(0.6) | 0,0,0,0|W,W,W,W| −1, −1, −1, −1

In the two input files shown above, each vertex is represented in the following format:

`VertexID list of adjacent vertices and weights | membership values target vertices | color decorations towards target vertices| immediate parent towards target vertices`

Take vertex 1 as an example; its membership values to the four target vertices (vertex 1, 2, 5, and 6) are initialized to be 1, 0, 0, 0, respectively. Its color decoration is set to be *gray* (*G*), *white* (*W*), *white* (*W*), *white* (*W*), respectively, where the first *G* means that more vertices belonging to the individual zone of the first target vertex need to be located starting from this vertex; the rest *W* means that, for other target vertices, no immediate action is needed from this vertex; and the other possible color value is *black* (*B*), which means no further development from the current vertex is needed for the corresponding target vertex. We further assume the ordinal among the three color values are $B > G > W$. Its immediate parent vertices toward each of the target vertices are initialized to be 0, −1, −1, and −1, respectively, where the first 0 means that this vertex itself is the first target vertex and the rest −1 mean that its parent vertex to the rest of the target vertices remains unknown for right now.

We further use the following data structure *vertex* to hold the information on each individual vertex:

```
Vertex
ID: int
NeighborsID: List<Integer>
MembershipOfNeighbors: List<Double>
MembershipToTargets: Array<Double>
ColorToTargets: Array<Char>
ParentToTargets: Array<Integer>
```

Then, the MapReduce algorithm can be described in Algorithm 1.

When this MapReduce job is executed on input File 1 and File 2, it will generate the following output, if $\alpha = 0{:}5$:

1. 2(0.7), 7(0.6) | 1, 0.7, 0, 0 | B, G, W, W | 0, 2, −1, −1
2. 1(0.7), 3(0.8), 6(0.6), 8(0.8) | 0.7, 1, 0, 0.6 | G, B, W, G | 1, 0, −1, 6
3. 2(0.8), 4(0.7), 5(0.8), 8(0.6), 6(0.9) | 0, 0.8, 0.8, 0.9 | W, G, G, G | −1, 2, 5, 6
4. 3(0.7), 5(0.5), 9(0.6) | 0, 0, 0.7, 0 | W, W, G, W | −1, −1, 5, −1
5. 3(0.8), 4(0.7), 8(0.7) | 0,0,1,0 | W,W,B,W | −1, −1, 0, −1

6. 2(0.6), 3(0.9) | 0,0.6,0,1 | W,G,W,B | −1, 2, −1, 0
7. 1(0.6), 8(0.6)| 0.6,0,0,0 | G,W,W,W | 1, −1, −1, −1
8. 7(0.6), 2(0.8), 3(0.6), 5(0.7), 9(0.7) | 0,0.8,0.7,0 | W,G,G,W | −1, 2, 5, −1
9. 8(0.7), 4(0.6) | 0,0,0,0 | W,W,W,W | −1, −1, −1, −1

Since the output records contain G color, the job counter numberOfIterations will be greater than 0. So we run above MapReduce job for another iteration by using the output of the first run as input. This process will continue until no record in output contains any G color. Then the output contains information on individual zones for the target vertices 1, 2, 5, and 6. In the same way, we are able to obtain individual zones for vertices 3, 4, 7, 8, and 9.

4.4.1 Algorithm 1: Mapper Part

```
Method Map(vertexID id, vertexRecord: r)
Instantiate Vertex v from id and r
for (int i = 0; i&amp;amp;lt;v.ColorToTargets.size(); i++)
{
 if (v.ColorToTargets[i] = 'G') {
 for(int j=0; j&amp;amp;lt;v.NeighborsID.size(); j++){
 double membershipToTarget =
 v.MembershipToTarges[i] *
 v.MembershipOfNeighbors[j];
 if (membershipToTarget > α){
 Instantiate Vertex vv for Neighbors[j], such
 that
 vv.ID = NeighborsID[j];
 vv.NeighborsID = Null;
 vv.MembershipOfNeighbors = Null;
 for (int k=0; k< v.ColorToTargets.size();
 k++){
 if (i==k)
 vv.MembershipToTargets[k] =
 membershipToTarget;
     vv.ColorToTargets[k] = G;
     vv.parentToTargets[k] = v.VertexID;
 else
 vv.MembershipToTargets[k]=0;
 vv.ColorToTargets[k] = W;
 vv.ColorToTargets[k] = -1;
 }
 Create record rr from vv in following format:
 list of adjacent vertices and weights | membership
```

```
values target vertices color decorations towards
target vertices | immediate parent towards
target vertices;
EMIT (vv.VertextID, rr);
}
}
v.ColorToTargets[i] = 'B'.
}
}
Create record r from v in the following format:
list of adjacent vertices and weights | membership values
target vertices | color decorations towards target verti-
ces| immediate parent towards target vertices;
EMIT(id, r);
```

4.4.2 Algorithm 1: Reducer Part

```
Method Reduce(vertexID id, [r1, r2, r3, ..., rl])
Instantiate Vertex v from id;
v.VertexID = id
v.NeighborsID = Null
v.MembersOfNeighbors=[0,0,0,...]
v.ColorToTargets=[W, W,W,...]
v.ParentToTargets=[-1,-1,-1,...]
for each ri in [r1, r2, r3, ..., rl] {
 Instantiate Vertex vv from id, and ri
 if (vv.NeighborsID != Null){
 v.NeighborsID = vv.NeighborsID;
 v.MembershipOfNeighbors=vv.MembershipOfNeighbors;
 }
  for  (int  j=0;  j&amp;amp;lt;vv.ColorToTargets.size();
j++){
 if (vv.ColorToTargets[j] > v.ColorToTargets[j]){
 v.ColorToTargets[j] = vv.ColorToTargets[j]
 }
 }
 for (int j=0; j&amp;amp;lt;vv.MembershipToTargets.size
(); j++){
 if (vv.MembershipToTargets[j] >
v.MembershipToTargets[j]){
 v.MembershipToTargets[j] =
vv.MembershipToTargets[j]
```

```
 }
 }
}
for (int i=0; i&amp;amp;lt;v.ColorToTargets.size(); i++)
{
 if (v.ColorToTargets[i] = G){
   Increment a predefined job counter called
numberOfIterations;
 }
}
Create record r from v in the following format:
list of adjacent vertices and weights | membership values
target vertices | color decorations towards target vertices
| immediate parent towards target vertices;
EMIT(id, r);
```

4.5 MapReduce Algorithm for Solving MIV

Using the graph shown in Fig. 4.2 as an illustration again, the output of algorithm 1 on this graph can be easily converted to the following format by a MapReduce process. Each record represents the α-cut individual zone for a vertex. Given a vertex, its influence is the sum of all the membership values in its α-cut individual zone. For example, the influence of vertex 1 is $0.7 + 0.6 + 0.56 + 0.56 = 2.42$. Now, the task is to find minimum-sized influential vertices (MIV) whose influence reaches $N^*x\%$:

1. 2(0.7), 7(0.6), 8(0.56), 3(0.56)
2. 1(0.7), 8(0.8), 6(0.6), 3(0.8), 4(0.56), 5(0.64), 9(0.56)
3. 2(0.8), 6(0.9), 8(0.64), 5(0.8), 4(0.7), 1(0.56)
4. 3(0.7), 5(0.5), 9(0.6), 2(0.56), 6(0.63)
5. 8(0.7), 3(0.8), 4(0.56), 2(0.64), 6(0.72)
6. 2(0.6), 3(0.9),8(0.54),5(0.72), 4(0.63)
7. 1(0.6), 8(0.6)
8. 7(0.6), 2(0.8), 3(0.6), 5(0.7), 9(0.7), 1(0.56), 6(0.54)
9. 8(0.7), 4(0.6)

We design a MapReduce-based greedy algorithm for computing this task. Let the minimum-sized set of influential vertices be denoted as S. We also introduce another set denoted as I, which includes all the vertices that are influenced by vertices in S as well as their maximum membership values toward all influential vertices in S. For instance, for the graph shown in Fig. 4.2, if $S = \{2, 8\}$, then $I = \{2(1), 8(1), 1(0.7), 6(0.6), 3(0.8), 4(0.56), 5(0.7), 9(0.7), 7(0.6)\}$, and the influence of S is $1 + 1 + 0.7 + 0.6 + 0.8 + 0.56 + 0.7 + 0.7 + 0.6 = 6.66$.

A MapReduce-based greedy algorithm for identifying minimum-sized influential vertices can be described in Algorithm 2:

4.5.1 Algorithm 2: Mapper Part

```
Method Map (vertexID id, vertexRecord: r)
I_temp = I;
if id is in I_temp then
  | reset its membership value in I_temp to be 1;
else
  | add id(1) to I_temp;
for each vertex v_i in r do
  if v_i is in set I_temp then
    let its membership value recorded in I_temp be mm;
    if m > mm then
      | replace mm with m in I_temp for v_i;
  else
    | add v_i(m) to I_temp;
sumInfluence = sum of all membership values in I_temp;
Emit(0, id | Itemp | sumInfluence);
```

4.5.2 Algorithm 2: Reducer Part

```
Method Reduce(Key k, [I_temp_0 | sumInfluence_0,... ,id_i
| I_temp_i | sumInfluence_i,...])
max = 0, id_max = null, I_max = null;
for each id_i | I_temp_i | sumInfluence_i do
  if sumInfluencei > max then
    max = sumInfluence_i;
    id_max = id_i;
    I_max = I_temp_i;
Add id_max to the set S;
I = I_max;
if max < N × x% then
  | Increment a predefined job counter called
  | numberOfIterations;
```

As shown in Algorithm 2, in the mapper part, the sum of the influences of each vertex is calculated and stored. In the reducer part, if the job counter numberOfIterations is greater than 0, then the algorithm 2 will run again to add the next most influential vertex to the set S, which is the influential set we are trying to construct.

4.6 Conclusion and Future Work

In this chapter, we proposed a fuzzy propagation model to simplify the description of how a vertex influences others in a large-scale weighted social network from big data perspective. A MapReduce-based algorithm was then designed to locate individual zone for each vertex of the network. The concept of individual zone approximates the influence propagated from a vertex by using an α-cut fuzzy set. Then, a MapReduce-based greedy algorithm was designed to identify MIV from all individual zones.

The MapReduce algorithms are well designed in the chapter. However, how to validate the proposed algorithms needs to be investigated in the future work. In the future, we plan to conduct both simulations and experiments (crawl real social data online) to validate and evaluate the performance of the proposed framework. Another research direction is to extend the influence diffusion model. In the chapter, we assume linear threshold model which is adding neighbors' social influences together to compare with the preset threshold. More practical diffusion model can be designed and proposed to solve the MIV problem, such as independent cascade model [21], susceptible/infected/susceptible (SIS) model [22], or voter model [23]. One interesting direction could be model social influences as positive and negative influences [23], since there are difference campaigns and opposite ideas competing for their influence in the social network.

References

1. Nail J (2004) The consumer advertising. Forrester Research and Intelliseek Market Research Report
2. Domingos P, Richardson M (2001) Mining the network value of customers. ACM SIGKDD, pp 57–66
3. Richardson M, Domingos P (2002) Mining knowledge-sharing sites for viral marketing. ACM SIGKDD, pp 61–70
4. Kempe D, Kleinberg J, Tardos E (2003) Maximizing the spread of influence through a social network. ACM SIGKDD, pp 137–146
5. Kwak H, Lee C, Park H, Moon S (2010) What is Twitter, a social network or a news media? WWW, pp 591–600
6. Goodrich MT (2010) Simulating parallel algorithms in the MapReduce framework with applications to parallel computational geometry, CORR
7. Lattanzi S, Moseley B, Suri S, Vassilvitskii S (2011) Filtering: a method for solving graph problems in MapReduce. SPAA, pp 85–94
8. Qin L, Yu J, Chang L, Cheng H, Zhang C, Lin X (2014) Scalable big graph processing in MapReduce. ACM SIGMOD, pp 827–838

9. White T (2012) Hadoop: the definitive guide, 3rd edn. O'Reilly, Beijing
10. Leskovec J, Krause A, Guestrin C, Faloutsos C, VanBriesen J, Glance N (2007) Cost-effective outbreak detection in networks. ACM SIGKDD, pp 420–429
11. Goyal A, Lu W, Lakshmanan LV (2011) Celf++: optimizing the greedy algorithm for influence maximization in social networks. WWW, pp 47–58
12. Kimura M, Saito K, Nakano R (2007), Extracting influential nodes for information diffusion on a social network. NCAI, vol 22, 1371–1380
13. Chen W, Wang Y, Yang S (2009) Efficient influence maximization in social networks. ACM SIGKDD
14. Chen W, Wang C, Wang Y (2010), Scalable influence maximization for prevalent viral marketing in large scale social networks. ACM SIGKDD
15. Liu X, Li S, Liao X, Wang L, Wu Q (2012) In-time estimation for influence maximization in large-scale social networks. ACM proceedings of EuroSys workshop social network systems, pp 1–6
16. Hu J, Meng K, Chen X, Lin C, Huang J (2013) Analysis of influence maximization in large-scale social networks. ACM sigmetrics Big Data analytics workshop
17. Kim J, Kim SK, Yu H (2013) Scalable and parallelizable processing of influence maximization for large-scale social networks. ICDCS
18. He J, Ji S, Beyah R, Cai Z (2014) Minimum-sized influential node set selection for social networks under the independent cascade model. ACM MOBIHOC 2014, Philadelphia, PA, USA, August 11–14 2014
19. Shi Q, Wang H, Li D, Shi X, Ye C, Gao H (2015) Maximal influence spread for social network based on MapReduce. Commun Comput Inf Sci 503:128–136
20. Liu X, Li M, Li S, Penng S, Liao X, Lu X (2014) IMGPU: GPU accelerated influence maximization in large-scale social networks. Trans Parallel Distrib Syst 25(1):136–145
21. He S Ji, Beyah R, Cai Z (2014) Minimum-sized influential node set selection for social networks under the independent cascade model. ACM MOBIHOC, August
22. Saito K, Kimura M, Motoda H (2009) Discovering influential nodes for SIS models in social networks. Discov Science 5808:302–316
23. Li Y, Chen W, Wang Y, Zhang ZL (2013) Influence diffusion dynamics and influence maximization in social networks with friend and foe relationships. WSDM

Part II
Big Data Modelling and Frameworks

Chapter 5
A Unified Approach to Data Modeling and Management in Big Data Era

Catalin Negru, Florin Pop, Mariana Mocanu, and Valentin Cristea

Abstract The emergence of big data paradigm and developments in other areas, such as cyber-infrastructures, smart cities, e-health, social media, Web 3.0, etc., has led to the production of huge volumes of data. Moreover, these data are often unstructured or semi-structured, with a high level of heterogeneity. Nowadays, information represents an essential factor in the process for supporting decision-making, and that is the reason that heterogeneous data must be integrated and analyzed to provide a unique view of information for many types of application. This chapter addresses the problem of modeling and integration of heterogeneous data that comes from multiple heterogeneous sources in the context of cyber-infrastructure systems and big data platforms. Furthermore, this chapter analyzes different heterogeneous data models in relation to heterogeneous sources such as the following: sensors, mobile users, web, and public open data sources (e.g., regulatory institutions). A CyberWater case study is also presented for the purposes of modeling, integration, and operation of these data in order to provide a unified approach and a unique view. The case study aims to offer support for different processes inside the CyberWater platform such as monitoring, analysis, and control of natural water resources, with the scope to preserve the water quality.

Keywords Big data • Data modeling • Data management • Cloud computing • Heterogeneous distributed systems • Cyber-infrastructure • Natural resources

5.1 Introduction

Big data paradigm represents an interesting and challenging research topic. Many scientific fields, such as cyber-infrastructures, smart cities, e-health, social media, Web 3.0 etc., try to extract valuable information from huge amounts of data, generated on a daily basis. Moreover these data are unstructured or semi-structured, having a high level of heterogeneity [1].

C. Negru • F. Pop (✉) • M. Mocanu • V. Cristea
Computer Science Department, Faculty of Automatic Control and Computers, University Politehnica of Bucharest, Bucharest, Romania
e-mail: florin.pop@cs.pub.ro

Z. Mahmood (ed.), *Data Science and Big Data Computing*,
DOI 10.1007/978-3-319-31861-5_5

In the case of smart cities, big data relates to urban data, basically referring to space and time perspective, which is gathered mostly from different sensors. Furthermore, the growth of big data changes the planning strategies from long-term thinking to short-term thinking as the management of the city can be made more efficient [2]. Moreover, this data can open the possibility for real-time analysis of city life and new modes of city administration and offer also the possibility for more efficient, sustainable, competitive, productive, open, and transparent city [3].

Healthcare systems are also transformed by big data paradigm as data is generated from different sources such as electronic medical records systems, mobilized health records, personal health records, mobile healthcare monitors, genetic sequencing, and predictive analytics as well as a large array of biomedical sensors and smart devices that rise up to 1000 petabyte [4].

People and the interaction between them produce, regarding the social media and Web 3.0, massive quantities of information. A huge effort is made to understand social media interactions, online communities, human communication, and culture [5].

The motivation for this chapter comes from the necessity of a unified approach of data processing in large-scale cyber-infrastructure systems as the characteristics of nontrivial scale cyber-physical systems (e.g., data sources, communication, and computing) exhibit significant heterogeneity. The main contribution of the chapter is the analysis of different heterogeneous sources and format of data collected from sensors, users, and the web from public open data sources in the context of the big data and cyber-infrastructure systems. First, we address the problem of data modeling for the water resource monitoring and management processes. Next, this analysis is extended considering the big data characteristics of the collected data. Furthermore, describe a unified approach for the water resource data models. Finally, we illustrate the applicability of the presented analysis on a real research project. The chapter is structured as follows.

The first part of the chapter analyzes existing data models for water resource monitoring and management considering computational methods (e.g., neural networks) for handling environmental data, using software engineering to deliver the computational solutions to the end user and developing methods for continuous environmental monitoring based on sensor networks as part of Internet of things.

In the second part, based on this analysis, we address the problem of data modeling considering a huge volume, high velocity, and variety of data collected from environments (another big data problem) and process in real time to be used for water quality support.

The third part presents the unified approach for water resource data models considering the context of observation and the possibility of usage to enhance the organization, publication, and analysis of world-distributed point observation data while retaining a simple relational format. This will enable development of new applications and services for water management that are increasingly aware of and adapt to their changing contexts in any dynamic environment. A context-aware approach requires an appropriate model to aggregate, semantically organize, and access large amounts of data, in various formats, collected from sensors or users, from public open data sources.

Finally, we present the case study of CyberWater project, a prototype cyber-infrastructure-based system for decision-making support in water resource management, as data modeling of water resources is a fundamental requirement. Cyber-infrastructure basically refers to a mix of advanced data acquisition through real-time sensor networks, big data, visualization tools, high-performance computational platforms, analytics, data integration, and web services.

5.2 Big Data: Heterogeneous Data

Big data refers to dealing with huge amounts of heterogeneous data. This requires a pipeline of processing operations in order to accomplish efficient analytics. The overall scope is to offer support in the decision-making process. An important challenge, besides the large volumes and high-speed production rates (e.g., velocity), is raised by the enormous heterogeneity (e.g., variety) of such data.

Research fields, such as environmental research, disaster management, decision support systems, information management in relief operations, crowdsourcing, citizen sensing, and sensor web technologies, need to make use of new and innovative tools and methods for big data. Ongoing research, challenges, and possible solutions have been presented through several important projects and in research papers that face with the previous presented challenges and try to overcome them.

A new and innovative approach is presented in Dancer FP7 project being oriented on developing new instruments and tools that will enhance environmental research and promote innovation in Danube Region, including the Danube Delta and the Black Sea. The project will undertake a critical analysis of what has been achieved so far in the region and will build upon results of achievements to date, to design innovative solutions and to strengthen knowledge transfer in this area.

In [6], the authors address the use of Geo-ICT in two stages of handling disasters, before and during occurrence, giving special attention to the real utilization of geo-information, e.g., risk maps, topographic maps, etc. They conclude that in both risk management and disaster management is a growing awareness of the importance of spatial information and the increasing types of spatial data are used for performing tasks within risk and disaster management; secondly a general understanding is building up about sharing information between the two domains.

A solution, called Sahana, is presented in [7] of a free and open-source software (FOSS) application. It aims to be a comprehensive solution for information management in relief operations, recovery, and rehabilitation. After presenting the architecture of the system, they conclude that there is much left to be done in the technical development of FOSS disaster management software, and of primary importance is the need to deal with a heterogeneity of data types (text, semi-structured, Web HTML and XML, GIS, tabular, and DBMS) and to develop standards and protocols for data sharing. The purpose of these solutions is to assist saving procedures for environment (by improving the quality of air, water, sediments, and soil) and people in the same time during a disaster. In periods without disasters, prevention and improvement actions, like people training, will be considered.

The authors of [8] survey the crowdsourcing, citizen sensing, and sensor web technologies for public and environmental health surveillance and crisis management and reinforce the roles of these technologies in environmental and public health surveillance and crisis/disaster. The data used by these services, collected from different sources (mobile smart devices, cameras, dedicated equipment, sensors, etc.), must be stored in an aggregated way in the repository.

The authors of [9] propose a cloud-based natural disaster management system. They conclude that it is essential to have a scalable environment with flexible information access, easy communication, and real-time collaboration from all types of computing devices, including mobile handheld devices (e.g., smartphones, PDAs, and tablets (iPads)). Also, it is mandatory that the system must be accessible, scalable, and transparent from location, migration, and resource perspectives.

An overview of all the extreme events that threaten people and what they value in the twenty-first century is presented by Keith Smith in the book named *Environmental Hazards: Assessing Risk and Reducing Disaster* [10]. It integrates cutting-edge material from the physical and social sciences to illustrate how natural and human systems interact to place communities of all sizes, and at all stages of economic development, at risk. It also explains in detail the various measures available to reduce the ongoing losses to life and property.

5.2.1 Characteristics, Promise, and Benefits

Big data is a business term that can also represent a buzzword for the advancing trends in the latest technologies. It describes a new approach in order to understand the process of data analysis for decision-making. Currently, data is growing at a rate of 50 % per year through new streams such as digital sensors, industrial equipment, automobiles, and electric meters (measuring and communicating all sorts of parameters like location, temperature, movement) [11].

In this section, some of the most important big data characteristics such as volume, variety, velocity, value and veracity, volatile, and vicissitude are analyzed.

Volume. represents the main challenge, as traditional relational database management systems did not succeed to handle volumes of data in terms of terabyte and petabyte levels. According to [12], it is estimated that 2.5 quintillion bytes of data are created each day, six billion people have cell phones, and 40 zettabytes will be created by 2020, which represents an increase of 300 times from 2005.

Variety. characteristic refers to different data formats and sources such as data from sensors, documents, emails, social media texts, mobile devices, etc.

According to [12], 400 million tweets are sent per day by about 200 million monthly active users, four billion hours of video are watched on YouTube each month, and more than 420 million wearable wireless health monitors exist.

Velocity. refers to the data acquisition rate, as data can be acquired at different speeds. For instance, in the case of a natural disaster or pollution accident, the data acquisition rate from the wireless sensor network rises exponentially, compared with normal circumstances. According to [12], the New York Stock Exchange captures 1 TB of trade information during each session, modern cars have close to 100 sensors that monitor items such as fuel level and tire pressure, and by 2016 it is projected that there will be 18.9 billion network connections, which represent almost 2.5 connections per person on Earth.

Value property illustrates the potential gain of data, obtained after some processing operations. Big data processing operations can help to uncover the fine interactions in data, allowing to manipulate heretofore hidden – often counter-intuitive – levels that directly impact different domains and activities.

Veracity. characteristic describes how accurate is data collected from different sources. For example, data gathered from a social media website have a specific degree of accuracy. In other words, veracity represents the degree of uncertainty in the data. According to [12], the poor data quality costs the US economy around $1.3 trillion per year, one of three business managers don't trust the information they use to make decisions, and 27 % of respondents in one survey were unsure about how much of their data was inaccurate.

Volatility. propriety of big data refers to the time of storage of the data after the processing step. Volatility has a direct impact on the other aspects of big data such as volume and veracity. So, data volatility policies have to be defined, in order to avoid the interference with other properties of even the damage of those.

Vicissitude. property refers to the challenge of scaling big data complex workflows. This property signifies a combination between the large volume of data and the complexity of the processing workflow, which prevent to gather useful insights for data [13].

There are three categories of promises related to big data analytics identified by the authors of [14]:

- Cost reduction – technologies like Hadoop and cloud-based analytics can provide significant cost advantages.
- Faster and better decision-making – this is obtained with the aid of frameworks such as Apache Storm and Apache Spark designed to run programs up to 100× faster than Hadoop. Also the possibility to analyze new sources of data can help in decision-making process. For example, healthcare companies try to use natural language processing tools to better understand customer's satisfaction.
- New products and services – this is possible due to new sources of data such as mobile phones. For example, a company called Verizon sells data about how often mobile phone users are in certain locations.

5.2.2 Data Models

Data models represent the building blocks of big data applications, having a major impact on the performance and capabilities of those applications. Moreover, the different tools that are used for data processing impose these data models. Next, we will discuss the different data models that exist in the context of big data: structured data, text file data, semi-structured data, key-value pair data, and XML data [15].

Structured data models refer to data that is contained in database or spreadsheet files. The sources of structured data in a cyber-infrastructure are water quality sensors (e.g., water parameters such as conductivity, salinity, total dissolved solids, resistivity, density, dissolved oxygen, pH, temperature, and so on), mobile phone sensors (e.g., location data), geographical information systems (e.g., geo-database that contains spatial data such as points, polygons, raster, annotations, and so on), and click-stream sources (e.g., data generated by human intervention in the case of reporting an incident related to a pollution event). Figure 5.1 presents a database table that contains information about a spatial dataset which has different attributes.

Data can be aggregated and queried across the entire database. Things get more complicated when we want to aggregate data from many tables because the problem becomes exponentially more complex. The reason behind this complexity is that each query requires the read of the entire dataset.

Text data models, on the other hand, are at the opposite end of structured data, as this type of data has no well-defined structure and meaning. The sources for this data are represented by different documents related to different regulations released by regulatory institutions in the field of cyber-infrastructure domains such as water management [15].

Semi-structured data represents data that has a structure but is not relational. Instrumentation equipment such as sensors generates this data. In order to be stored in a relational database, data must be transformed. A major advantage of semi-structured data is represented by the fact that can be loaded directly into a Hadoop HDFS file system and processed in raw from there [16].

Key-value pair data represents the driver for performance in MapReduce programming model. This model has a single key-value index for all data being similar to *memcached* distributed in memory cache. This type of data is stored in key-value stores and in general provides a persistence mechanism and additional functionality as well: replication, versioning, locking, transactions, sorting, and/or other features [17].

OBJECT_ID	Zone	SHAPE Length (m)
1	RightBank	54.021,952344830665
2	MainStream	37.099,640979200485
3	LeftBank	40.974,860305322997
5	MainStream	21.167,611441057656
6	LeftBank	23.707,147651820300

Fig. 5.1 Geographical data example

5.2.3 Data Gathering

Environmental data gathering and analysis, resulting from continuous monitoring, pose unprecedented challenges to the design and analysis of methods, workflows, and interaction with datasets. A big part of these data that resulted from monitoring is made available by public institutions, private companies, and scientists [18]. Furthermore, data collected from social media sites such as geo-tagged photos, video clips, and social interactions may contain information related to environmental conditions. Moreover this process of data gathering must be energy efficient and cost aware in the context of cloud computing services. New methods for gathering and integration for these types of data for further processing stages must be designed.

Geospatial sensor web [19] has been used widely for environmental monitoring. Different from a sensor network, in a sensor web infrastructure, the device layers, the network communication details, and the heterogeneous sensor hardware are hidden [20]. The Sensor Web Enablement (SWE) of the Open Geospatial Consortium (OGC) defines a sensor web as an infrastructure that enables access to sensor networks and archived sensor data that can be discovered and accessed using standard protocols and interfaces [21]. In [22] the authors propose a sensor web heterogeneous node meta-model discussing development of five basic metadata components, the design of a nine-tuple node information description structure as shown in Fig. 5.2.

In [23], the authors describe the development of a service-oriented multipurpose SOS framework in order to access data in a single method approach, integrating the sensor observation service with other OGC services. The proposed solution includes few components such as extensible sensor data adapter, OGC-compliant

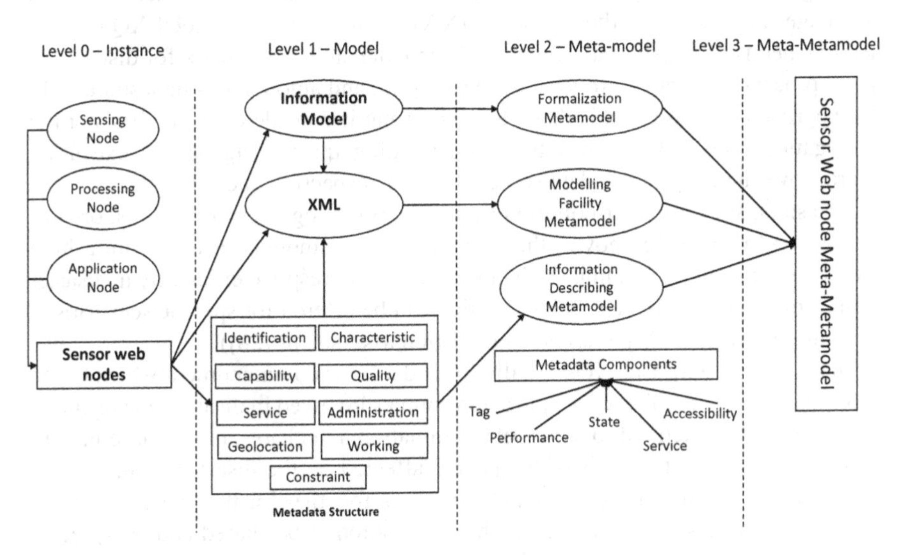

Fig. 5.2 Sensor web meta-model

geospatial SOS, geospatial catalogue service, a WFST, and a WCS-T for the SOS, and a geospatial sensor client. Data from live sensors, models, and simulation is stored and managed with the aid of extensible sensor data adapter.

There is an important need for data gathering and content analysis techniques from social media streams, such as Twitter and Facebook. These have become essential real-time information resources with a wide range of users and applications [24]. Gathering valuable data from social media represents a good opportunity for environmental monitoring. Also social media can play an important role in pollution accidents or natural disasters as an information propagator.

5.2.4 Open Issues and Challenges

The challenges addressed in the chapter were also presented in the literature in various forms: methodologies, applications, etc. The authors of [25] review state-of-the-art research issues in the field of analytics over big data. Among the open problems and research trends in big data analytics identified by the authors are the following:

- Data source heterogeneity and incongruence
- Filtering-out uncorrelated data
- Strongly unstructured nature of data sources
- High scalability
- Combining the benefits of RDBMS and NoSQL database systems
- Query optimization

Integration of distributed heterogeneous web data sources is another important challenge. In [26], the authors propose DeXIN (Distributed extended XQuery for heterogeneous data INtegration), which is an extensible framework for distributed query processing over heterogeneous, distributed, and autonomous data sources. In this approach the base idea is that one data format is considered as the basis or the aggregation model. It extends the corresponding query language to executing queries over heterogeneous data sources in their respective query languages.

Disasters, such as water pollution accidents, bring together institutions, organizations, and people. Moreover, these produce large amounts of data with a high level of heterogeneity that must be handled in order to respond efficiently in disaster management operations (a special attention can be offered for specific scenarios in agriculture disasters [27] and land degradation assessment [28]).

In the case of a disaster, it is usually formed a crisis commandment which has to coordinate a large number of people and resources involved in disaster management operations. This often led to a lack of coordination and chaos among these institutions, as they track the needs of the people affected by the disaster. Many autonomous actors control information and may work together for the first time, so the development of a system that enable the information to be shared and analyzed to target resources is fundamental to build a better response capacity [29].

With the aid of big data solutions, we can acquire better results in response time, coordination, and distribution of efforts, tracking displaced and vulnerable populations; logging the damage to housing, infrastructure, and services; dealing with the sudden influx of humanitarian supplies; and coordinating the work of dozens and even hundreds of responding agencies [29]. However, there is no current solution that provides information system or data management support for the basic functionality of disaster management such as registering organizations, locating missing persons, and requesting assistance [30].

So, a major challenge is to provide an integrated and interoperable framework for environmental monitoring and data processing that include development of sound solutions for a sustainable crisis management in order to support human life in the case of disasters. For instance, the authors of [31] propose an approach to integrate heterogeneous data with uncertainty in emergency systems.

Based on ICT (information and communications technology) support and big data technologies, the response to natural disasters needs to be both rapid and carefully coordinated. Saving lives and providing shelter for those displaced are two key priorities for first responders. Helping citizens post disaster is a key part of the mandate for damage assessors. This is by measuring and quantifying a disaster's impact on the community and then providing assistance to individuals. Mobile and cloud-based GIS offers many potential benefits to improve disaster management. This is the position of WebMapSolutions Company, which is implementing state-of-the-art disaster management GIS solutions.

Another important challenge is to combine mobile applications, existing electronic services, and data repositories in an architecture based on cloud solutions and existing big data approaches. Also we have to contribute to the development of INSPIRE compliant solutions [32], based on the ISO 19156 standard on Observations and Measurements (O&M) and SOS (Sensor Observation Service), SES (Sensor Event Service), and SAS (Sensor Alert Service) as a main contribution to standardization.

Incorporating and extending the technology presented in the literature to create a new one that will transform data into valuable information is another big challenge. Even when faced with the big data problem, solutions based on cloud computing and HPC infrastructure to process data must be used.

The main challenges when dealing with systems for water quality management are the following: storage, processing, and management of data. For the storage part, scale-out architectures have been developed to store large amount of data, and also purpose-built appliance has improved the processing capability.

The major challenge is represented by the management of big data throughout its entire life cycle, from acquisition to visualization in order to get valuable information that helps in decision-making processes. Further challenges are in the field of real-time control and action prediction for people and authority's assistance and setting up correlations between related phenomena, based on the available information.

5.3 Unified Approach to Big Data Modeling

In this part of the chapter, we present a unified approach for water resource data models considering the context of observation and the possibility of usage to enhance the organization, publication, and analysis of world-distributed point observation data while retaining a simple relational format. This will enable development of new applications and services for water management that are increasingly aware of and adapt to their changing contexts in any dynamic environment.

According to the National Oceanic and Atmospheric Administration (NOAA) that collects, manages, and disseminates a wide range of climate, weather, ecosystem, and other environmental data, there are nine principles for effective environmental data management:

- Environmental data should be archived and made accessible.
- Data-generating activities should include adequate resources to support end-to-end data management.
- Environmental data management activities should recognize user's needs.
- Effective interagency and international partnerships are essential.
- Metadata are essential for data management.
- Data and metadata require expert stewardship.
- A formal, ongoing process, with broad community input, is needed to decide what data to archive and what data not to archive.
- An effective data archive should provide for discovery, access, and integration.
- Effective data management requires a formal, ongoing planning process.

A matrix represents the basic structure of an environmental dataset, where usually the rows correspond to the individual objects (measurements, time units, or measuring spots) and the columns contain the series of reading for the corresponding variable. The units in the columns may be logical characters (true = 1, false = 0), ordered or unordered categories, integers (count data), or reals (measurements); they may also contain a time information or a coding of the measurement spots. The coding of missing data and of censored data (for extreme values) has to be fixed. Of course some describing or classifying text may be contained as well in the rows.

In a geographical information system (GIS), there are four basic data structures, viz., vectors, rasters, triangulated irregular networks, and tabular information (table of attributes). For example, a virtual representation of the Earth mostly contains data values that are observed within the physical Earth system. On the other side, data models are required to allow the integration of data across the silos of various Earth and environmental science domains. Creating a mapping between the well-defined terminologies of these silos is a challenging problem. A generalized ontology for use within Web 3.0 services, which builds on European Commission spatial data infrastructure models, that acknowledge that there are many complexities to the description of environmental properties which can be observed within the physical Earth system is presented in [33]. The ontology is shown to be flexible and robust enough to describe concepts drawn from a range of Earth science disciplines, including ecology, geochemistry, hydrology, and oceanography.

5.3.1 Unified Data Representation and Aggregation

When it comes down to big data representation and aggregation, the most important question that has to be answered is how to represent and aggregate relational and non-relational data in the same storage engine. Moreover, this data must be queried in an efficient way so that it offers relevant results across all data types.

An important aspect in wireless sensor networks is end-to-end data aggregation without degrading sensing accuracy that can prevent network congestion to occur [34]. With the aid of sensor web technologies, environmental monitoring observations are published. The integration of heterogeneous observations in different applications poses important challenges as these differ in spatial-temporal coverage and resolution. In [35], the authors present an approach for spatial-temporal aggregation in the sensor web using the geo-processing web by defining a tailored observation model for different aggregation levels, a process model for aggregation processes, and a spatial-temporal aggregation service.

Also, context awareness represents a core function for the development of modern ubiquitous systems. This offers the capacity to gather and deliver to the next level any relevant information that can characterize service-provisioning environment, such as computing resources/capabilities, physical device location, user preferences, time constraints, and so on [36].

There are numerous applications that need to query multiple databases, with heterogeneous schemas. The semantic integration approaches focus on heterogeneous schemas, with homogenous data sources. In [37], the authors propose a new query type, called decomposition aggregate query, to integrate heterogeneous data source domains. This approach is based on three-role structure. Decomposing compounds into components and translating non-aggregate queries over compounds into aggregate queries answerable by other data sources are achieved by a type of data sources called *dnodes*. It is in order to mention that this is a solution mainly designed for database management systems.

5.3.2 Data Access and Real-Time Processing

A monitoring platform for water data management needs to access distributed data sources (e.g., sensor networks, mobile systems, data repository, social web, and so on). Next, this data has to be processed in real time in order to prevent natural disasters such as water pollution and more importantly to alert the possible affected people.

Apache Kafka is a solution that proposes a unified approach to offline and online processing by providing a mechanism for parallel load in Hadoop systems as well as the ability to partition real-time consumption over a cluster of machines. Also it provides a real-time publish-subscribe solution, which overcomes the challenges of real-time data usage for consumption, for data volumes that may grow in order of magnitude, larger than the real data [38]. Figure 5.3 presents big data aggregation-and-analysis scenario supported by the Apache Kafka messaging system.

There are two main categories of data: one which has a value at the given moment in time such as prediction data and the other whose value remains forever such as the maximum value possible for pollutants or sensor data which represents historical data. Mining the instantaneously valued data requires a real-time platform. In [39], the authors propose a method of dynamic pattern identification for logically clustering log data. The method is a real-time and generalized solution to the process of log file management and analysis.

Computing frameworks such as MapReduce [40] or Dryad [41] are used for large-scale data processing. In this paradigm users write parallel computations with the aid of high-level operators, without paying attention to data distribution or fault tolerance. The main drawback of these systems is represented by the fact that these are batch-processing systems and are not designed for real-time processing.

Storm [42] and Spark [43] represent a possible solution for real-time data streaming processing. Storm is used currently at Twitter for real-time distributed processing of stream data. The most important proprieties of Storm are scalability, resiliency, extensibility, efficiency, and easy administration [44].

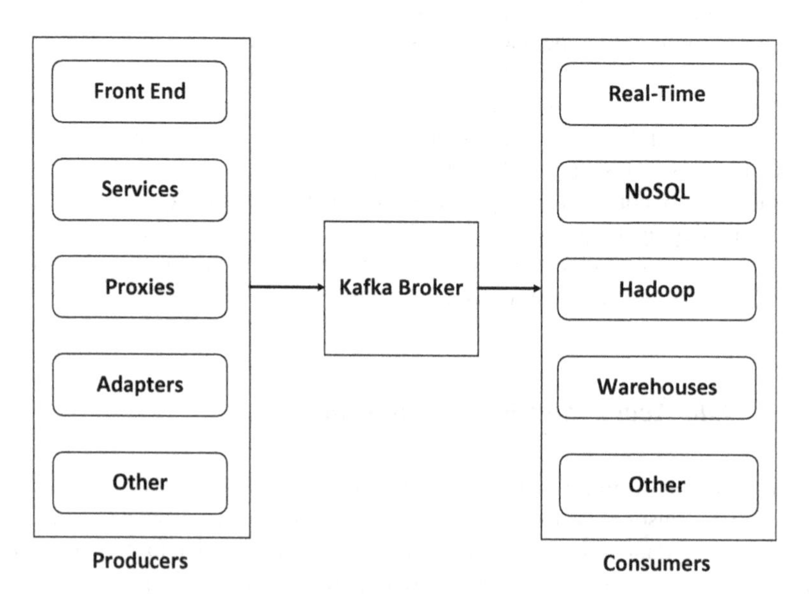

Fig. 5.3 Big data aggregation and analysis in Kafka

5.4 Uniform Data Management

In relation to the uniform data management, a context-aware approach requires an appropriate model to aggregate, semantically organize, and access large amounts of data, in various formats, collected from sensors or users, from public open data sources. When it comes down to uniform data management, the most important question that has to be answered is how to handle relational and non-relational data in the same storage engine. Moreover, this data must be queried in an efficient way so that it offers relevant results across all data types.

Data handling methods have been applied in several areas including water network data analysis and modeling, water quality analysis, energy-water relationship, efficiency modeling of regional water resource under disaster constrains, etc. Relational databases can handle various types of data, for example, sensor data or GIS data. Every query has the same kind of data – location, water parameters, map, and so on. These are all stored in a table with one column for each piece of data.

On the other hand, we have multimedia files such as images or videos that can be attached to an event reporting action. Multimedia files cannot be represented in the same way as a series of columns. What can be stored in a relational way is represented by the data about the files, which is in fact metadata. Alongside with multimedia files, we have social media objects such as blogs, tweets, and emails, which can be categorized in the category of non-relational data.

One approach is to modify existing relational database management systems in a way that can store non-relational data. For example, multimedia files can be stored in Blob data type. In this way, data can be retrieved. The worst part is represented by the fact that data stored in this way cannot be processed, for example, the image cannot be scanned in order to find useful information.

Another approach is to design new engines for database systems that can handle all these big data challenges. So, for example, the new engine can have functions for parsing multimedia files in order to find information about a pollution event.

The authors in [45] propose a hybrid system called HadoopDB, which combines the parallel database systems with MapReduce-based systems, in order to benefit from the performance and efficiency of the first type of systems and scalability, fault tolerance, and flexibility of the second ones. The main idea of this solution is to connect multiple single-node database systems using Hadoop as the task coordinator and to communicate through network layer. In this way queries can be parallelized with MapReduce. The drawback of this approach is that HadoopDB does not match the performance of parallel database systems.

Another approach is presented in [46]. This proposes a novel definition of a declarative language that has to be able to map, with precision, an ontology into queries for a set of data sources. This method is mainly designed for the case of integration of multiple heterogeneous relational database systems. The authors introduce the concepts of semantic identifier and semantic join. The first represents a solution for the problem of entity resolution, and the second one is designed to help in the problem of record linkage. Although, it is an interesting approach, this has to be modified in order to be used for multiple heterogeneous data sources in the context of big data. Still, we cannot know for sure how accurate the mapping phase can be when dealing with these types of data sources.

Tasso Argyros and Mayank Bawa from Teradata propose also an interesting approach. This supposes the design of a management system composed of three elements, storage engine, processing layer, and function library [47]. In the storage engine, relational data is stored in database tables and non-relational data is stored as de-serialized objects that are similar to Blobs. An extended SQL engine that includes MapReduce functions represents the processing layer. In this way relational data is queried with SQL and non-relational data is queried with MapReduce functions. The layer of function library is the core element. In this layer functions written by users permit the manipulation and query non-relational data. Next these functions are stored in a library and also the results of the functions are stored in database tables. This is due to the closure principle from relational database model. Basically this principle states that any query against a table or tables of data must return the answer in the form of a table, permitting in this way the chaining of queries.

So, in order to overcome the data heterogeneity in big data platforms and to provide a unified and unique view of heterogeneous data, a layer on top of the different data management systems, with aggregation and integration functions, must be created.

5.5 CyberWater Case Study

Water resource monitoring implies a huge amount of information with different levels of heterogeneity (e.g., spatial data, sensor data, multimedia data), availability (e.g., data must have a minimum degree of redundancy), and accessibility (e.g., methods for data access such as REST, SOAP, WSDL) [48]. It is very important to acquire, store, transmit, and analyze data in order to respond in real time to possible threads or pollution accidents, for instance.

In this section, we present the case study of CyberWater project. This is a prototype cyber-infrastructure-based system for decision-making support in water resource management. The main reason we present this case study is to highlight the need for heterogeneous data modeling in big data era, especially the context of cyber-infrastructure-based systems.

Water resource data modeling is a fundamental process for cyber-infrastructures. These are characterized by a mix of technologies such as advanced data acquisition through real-time sensor networks, big data, visualization tools, high-performance computational platforms, analytics, data integration, and web services.

Prototype Cyber Infrastructure-based System for Decision-Making Support in Water Resources Management (CyberWater) project [49], a national project, aims to create a prototype platform using advanced computational and communication technology for implementation of new frameworks for managing water and land resources in a sustainable and integrative manner. The main focus of this effort will be on acquiring diverse data from various disciplines in a common digital platform that is subsequently used for routine decision-making in normal conditions and for providing assistance in critical situations related to water, such as accidental pollution flooding. So, one of the main challenges of this project will be to integrate the heterogeneous data sources into a unique and unified view.

In Fig. 5.4, we present a layered architecture of CyberWater monitoring platform. This is focused on three levels: data level, storage level, and level of access, management, and the data processing. Between data and storage levels, there exists the interface for monitoring, analysis, and processing rules. Further, between storage and management level is placed the data access interface.

The *data level* consists of various heterogeneous data sources used by the platform, such as sensor network, data suppliers (e.g., GIS, water treatments plants, etc.), and third-party services (e.g., ANAR, ApaNova, and other Romanian institutions).

At the *storage level*, data collected from heterogeneous data sources is stored, in order to offer aggregation services, such as workflow execution, pollution propagation modeling, pollution prediction methods, and platform configuration. Also, the aggregation service is connected with a knowledge management service and model parameter estimator service, in order to offer support for the offered services.

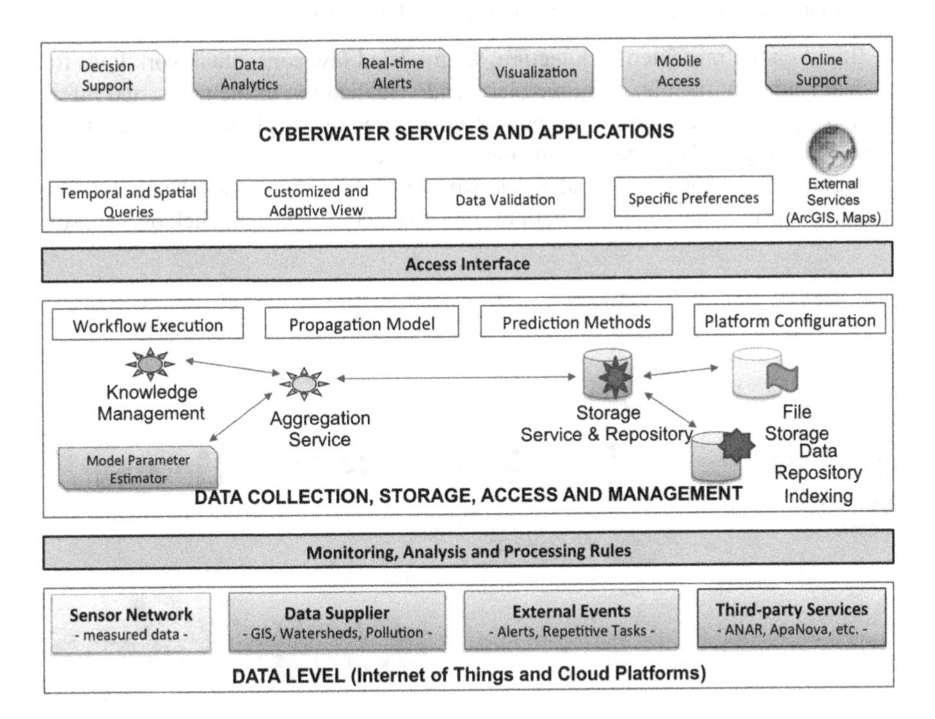

Fig. 5.4 CyberWater-layered architecture

The top layer of the architecture is the *level of access, management, and the data processing*. Here, we have services such as spatial and temporal query services, customized view of data, data validation, or specific preferences providing functionality for applications like decision support systems, data analysis, real-time alerts, video, mobile access, and online support.

The interface for monitoring, analysis, and processing rules has the role to store data from data level (e.g., sensor network, data supplier, external events, and third-party services) in a flexible and efficient manner.

Data access interface ensures the access to data for CyberWater services and applications and also for third-party services which need access to data, offering access methods such as REST, SOAP, and WSDL.

A sensor gathers the following parameters from water resources: temperature, pH, specific conductivity, turbidity, and dissolved oxygen.

Data obtained from third-party suppliers is received in heterogeneous formats, in a form of files such as pdf, csv, txt, etc. For example, data obtained from ApaNova is in a form of a pdf file, called analysis bulletin, and contains information about taste, color, odor, turbidity, pH, conductivity, free residual chlorine, ammonium, nitrites, nitrates, iron, total hardness, and aluminum. From this format, we can identify three main components:

- Limitations specification – in a fixed format
- Measured values
- Semantics – which refers to explanation of the measures

Based on the presented architecture, we modeled few conceptual workflows for decision support module, real-time alerts and visualization of the maps, and information's presented on them. The applications offered by the platform need a unique view and integrated approach of the data.

Figure 5.5 presents the visualization workflow. A user accesses the application and can do one of the following actions: can do a zoom in action, click on a sensor,

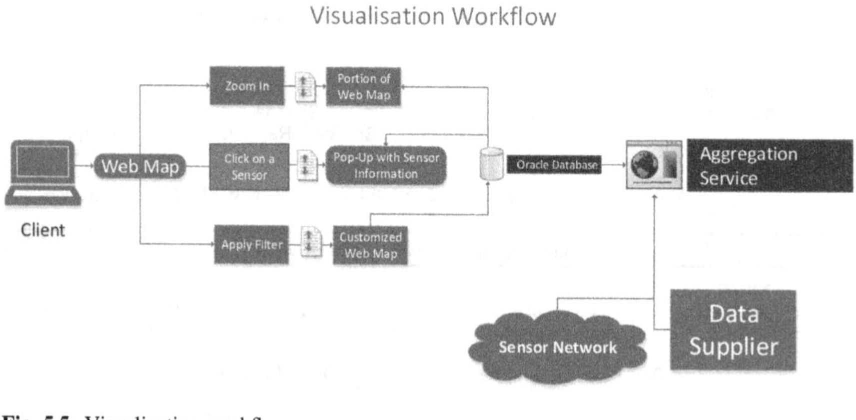

Fig. 5.5 Visualization workflow

or apply a filter. In the case of the first action, a detailed portion of a map is visualized and data are brought from storage system by the aggregation service. In this module can be identified the following components: analytics component, data filtering component, and access to distributed databases. Analytics component refers to all the activities in the workflow, such as zoom in, click on a sensor, and apply a filter, and is important because it is the most accessed and utilized part of the platform, all the users having access to this component.

Data filtering component occurs when users want to find specific data about an event. So, in order to get data, the user must apply a series of filters, which will be used as an input to query the storage system. Access to distributed storage supposes that data gathered from sensors are stored in the near storage site and all the data must be synchronized and processed.

Figure 5.6 presents the conceptual model for decision support module. First, the client accesses decision support module from the web application, by making one of the two requests: data validation or temporal and spatial queries. Next, the aggregation service will interrogate the storage and service repositories in order to calculate the propagation model or prediction methods. There are two important components: analytics component and automatic control component.

Analytics component can be identified in the sub-workflow, which is responsible with temporal and spatial queries. Data is aggregated from the storage system in order to provide the application with historical data. In this case the cost for data aggregation is very important in order to optimize the overall cost of the entire system.

Another important service offered by the CyberWater project is real-time alerts. Figure 5.7 presents the workflow of the service. The clients or administrators will make preferences for platform configuration, and the service will perform spatial and temporal queries, aggregating data from storage and service repositories. After this step data is validated against third-party data suppliers and passed to alert

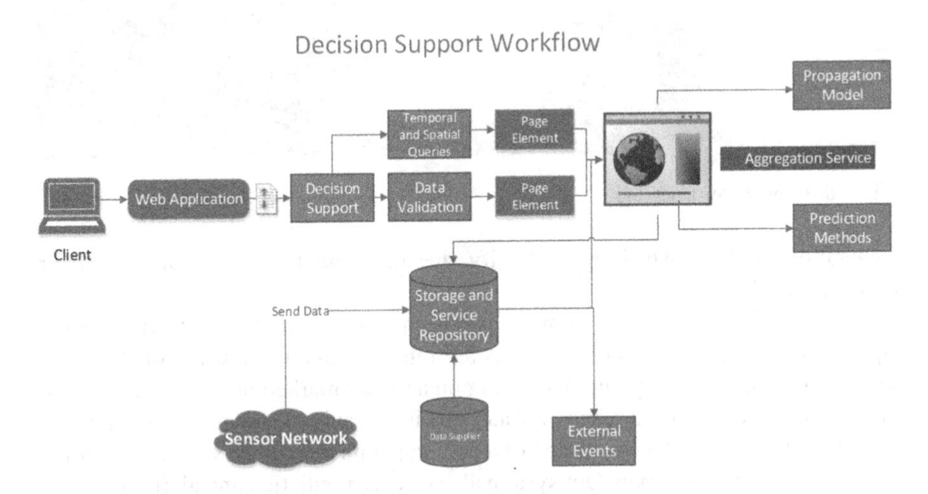

Fig. 5.6 Decision support workflow

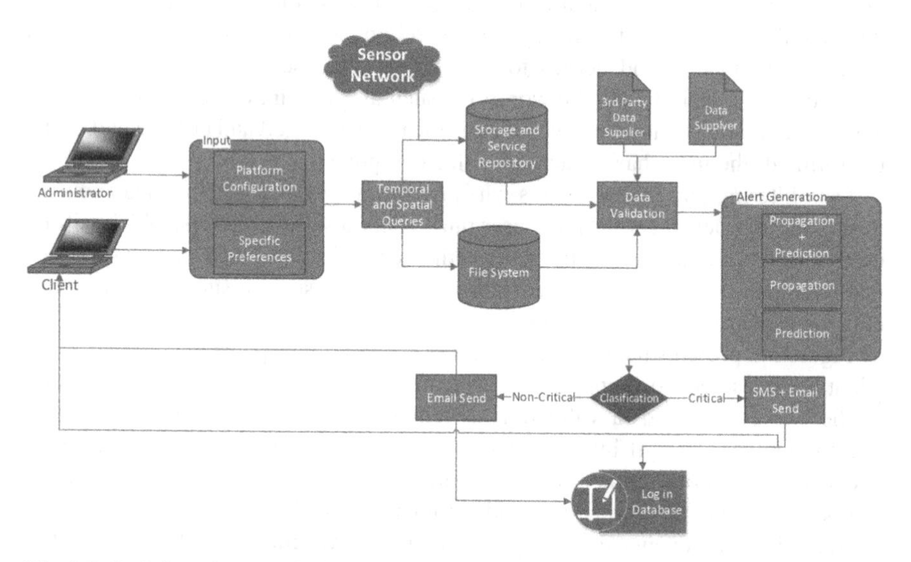

Fig. 5.7 Real-time alert workflow

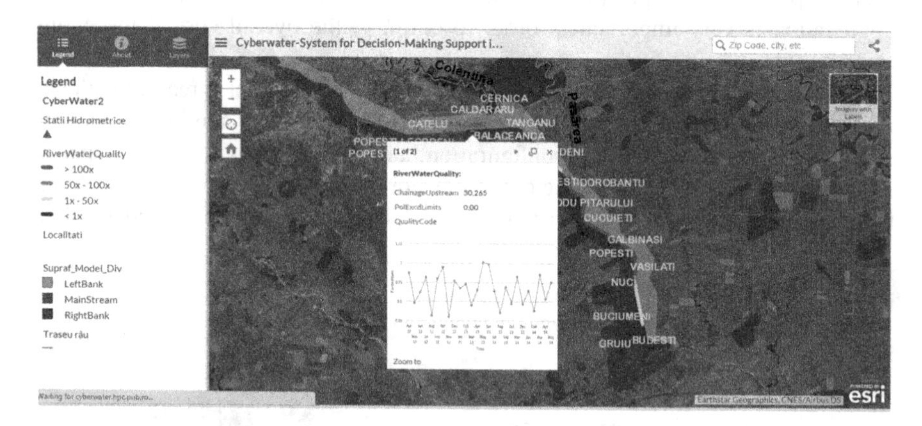

Fig. 5.8 CyberWater front-end

generation module, which will classify the alert and take necessary steps for sending the alert.

All workflows presented above have been encapsulated in a web mapping application presented in Fig. 5.8. Users are informed about the water quality on a certain river in a certain point by clicking on a sensor marked on river course. Also the watercourse is colored based on water quality (blue is a good quality and red bad quality of the water). The evolution of a certain parameter can be viewed in a graph displayed on top of the map. Our system also offers a prediction and alert service for water quality in the case of a pollution event.

Users can report an incident by clicking on the map and filling a web form that describes the reported incident, can upload multimedia files related to the incident, and can share the reported event on social networks.

5.6 Conclusion

In this chapter, we analyzed the problem of modeling and integration of heterogeneous data sources in the context of cyber-infrastructure systems and big data as it is a significant research issue. In this type of systems, information represents a key factor in the process of decision-making. Hence, there is an urgent need to integrate and operate heterogeneous data, in order to provide a unique and unified view of information. Benefits such as cost reduction, faster and better decision-making, and the development of new products and services can be obtained.

The unified approach for big data in environmental monitoring relies on a separation of heterogeneous information into data, metadata, and semantics. In terms of data management, operations such as integration, reduction, querying, indexing analysis, and mining, must be performed. Every operation corresponds to a specific layer and represents an important stage in the processing pipeline.

Data integrity check and validation also play an important role as there are often many errors and missing data in environmental datasets. Visualization tools also play a central role as front-end interface can offer a unified and uniform view of the entire system permitting users to get useful insights from data and take the necessary actions.

A key research issue nowadays is to find an approach that combines the relational database management systems and NoSQL database systems in order to benefit from the two paradigms. Equally important is to have new methods for the query optimization.

The CyberWater case study presented in the chapter highlights the need for methods and tools that integrate heterogeneous data sources. These need to be integrated in order to build a robust and resilient system that offers support in the decision-making process in the case of water resource management.

Although there are many tools for integrating various heterogeneous data sources, these do not provide the performance needed by a real-time system. New scalable and resilient methods and tools to integrate data source heterogeneity to filter our uncorrelated data need to be developed.

References

1. Jagadish HV, Gehrke J, Labrinidis A, Papakonstantinou Y, Patel JM, Ramakrishnan R, Shahabi C (2014) Big Data and its technical challenges. Commun ACM 57(7):86–94. doi:10.1145/2611567

2. Kitchin R (2013) Big Data and human geography opportunities, challenges and risks. Dialogues in Hum Geogr 3(3):262–267. doi:10.1177/2043820613513388
3. Kitchin R (2014) The real-time city? Big Data and smart urbanism. GeoJournal 79(1):1–14. doi:10.1007/s10708-013-9516-8
4. Liu W, Park EK (2014) Big Data as an e-health service. In: International conference on Computing, Networking and Communications (ICNC), February 2014, IEEE, pp 982–988. doi:10.1109/ICCNC.2014.6785471
5. Boyd D, Crawford K (2012) Critical questions for Big Data: provocations for a cultural, technological, and scholarly phenomenon. Inf Commun Soc 15(5):662–679. doi:10.1080/1369118X.2012.678878
6. Zlatanova S, Fabbri AG (2009) Geo-ICT for risk and disaster management. In: Geospatial technology and the role of location in science, Springer, Dordrecht, pp 239–266. doi:10.1007/978-90-481-2620-0_13
7. Careem M, De Silva C, De Silva R, Raschid L, Weerawarana S (2006) Sahana: overview of a disaster management system. In: International conference on Information and Automation, 2006. ICIA 2006, IEEE, pp 361–366. doi:10.1109/ICINFA.2006.374152
8. Boulos MNK, Resch B, Crowley DN, Breslin JG, Sohn G, Burtner R, Chuang KYS (2011) Crowdsourcing, citizen sensing and sensor web technologies for public and environmental health surveillance and crisis management: trends, OGC standards and application examples. Int J Health Geogr 10(1):67. doi:10.1186/1476-072X-10-67
9. Habiba M, Akhter S (2013) A cloud based natural disaster management system. In: Grid and pervasive computing. Springer, Berlin/Heidelberg, pp 152–161. doi:10.1007/978-3-642-38027-3_16
10. Smith K (2013) Environmental hazards: assessing risk and reducing disaster. Routledge, New York
11. Lohr S (2012) The age of Big Data. New York Times, http://www.nytimes.com/2012/02/12/sunday-review/big-datas-impact-in-the-world.html?_r=0. Accessed 25 Mar 2015
12. Bizer C, Boncz P, Brodie ML, Erling O (2012) The meaningful use of Big Data: four perspectives-four challenges. ACM SIGMOD Rec 40(4):56–60. doi:10.1145/2094114.2094129
13. Ghit B, Capota M, Hegeman T, Hidders J, Epema D, Iosup A (2014) V for Vicissitude: the challenge of scaling complex Big Data workflows. In: 2014 14th IEEE/ACM international symposium on Cluster, Cloud and Grid Computing (CCGrid), IEEE, May, pp 927–932. doi:10.1109/CCGrid.2014.97
14. Davenport T (2014) Three big benefits of Big Data analytics. http://www.sas.Com/tr_tr/news/sascom/2014q3/Big-data-davenport.html. Accessed 25 Mar 2015
15. Pop F, Cristea V (2015) The art of scheduling for big data science. In: Kuan-Ching Li, Hai Jiang, Yang LT, Cuzzocrea A (eds) Big data: algorithms, analytics, and applications. Chapman & Hall/CRC Big Data Series, pp 105–120, ISBN 978-1482240559
16. Freitas A, Curry E, Oliveira JG, O'Riain S (2012) Querying heterogeneous datasets on the linked data web: challenges, approaches, and trends. IEEE Internet Comput 16(1):24–33. doi:10.1109/MIC.2011.141
17. Calì A, Calvanese D, De Giacomo G, Lenzerini M (2013) Data integration under integrity constraints. In: Seminal contributions to information systems engineering. Springer, Berlin/Heidelberg, pp 335–352. doi:10.1007/3-540-47961-9_20
18. Buytaert W, Vitolo C, Reaney SM, Beven K (2012) Hydrological models as web services: experiences from the environmental virtual observatory project. In: AGU fall meeting abstracts, vol 1, p 1491
19. The Open Geospatial Consortium (OGC) Why is the OGC involved in sensor webs? http://www.opengeospatial.org/domain/swe. Accessed: 15 Apr 2015
20. Bröring A, Echterhoff J, Jirka S, Simonis I, Everding T, Stasch C, Lemmens R (2011) New generation sensor web enablement. Sensors 11(3):2652–2699. doi:10.3390/s110302652

21. Reed C, Botts M, Davidson J, Percivall G (2007) OGC® sensor web enablement: overview and high level architecture. In: Autotestcon, 2007 IEEE, IEEE, pp 372–380. doi:10.1007/978-3-540-79996-2_10
22. Chen N, Wang K, Xiao C, Gong J (2014) A heterogeneous sensor web node meta-model for the management of a flood monitoring system. Environ Model Softw 54:222–237. doi:10.1016/j.envsoft.2014.01.014
23. Chen N, Di L, Yu G, Min M (2009) A flexible geospatial sensor observation service for diverse sensor data based on web service. ISPRS J Photogramm Remote Sens 64(2):234–242. doi:10.1016/j.isprsjprs.2008.12.001
24. Gao Y, Wang F, Luan H, Chua TS (2014) Brand data gathering from live social media streams. In: Proceedings of international conference on multimedia retrieval, ACM, April, p 169. doi:10.1145/2578726.2578748
25. Cuzzocrea A, Song IY, Davis KC (2011) Analytics over large-scale multidimensional data: the Big Data revolution! In: Proceedings of the ACM 14th international workshop on Data Warehousing and OLAP, ACM, October, pp 101–104. doi:10.1145/2064676.2064695
26. Ali MI (2011) Distributed heterogeneous web data sources integration: DeXIN approach. LAP Lambert Academic Publishing, Saarbrücken
27. Riley J (2001) The indicator explosion: local needs and international challenges. Agric Ecosyst Environ 87:119–120. doi:10.1016/S0167-8809(01)00271-7
28. Wessels KJ, Van Den Bergh F, Scholes RJ (2012) Limits to detectability of land degradation by trend analysis of vegetation index data. Remote Sens Environ 125:10–22. doi:10.1016/j.rse.2012.06.022
29. Amin S, Goldstein MP (eds) (2008) Data against natural disasters: establishing effective systems for relief, recovery, and reconstruction. World Bank-free PDF
30. Alazawi Z, Altowaijri S, Mehmood R, Abdljabar MB (2011) Intelligent disaster management system based on cloud-enabled vehicular networks. In: 2011 11th international conference on ITS Telecommunications (ITST), IEEE, August, pp 361–368. doi:10.1109/ITST.2011.6060083
31. Huang W, Chen KW, Xiao C (2014) Integration on heterogeneous data with uncertainty in emergency system. In: Fuzzy information & engineering and operations research & management. Springer, Berlin/Heidelberg, pp 483–490. doi:10.1007/978-3-642-38667-1_48
32. van Loenen B, Grothe M (2014) INSPIRE as enabler of open data objectives. In: INSPIRE conference: INSPIRE for good governance
33. Leadbetter AM, Vodden PN (2015) Semantic linking of complex properties, monitoring processes and facilities in web-based representations of the environment. Int J of Digital Earth 9(3):1–38. doi:10.1080/17538947.2015.1033483
34. Sicari S, Grieco LA, Boggia G, Coen-Porisini A (2012) DyDAP: a dynamic data aggregation scheme for privacy aware wireless sensor networks. J Syst Softw 85(1):152–166. doi:10.1016/j.jss.2011.07.043
35. Stasch C, Foerster T, Autermann C, Pebesma E (2012) Spatio-temporal aggregation of European air quality observations in the sensor web. Comput Geosci 47:111–118. doi:10.1016/j.cageo.2011.11.008
36. Bellavista P, Corradi A, Fanelli M, Foschini L (2012) A survey of context data distribution for mobile ubiquitous systems. ACM Comput Surv (CSUR) 44(4):24. doi:10.1145/2333112.2333119
37. Xu J, Pottinger R (2014) Integrating domain heterogeneous data sources using decomposition aggregation queries. Inf Syst 39:80–107. doi:10.1016/j.is.2013.06.003
38. Garg N (2013) Apache Kafka. Packt Publishing Ltd., Birmingham
39. Moharil B, Gokhale C, Ghadge V, Tambvekar P, Pundlik S, Rai G (2014) Real time generalized log file management and analysis using pattern matching and dynamic clustering. Int J Comput Appl 91(16):1–6. doi:10.5120/15962-5320
40. Dean J, Ghemawat S (2008) MapReduce: simplified data processing on large clusters. Commun ACM 51(1):107–113. doi:10.1145/1327452.1327492

41. Isard M, Budiu M, Yu Y, Birrell A, Fetterly D (2007) Dryad: distributed data-parallel programs from sequential building blocks. In: ACM SIGOPS operating systems review, vol 41, no. 3, pp 59–72, March ACM. doi:10.1145/1272998.1273005
42. Ankit T, Siddarth T, Amit S, Karthik R, Jignesh MP, Sanjeev K, Jason J, Krishna G, Maosong F, Jake D, Nikunj B, Sailesh M, Dmitriy R (2014) Storm@twitter. In: Proceedings of the 2014 ACM SIGMOD international conference on Management of data (SIGMOD '14). ACM, New York, pp 147–156. doi:10.1145/2588555.2595641
43. Matei Z, Mosharaf C, Michael JF, Scott S, Ion S (2010) Spark: cluster computing with working sets. In: Proceedings of the 2nd USENIX conference on Hot topics in Cloud Computing (HotCloud'10). USENIX Association, Berkeley, CA, USA, pp 10–10
44. Toshniwal A, Taneja S, Shukla A, Ramasamy K, Patel JM, Kulkarni S, Ryaboy D (2014) Storm@twitter. In: Proceedings of the 2014 ACM SIGMOD international conference on Management of data, ACM, June, pp 147–156. doi:10.1145/2588555.2595641
45. Abouzeid A, Bajda-Pawlikowski K, Abadi D, Silberschatz A, Rasin A (2009) HadoopDB: an architectural hybrid of MapReduce and DBMS technologies for analytical workloads. Proceedings of the VLDB Endowment 2(1):922–933. doi:10.14778/1687627.1687731
46. Leida M, Gusmini A, Davies J (2013) Semantics-aware data integration for heterogeneous data sources. J Ambient Intell Humaniz Comput 4(4):471–491. doi:10.1007/s12652-012-0165-4
47. Whitehorn M. Aster Data founders explain unified approach to data big and small. http://www.computerweekly.com/feature/Aster-Data-founders-explain-unified-approach-to-data-big-and-small. Accessed: 15 Apr 2015
48. Singh VK, Gao M, Jain R (2012) Situation recognition: an evolving problem for heterogeneous dynamic big multimedia data. In: Proceedings of the 20th ACM international conference on Multimedia, ACM, October, pp 1209–1218. doi:10.1145/2393347.2396421
49. Ciolofan SN, Mocanu M, Pop F, Cristea V (2014) Improving quality of water related data in a cyberinfrastructure. In: Third international workshop on cyber physical systems. doi:10.13140/2.1.1380.4803

Chapter 6
Interfacing Physical and Cyber Worlds: A Big Data Perspective

Zartasha Baloch, Faisal Karim Shaikh, and Mukhtiar A. Unar

Abstract With the increase in utilization and pervasiveness of smart gadgets, there is a rise in new application domains. For that reason, computational technologies are progressing very rapidly, and computations are becoming an essential part of our life. Cyber-physical systems (CPSs) are a new evolution in computing that are integrated with the real world along with the physical devices to provide control in real-time environments. CPS generally takes input through sensors and controls the physical system through cyber systems using actuators. Such systems are really complex and challenging as they control real environments. This necessitates a proper interfacing of physical and cyber domains. To this end, the data generated by physical devices is getting bigger and bigger that is collectively acknowledged as big data. The real challenge in interfacing cyber and physical domains is the efficient management of big data. Accordingly, this chapter discusses big data sources and the relevant computing paradigms. It also classifies and discusses the main phases of data management for interfacing CPS, viz., data acquisition, data preprocessing, storage, query processing, data analysis, and actuation.

Keywords Big Data • Cyber-physical systems • Cloud computing • Data analytics • Decision support systems • Data management • Big data sources

6.1 Introduction

The computing paradigm has evolved in line with the development of the latest and newer technologies. With these advancements, there is a perception that 1 day computing will become the fifth utility (after water, gas, electricity, and telephone) which will be essential for everyday needs of the society [1]. Cyber-physical

Z. Baloch (✉) • M.A. Unar
IICT, Mehran University of Engineering and Technology, Jamshoro, Pakistan
e-mail: zartasha.baloch@faculty.muet.edu.pk

F.K. Shaikh
IICT, Mehran University of Engineering and Technology, Jamshoro, Pakistan

TCMCORE, STU, University of Umm Al-Qura, Mecca, Saudi Arabia

Z. Mahmood (ed.), *Data Science and Big Data Computing*,
DOI 10.1007/978-3-319-31861-5_6

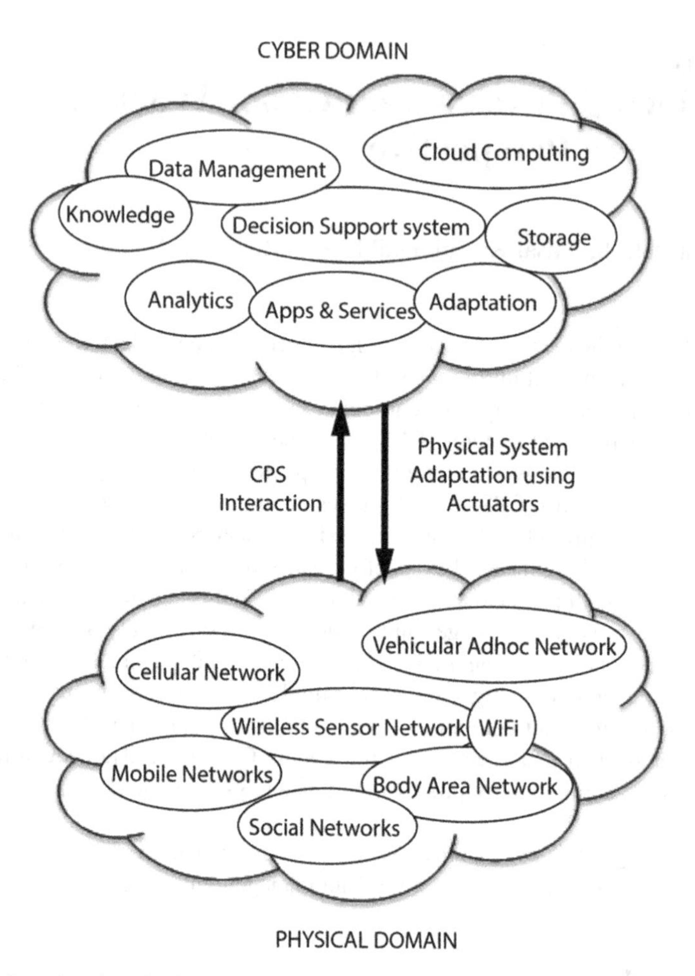

Fig. 6.1 Generic cyber-physical system

systems (CPSs) may support a new wave of computing by actively engaging it with the real world in real time [2]. Cyber-physical system is a *new generation of systems with integrated computational and physical capabilities that can interact with humans through many new modalities* [3]. It is a bridge between the cyber world and the physical world [4], where the physical world is simply the real world and the cyber world comprised of computing paradigms.

A cyber-physical system is the integration of the physical world with the cyber world to monitor and control physical entities by using feedback loops. It is an emerging technology which provides computing and communication facilities to the real-world systems and adds intelligence to the physical entities (see Fig. 6.1). CPS uses digital capabilities of computing to control analog physical systems.

In cyber-physical systems, multiple static or mobile sensors and actuators may be used that are integrated with intelligent decision support systems [5, 6]. The sensors are constrained due to low energy, low computational power, and less storage capacity. Also a sensor does not have enough storage capacity to accommodate huge datasets. Cloud computing is a solution to some of these issues related to sensors. The combination of sensors and cloud is known as sensor cloud [7]. The sensor cloud infrastructure is a vital part of CPS, where the cloud performs computing (cyber) activities and sensor supports physical activities [7].

Kim et al. [8] proposed a generic framework for design, modeling, and simulation of CPS. The paper highlights many important features that need to be part of that framework. The features include heterogeneous application support, physical modeling environments that support mathematical expressions, scalability that helps to increase in number of sensors deployed, support to connect with existing simulation tools, and software reusability, and all the proprietary solutions and open standards should be integrated into generic framework [8].

Due to the increase in the use of smart computing devices, a huge amount of data is generated across the physical world. The term big data is used for those huge datasets and is defined as *a massive volume of both structured and unstructured data that is so large that it is difficult to process using traditional database and software techniques* [9]. There are many sources of big data at the physical world; there may be wireless sensor networks, social networks, wireless body area networks, mobile networks and vehicular ad hoc networks, etc. The data are physically managed through some data management frameworks, and then that captured data is sent to the cyber world for analytics.

The cyber world may include big data, big data management, cloud storage, data analytics, and decision support systems. The continuous data growth poses many challenges. The major issues are storing that data and extracting valuable information from such a large amount of data. The data is not limited, but it is increasing exponentially so there is a major issue to store this data efficiently and in a cost-effective manner. The cloud storage provides a cost-effective way to facilitate the users with ease of computing, storing, and networking resources. As the big data is in large quantity and all of that data is not important, there is a need to extract valuable data through data analytics. Big data analytics is the process of capturing, arranging, and analyzing huge sets of data to identify patterns and valuable information [10]. The analyzed data will be sent back to the physical world. Big data is a buzz word today, so it provides wide space for research in this field. This chapter presents the review of various technical aspects of big data for cyber-physical systems.

The remaining chapter is organized as follows. Section 6.2 discusses various sources of big data. Section 6.3 briefly describes data management at cyberspace that includes cloud computing and decision support systems. Interfacing cyber and physical worlds is discussed in Sect. 6.4. Section 6.5 identifies the main challenges of cyber physical systems in terms of big data, and Sect. 6.6 concludes the chapter.

6.2 Data Generation by Physical Systems: Big Data Sources

The first step in big data scenario is data generation. There are many sources of big data which are generating highly diverse and complex datasets. These sources include wireless sensor networks, mobile ad hoc networks, social networks, vehicular networks, RFIDs, web servers, online transactions, etc.

Big data can be structured, unstructured, and semistructured. The data, which are well organized and are based on some data model, are referred to as structured data. On the other hand, the unstructured data does not follow any data model. The semistructured data is the combination of structured and unstructured. It is a type of structured data, but somehow it lacks the data model structure and uses markers or tags to mark specific data elements. For example, emails contain unstructured data, but it has some fields like date, time, sender, recipient, etc. which are considered to be as structured data. Generally, big data is considered as unstructured.

There are three main characteristics of big data: volume, variety, and velocity [11]. The *volume* characteristic is defined as the amount of data, *variety* as different formats of data/data sources, and *velocity* is the speed at which the data is growing [12]. The data is not just large in volume, but there is variety of complex datasets. The real challenge is to handle that diversity and variety. We can categorize the data growth as business application data, personal data, and machine data [13]. The data generated by business applications is moderate in volume, variety, and velocity. This type of data is highly structured data. It includes online transactions. The personal data includes web logs, documents, emails, social media, etc. It is highly unstructured data, and it is moderate in variety but high in volume and velocity. The data growth is two times more than business application data. The third category is machine data which include sensors, machine logs data, audio and video recordings, bio-informatics, etc. This type of data is highly structured, and it is high in volume, variety, and velocity [14]. The growth is three times more than business application data.

In this section, we discuss a number of common data sources such as wireless sensor networks, social networks, body area networks, and vehicular ad hoc networks.

6.2.1 Wireless Sensor Networks

In the past few years, the applications of wireless sensor networks (WSNs) have been increasing rapidly, such as monitoring, event detection, surveillance, etc. Wireless sensor network is a wireless network of many small devices which are capable of sensing, computation, and communication. A sensor network consists of multiple sensor nodes, which are small and lightweight. The sensor nodes are generally dispersed in a sensor field as shown in Fig. 6.2. Every sensor node contains a transducer, microcomputer, transceiver, and a power source [15, 16]. When a sensor node senses a physical phenomenon, an electrical signal is

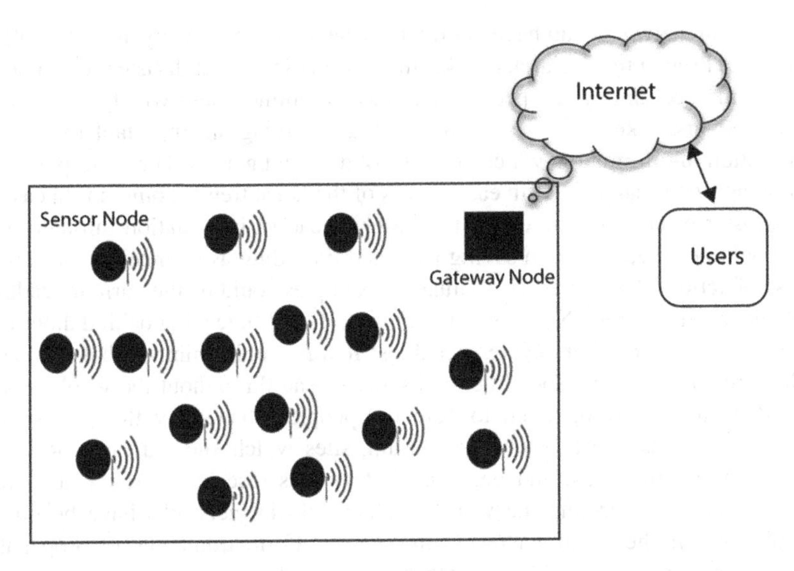

Fig. 6.2 Generic wireless sensor network model

generated by the transducer, which is processed and stored by microcomputer. The collected data will then be sent to the sink/gateway which sends it back to the end user via Internet/satellite or any other means of communication [17].

The traditional technologies for data processing, storing, and reporting provide limited support for analyzing WSN data. These technologies have become prohibitively expensive, while dealing with sensor-generated big data [18]. Even then, they cannot handle the processing requirements for real-time processes such as fire detection, natural disasters, and traffic control [18]. Thus, the research is directed toward new technologies for processing big data. There are many attempts in combining big data and WSN. Jardak et al. [19] proposed a data model for structuring the stored data by allowing a wide range of analysis by using Bigtable, Hadoop, and MapReduce algorithms. A Hadoop-based cloud storage solution for WSN data is presented by Fan et al. [20]. Similarly, Ahmed et al. [21] proposed an infrastructure for integrating cloud computing with WSNs.

6.2.2 Social Networks

A social network is a Web-based service that connects people with each other. The user can make their profile and share information, experiences, ideas, etc. The most popular social networking sites are Twitter [22], Facebook [23], Pinterest [24], Google+ [25], Instagram [26], and many more. These social sites are very popular among youngsters. Data is being shared on these networks, which include billions of photos, videos, and other information. Most of the data is unstructured with high volume and velocity.

The social networks can have enormous benefits for the society as it can help in disasters. Through the social networks, the information about disasters can quickly be disseminated among the people. The most prominent and widely used social media networks like Twitter and Facebook are playing an important role in the propagation of information which could be of different genres. The widespread use of hashtag trends can help with easy access of the latest trends going on. In case of any disaster or catastrophic crisis, the faster spread of information through these sites could be an epidemic in saving lives and providing assistance for the further course of action. One of the crisis situation examples could be the current deadliest earthquake which struck Nepal in April 25, 2015 [27, 28], leaving behind thousands of people dead and other severe casualties. It was within minutes that this news broke through the whole social networks and spread throughout the whole world. Immediate actions were taken to help the people affected by this devastating tragedy. This was due to social networking sites which showed the world how severe the situation was, and because of it various rescue and relief aids were instantly sent from around the world to Nepal. Social networks have become a binding force in the world where within seconds information could be propagated from one corner of the world to the other. The only disadvantage is that we cannot verify the credibility of the information being generated on the social networking sites. Furthermore, social media analysis can be helpful for the organizations to redesign their policies to address the public issues [29]. Social networks are leading toward a new generation of crowd sourcing applications [30], which will help in analyzing in-depth physical environments.

6.2.3 Vehicular Ad Hoc Networks

The research on integrating communication technologies with vehicles has begun since long ago. The communication between vehicles by using ad hoc networks is known as vehicular ad hoc networks (VANETs) [31, 32] as shown in Fig. 6.3. It is a subcategory of intelligent transport systems and mobile ad hoc networks. The vehicles can share necessary information with each other (referred to as vehicle

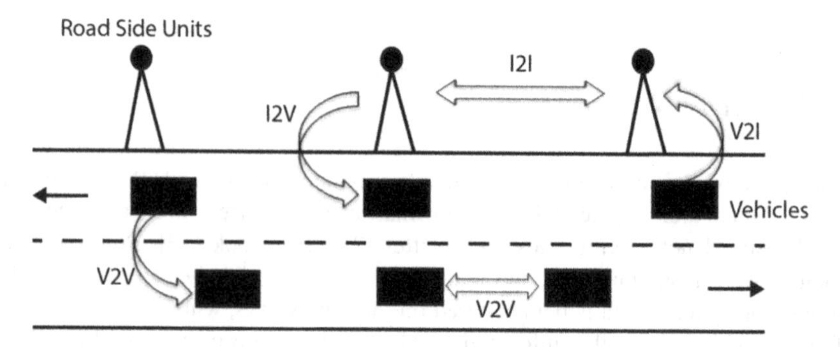

Fig. 6.3 Generic vehicular ad hoc network

to vehicle or V2V), such as traffic information (traffic jam or accident), emergency warnings, weather information, road condition warnings, etc. Furthermore, the data can also be shared with the data center to the passing vehicles (referred as infrastructure to vehicle or I2V and vice versa). It is not only important to pass along the latest information but also to remove the outdated data [33, 34]. Since there are many vehicles passing on the roads, they may consume the total bandwidth in data dissemination. Therefore, it is important to efficiently transmit the data by using limited bandwidth [35].

6.2.4 *Wireless Body Area Networks*

The recent development in wireless networks and microelectronics has resulted in wireless body area networks (WBANs). A WBAN may consist of miniature lightweight sensor nodes with low power and is used for healthcare application to monitor physiological status of the human body, such as blood pressure, blood sugar, ECG, pulse rate, etc. [36].

Figure 6.4 shows a generic architecture for WBAN-based health monitoring system where sensor nodes senses medical data and sends it to the base station (BS).

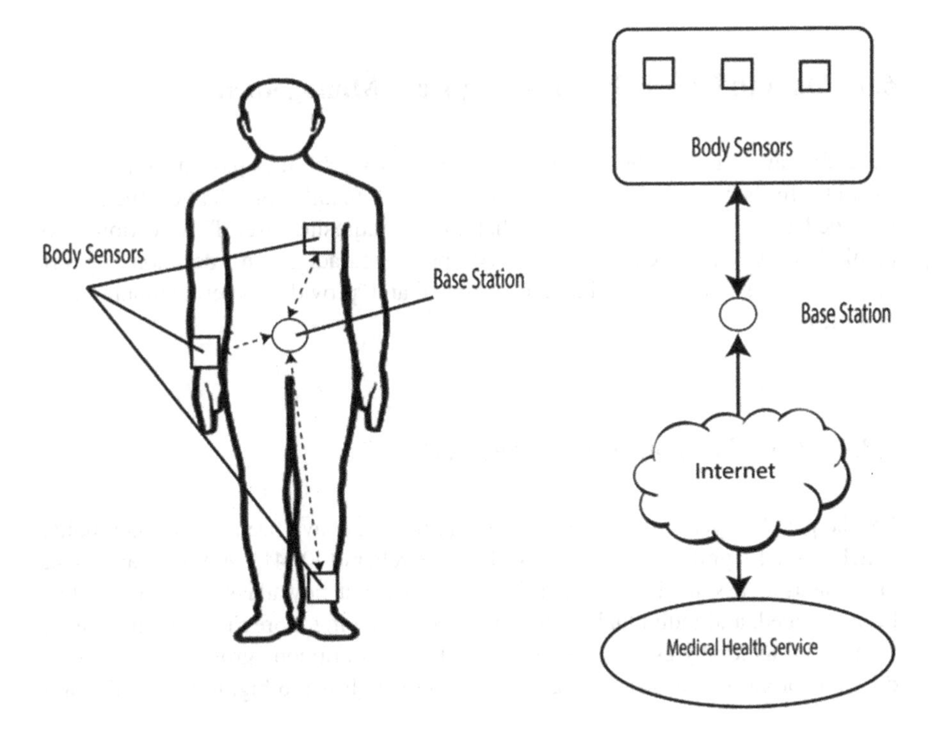

Fig. 6.4 Generic WBAN scenario

Table 6.1 Comparison of big data sources

Big data sources	Volume	Velocity	Variety	Structured/unstructured
WSN	High	High	Low	Structured
Social networks	Moderate to high	High	Low to moderate	Unstructured
VANET	Moderate to high	Moderate	Low to moderate	Structured
WBAN	High	High	High	Unstructured

The BS then transmits the data to the doctor for real-time diagnosis, to a database for keeping medical records or to a particular device that generates emergency alerts via medical health service (MHS) center [37, 38].

The common factor in all the emerging data sources is that they all are continuously generating data. The data is high in volume, with lots of variety and with dynamic velocities [39]. In general, the characteristics of big data for these data sources are summarized in Table 6.1. The data from these four sources are mostly high in volume and velocity. As the WSN, VANET, and WBAN data are sensor-generated data, so they have a vast data variety depending on the nature of sensors used and are mostly highly structured. The social network data is moderate to high in volume but does not have much data variety. This type of data is mostly unstructured, as it mostly contains images and audio and video streams.

6.3 Data in Cyber Systems: Big Data Management

Once the raw data is collected from the physical world, it is passed to the cyber world for further processing. In this section, we will mainly consider how the data is managed in the cyber world and what the prerequisites are. This section also highlights the role of existing cloud computing paradigm and decision support systems that solves the problem of storage and provides other computational facilities.

6.3.1 Cloud Computing Paradigms

For the past few years, cloud computing and big data have been the two key fields which gained significant attention of the researchers [40, 41]. A few years back, large datasets assumed to be of a few terabytes, but nowadays this concept has been changed, and individual applications are producing more than that: new units being used as terabytes and petabytes. With this continuous growth of data, it is difficult for an organization to handle the data which is too big, too versatile, and too fast because the traditional storage methods are not designed for such a huge data. One solution is cloud storage. Cloud computing is the emerging technology that provides users to perform complex computations without maintaining expensive hardware and software. Although cloud computing is currently used by almost all the leading companies, still there is no universally agreed definition [42]. Gartner [43] defines cloud computing as a computing style which provides scalable and elastic IT-enabled capabilities as a service by using Internet technologies. It provides many computational services to the users such as infrastructure as a service (IaaS) [44], platform as a service (PaaS) [44], and software as a service (SaaS) [44]. The IaaS offers storage and processing infrastructure as a service. The user is provided virtualized infrastructure without worrying about hardware resources [45]. The PaaS provides a platform for the software developers to write and upload their application code [45]. The SaaS is the most common layer of cloud computing. It provides software application as a service to users on pay-as-per-use basis [45]. Cloud computing has many advantages over other computational services which includes parallel processing, security, scalable data storage, and resource virtualization [46]. It also reduces maintenance of infrastructure. Cloud computing supports virtualization; through virtualization software, a simple computer can behave like a supercomputer at an affordable cost [46].

There is a big challenge for researchers to design an appropriate platform for cloud computing that handles it and performs data analytics. There are many cloud service providers, whenever an enterprise tend to migrate from IT system to the cloud, the decision can be difficult, as the enterprise want to evaluate cost, benefits, and risks of using cloud computing [47]. Hashem et al. [47] presented two decision support tools for migration to the public IaaS cloud. These tools help an enterprise to make cloud migration decisions. The first tool is a cost modeling tool [47], which can be used in modeling the requirements of enterprise data, applications, and infrastructure along with the usage patterns of computational resource. This tool can also be used to compare the cost of cloud services from different cloud service providers with different deployment options and usage. The second tool, which is a spreadsheet, shows the usage benefits of IaaS cloud, and it also provides beginning point for risk assessment [47].

The data management applications in the cloud include two main data management components: transactional data management and analytical data management [48]. Transactional data management is the database related with transactions like banking and online reservations [49]. Shared nothing is simply a distributed architecture in which each node consists of a processor, main memory, and disk, and the nodes communicate with each other via interconnecting network [50]. For implementation of transactional data management, the usage of shared nothing architecture may result in the complex distributed locking and commit protocols [48]. There is an extensive risk of storing transactional data on untrusted host because of the sensitive information which includes credit card numbers and pin codes. Analytical data management is for the applications that query a data store for business intelligence and decision making purposes [48]. Its scale is larger than transactional database management. The analytical data management systems are best suited for execution in a cloud environment. Generally, for implementation of analytical data management, the use of shared nothing architecture is well suited. The atomicity, consistency, and isolation are easy to obtain as compared to transactional data.

6.3.2 Service-Oriented Decision Support Systems

Decision support system (DSS) is a computer application that analyzes business data and presents it in a way so that users can make business decisions more easily [51]. It may use artificial intelligence for analyzing data. DSS finds certain patterns in data which helps humans to take decisions. For example, DSS helps doctors to diagnose the disease on the basis of symptoms.

There are three service models for service-oriented DSS [52], namely, data as a service (DaaS), information as a service (IaaS), and analytics as a service (AaaS). These are discussed in the following subsections.

6.3.2.1 Data as a Service

The service-oriented architecture provides access to the data from anywhere, independent of the platform. The data as a service provides the business applications, a facility to access the data wherever it resides [53]. With the provision of DaaS, the data quality can be maintained at central place, i.e., at cloud. According to Demirkan et al. [52], the data cleansing and data enriching can be done by two solutions, namely, master data management (MDM) and customer data integration (CDI), where customer data can be placed anywhere and can be accessed as a service through any application that has service provision.

6.3.2.2 Information as a Service

Sometimes the information repositories at organizations are not efficiently designed to transmit information to the required destinations; this is due to the increased complexity of processes and architectures. Demirkan et al. [52] defines information as a service as an idea to make information quickly available to users, processes, and applications in an organization. This shares real-time information with emerging applications and hides complexities. It increases availability with virtualization. It also provides master data management (MDM), content management services, and business intelligence services [52].

6.3.2.3 Analytics as a Service

Analytics as a Service can be defined as the combination of cloud computing and big data analytics [53]. It enables data scientists to access datasets that are centrally managed by cloud providers. The business analysts can make decisions effectively and delivers successful outcomes. AaaS is a cloud-based analytical platform, where several data analytical tools are available, which can be configured by end users to process and analyze large amounts of heterogeneous data.

6.4 Interfacing Cyber World with Physical World

The error-free interaction between cyber and physical worlds is not an easy task because the physical world is mostly unpredictable. The most critical element is that human lives are dependent on the system. Thus, availability, connectivity, predictability, and repeatability are very much important for the cyber-physical interface [54].

In general, CPS can be divided into three parts: physical world, cyber world, and interfacing physical and cyber worlds. The sensors sense physical characteristics, then the sensed data is passed to cyber world to perform computations, and finally, response is generated through some actuators (see Fig. 6.1).

The physical components may include power sources, energy storage, and physical transducers that perform energy conversions in physical domains [55]. The cyber components may include data stores, computation, and I/O interfaces. The interface contains both physical and cyber components and adds a few more components to connect them. Rajhans et al. [56] presented two connector types for modeling the interface between cyber and physical worlds. The connectors are physical-to-cyber (P2C) and cyber-to-physical (C2P) connectors. Simple sensors can be used for physical-to-cyber connector type and actuators can be used for cyber-to-physical connector type. For the complex interfaces in CPS, physical-to-cyber transducer and cyber-to-physical transducer may also be used, which have ports to cyber and physical components on each side [56].

Applications of CPS apparently have potential to overshadow the twentieth-century IT advancements. CPS applications include many components that cooperate through an unpredictable physical environment. In this regard, reliability and security are major issues to be resolved. The CPS applications include transportation, defense, energy and industrial automation, health and biomedical, agriculture, and other critical infrastructures [57].

Cyber-physical cloud computing is the integration of CPS and cloud computing. The CPCC architectural framework is "a system environment that can rapidly build, modify, and provision cyber physical systems composed of a set of cloud computing based sensor, processing, control, and data services" [58]. The customers can access available resources through Internet independent of location and devices.

CPCC automatically manages all the resources. CPCC can benefit many systems such as traffic management, intelligent power grids, disaster management systems, healthcare, etc. [58].

Despite the fact that data management is the focal point of interest for many researchers, there is still lack of an agreed-upon definition for data management. It includes many phases. The data management phases are defined by many researchers; these steps depend on the nature of data to be managed. Some phases can be added or removed accordingly. We will discuss some of them in this section. In TDWI report [39], the data management is defined as data collection, storage, processing, and delivery, and it considers data management as a broad practice that includes many data disciplines such as data quality, data integration, data warehousing, event processing, database administration and content management, etc. According to Mokashi et al. [59], data management includes data collection, data storage, and query processing. Padgavankar et al. [60] consider data management as a four-step process, i.e., data generation, big data acquisition, big data storage, and data analysis. Furthermore, Sathe et al. [61] use four tasks for WSN data management, i.e., data acquisition, data cleaning, query processing, and data compression.

Accordingly, we classify the main phases of data management for interfacing CPS as data acquisition, data preprocessing, storage, query processing, data analysis, and actuation.

6.4.1 Data Acquisition

Data acquisition is simply data gathering or data collection. As the physical world is generating huge datasets, the data acquisition process determines which data should be collected along with minimal energy consumption. This is challenging task, because of the data uncertainty due to natural errors, noise, and missed readings in sensor data [62]. It is responsible for efficiently collecting samples from sensors in CPS. In sensor data acquisition, the main objective is to achieve energy efficiency because sensors are battery powered and located mostly at unreachable locations. Sathe et al. [61] presented model-based data acquisition techniques that are designed to handle challenges such as minimal energy consumption and communication cost.

6.4.2 Data Preprocessing

In the data acquisition phase, the raw data gets collected. The acquired sensor datasets may sometimes contain erroneous or redundant data, which will definitely occupy more storage space and will affect data analysis [60]. Therefore, before data storage, preprocessing may be applied which may include data cleaning, data fusion, and data compression [60].

6.4.2.1 Data Cleaning

It is a process of finding incomplete, inaccurate, and unreasonable data and then correcting the errors to improve data quality [63]. In the data cleaning process, the errors will be removed from raw sensor data. For incomplete datasets, regression or interpolation models can be used to reconstruct missing data. Alonso et al. [64] proposed an extensible receptor stream processing (ESP) framework for online data cleaning of the acquired sensor data streams.

6.4.2.2 Data Fusion

As the name shows, data fusion combines data from various data sources. Sensor fusion is a technique to merge data from many sources to provide accurate and comprehensive information [65]. It is a technique to address sensor impairments. Some other terms are also used in the literature that are related to data fusion such as decision fusion, multisensor fusion, and information fusion.

Data fusion is a technique of combining data from various sensors and information from related databases to attain accuracy and more specific inferences than by using a single sensor [66]. The data fusion techniques can be categorized into three categories, that is, data association, state estimation, and decision fusion [67].

6.4.2.3 Data Compression

As the sensors continuously generate huge datasets, and sometimes the collected data contains redundant data, this redundancy is common in environmental monitoring. The data compression techniques can help to reduce the redundancy which helps in reducing storage space [60]. Data transmission is more energy consuming than computation; thus, reduced data size, before data transmission, will minimize the overall energy consumption [68]. Different data compression schemes have been discussed by Kimura et al. [68]. Sathe et al. [61] also discussed many data compression techniques such as linear approximation, model approximation, and orthogonal transformation. Marcelloni et al. [69] introduces a new lossless compression algorithm that is suitable for reduced computational and storage resources of a WSN node. Many other techniques are proposed in the literature, some of them are derived from signal processing [70] and some has used correlations between sensor data to compress the data streams [71–73].

6.4.3 *Data Storage*

Due to the limitations of sensors, it is important to store the sensor data efficiently elsewhere [74], to improve the data retrieval and analytics processes. As the sensors generate huge datasets, the question arise: do all the generated data is required to be stored? For different applications, the answer could be different. For example, in real-time applications, mostly, the recent data is important [75], so there is no need to keep all the data for long periods. In such cases, live data streaming will be an appropriate approach. In some applications, the historical data need to be stored for future analysis, in those cases; the historical storage approach will be appropriate [76]. For a variety of applications, both the storage approaches can be combined to make an efficient storage system, but this could be very challenging [75].

The second important issue is determine where to store that data. Many researchers have done work in this direction. Three data storage methods for WSN have been discussed by Xing et al. [74]; these are local storage, external storage, and data centric storage (DCS). In local storage, only short-lived data is stored in the sensor node. In external storage, data is stored at an external point for further processing. While in data-centric storage, data is stored along with the name or location. In DCS, the related data is classified and named according to its meaning. The data with the same name will be stored in the same sensor node. For a particular name, the user queries will be sent directly to that particular node which holds that named data [59].

For live querying data, many data management techniques have been proposed, whereas for querying historical data, only a few data management solutions have been developed. Those techniques are discussed by Diao et al. [76].

With the technology advancements, the storage devices are becoming more energy efficient and cheaper in price. Thus the sensor networks are transforming from communication-centric to storage-centric perspective which provides a network that efficiently stores data from sensors [76, 77]. The data can be batched or accessed later. Energy efficiency can be improved by batching sensed data. In storage-centric sensor network, the applications must be delay tolerant because the data is not transmitted immediately. For the applications where immediate response is needed, delay cannot be tolerated, in such cases communication-centric approach is appropriate [77].

6.4.4 *Query Processing*

Another important component of data management is data retrieval or query processing. Many important model-based query processing techniques, which aim to process queries by retrieving minimum amount of data, are presented by Sathe et al. [61]. Apart from that, these techniques also handle missing data and create an abstraction layer on sensor network by using these models [78, 79]. Some of the techniques are based on hidden Markov model (HMM) [80] or dynamic probabilistic model which is for spatiotemporal evolution of the data from sensors [79].

For querying the real-time applications of CPS, different researchers have developed many tools that can be named as information flow processing (IFP) systems [81]. Information flow processing is an application domain where users are required to collect data from various data sources to process it within due time [81]. After processing data, the collected data is generally discarded, except some critical applications where historical analysis is important. This is done using two popular models that are data stream processing [82] and complex event processing models [83]. The data stream processing model is processing data from various sources to produce output data streams. The data stream management system (DSMS) is also based on database management systems (DBMS) with a few differences such as DBMSs deals with data that is not updated constantly, whereas DSMSs are specially designed to deal with data that is updated continuously [81]. Apart from few differences, there are more similarities in between both of them.

The recent developments in DSMSs are reviewed by Golab et al. [84]. The complex event processing model considers information flow items as notification of events of the physical world, which will be filtered and combined to visualize what is happening in the form of high-level events [81]. The approach mainly focuses to detect patterns of low-level events that will eventually be combined to represent the high-level events that will be notified to the parties that are interested. An architecture for real-time analysis and processing of complex-event streams of sensor networks, which is based on semantically rich event models, is presented by Dunkel et al. [85].

6.4.5 Data Analysis

The data analysis is the most significant phase of data management for CPS interfacing. As CPS data is larger in magnitude, the real challenge is to extract the insight value from it which is valuable. The purpose of data analysis is to sift valuable information. It helps organizations to better cope with the needs of their customers and to make better decisions [86].

The traditional analytical methods, based on statistics and computer science, may still be used for big data analysis, such as cluster analysis, factor analysis, correlation analysis, regression analysis, real-time analysis, offline analysis, memory level analysis, business intelligence (BI) analysis, and massive analysis [60]. Some advanced analysis techniques are also required to handle the complexity of real-world heterogeneous datasets. Big data analytics is the set of modern techniques that are designed to operate on heterogeneous data with large magnitudes [87]. The intelligent quantitative methods, such as artificial intelligence, robotics, artificial neural networks, or machine learning, can be used to explore and to identify hidden patterns and their relationships [87].

In a typical CPS for environment monitoring, most of the collected data is considered as regular, but some of them may be irregular; such data is known as atypical data [88]. Atypical data is extremely crucial as it identifies a change in environmental condition; therefore, such data need to be analyzed. Different approaches have been discussed in the literature [88–90] to analyze atypical data in the CPS. Tang et al. [91] proposed a method named as Tru-alarm, which finds out trustworthy alarms for the cyber-physical systems. It uses data analysis to eliminate noisy data that can cause false emergency alarms.

6.4.6 Actuation

Actuation is the most crucial element in CPS because it controls the environment. In many CPS applications, sensing data is not just sufficient, but a response is also required to show how the system reacts in a particular situation [92]. For example, in fire alarm systems, the actuators may be deployed to shower water on the fire. Another example could be of an agricultural environment where crops can be monitored and pesticides can be sprinkled by an actuation process if needed.

Data actuation is the process in which the processed data is sent to actuators to perform some action. It transfers data back to the physical systems. Thouin et al. [93] discussed different actuation strategies to acquire desired actions to be performed on physical devices. A dynamic actuation strategy is group of decision rules to find the actuation nature which will be executed throughout the course of operations in a wireless sensor and actuator networks [93].

6.5 Future Challenges and Opportunities

The CPS is a multidisciplinary technology, which involves communication and networks, embedded systems, and semantic technologies. To take the maximum benefit from CPS and to handle big data that flows in between the cyber and physical worlds, there are many challenges to be addressed. A few challenges of CPS in big data perspective are given below.

Volume As discussed earlier, CPS data is enormous and keeps growing continuously, processing that huge dataset can lead to many challenges. Data abstraction, i.e., summarizing the data and making it human comprehensible, is one of the biggest challenge for big data generated across CPS. Another challenge could be efficient use of distributed processing to scale the CPS computations. The simple computations can become complex when scaling from terabytes to petabytes. Even sequential scans to petabytes data takes too much time. The indexing techniques are also very challenging while scaling to huge volume.

Variety There is a huge variety of datasets with different data formats, which need to be integrated together. As the data is collected from distinct sources, the structure of data can be complex and data processing can also be very complex. Thus, efficient techniques are needed to cope with the increasing variety of data.

Velocity With these fast growing datasets, it is challenging to focus on the data trends and the correlations between data. There is a great need of robust and real-time techniques to cope with velocity of the data generation and processing.

Veracity Sometimes, the sensors in a CPS generate erroneous data, or some data is missed due to erroneous communication. Therefore, it is challenging to find trustworthiness of the data.

Value The main challenge of interfacing CPS is to transform the collected raw data into useful information in order to facilitate the decision making process. The efficient transformation techniques are needed to provide the accurate value of the information.

Query Load Generally, the query loads vary and are unpredictable. Due to lack of flexibility, it is complex to handle these variations. Conti et al. [94] proposed a new term data vitalization to sense the query load variations. There is still lot of work needed to optimize the query load in accordance with the needed information and available resources.

Quality of Service The methodologies to precisely capture and communicate information and the quality needs of an application should be researched. Due to increase in scalability and complexity of data, the computational techniques and their results are very complex to reproduce. Thus, the relationship between data from information producing systems and the operational systems need to be studied such that application's quality of service requirements are fulfilled efficiently.

Knowledge Association The constant sensor data streams are required to be processed by CPS. These streams need to be efficiently associated with the existing knowledge [6, 58]. For the complex and uncertain data, the temporal and spatial correlations must be used with data mining tools to retrieve valuable knowledge [6]. There is very little work in this direction, and more research is needed for efficient knowledge association across CPS.

Open CPS Architecture A new open architecture is required, which can be customized in different situations by different application scenarios. The physical components are mostly unreliable; tools are needed to build a reliable CPS that should be resilient to tolerate malicious attacks on the data [57]. For the complex design of CPS, new modeling and analytical tools are essential to be utilized [95].

6.6 Conclusion

In the era of advanced computing, there is an emergent and rapid technological enhancements in the fields of embedded systems, human computer interaction, cloud computing, data analysis, cyber-physical systems, and many other computing aspects. Cyber-physical systems, a new wave of computing, have enabled many applications that were not practical before. The data from cyber-physical systems is enormous and growing constantly which poses many challenges in this field. This chapter discusses the state-of-the-art of the cyber-physical systems from big data perspective. The data generation sources, cyberspace paradigms, and interfacing them with the physical and cyber world have been discussed. From data generation to its storage, different phases of data management for interfacing the two worlds have also been elaborated. The main issues are efficient storage and processing of cyber-physical systems big data. The cyber-physical system cloud computing infrastructure has also been discussed which provides the framework to interface with the computing devices. Furthermore, the research issues related to big data in cyber-physical systems have been highlighted. The cyber-physical systems are in its way of development; therefore, significant issues and challenges must be addressed by researchers for long-term success.

References

1. Buyya R, Yeo CS, Venugopal S, Broberg J, Brandic I (2009) Cloud computing and emerging IT platforms: vision, hype, and reality for delivering computing as the 5th utility. Futur Gener Comput Syst 25(6):599–616
2. Wolf W (2009) Cyber-physical systems. Computer 3:88–89
3. Baheti R, Gill H (2011) Cyber-physical systems. Impact Control Technol 12:161–166
4. Rajkumar R, Lee I, Sha L, Stankovic J (2010) Cyber-physical systems: the next computing revolution. In: Proceedings of the 47th design automation conference. ACM, pp 731–736
5. Shaikh FK, Zeadally S (2015) Mobile sensors in cyber-physical systems. Book Chapter in cyber physical system design with sensor networking technologies, IET, 2015 (to appear)
6. Wu FJ, Kao YF, Tseng YC (2011) From wireless sensor networks towards cyber physical systems. Pervasive Mob Comput 7(4):397–413
7. Haque AS, Aziz SM, Rahman M (2014) Review of cyber-physical system in healthcare. Int J Distrib Sens Netw 2014:1–20
8. Kim JE, Mosse D (2008) Generic framework for design, modeling and simulation of cyber physical systems. ACM SIGBED Rev 5:1
9. Bloomberg J (2013) The big data long tail. http://www.devx.com/blog/the-big-data-long-tail.html. Accessed 17 Jan 2015
10. Kambatla K, Kollias G, Kumar V, Grama A (2014) Trends in big data analytics. J Parallel Distr Com 74(7):2561–2573
11. Madden S (2012) From databases to big data. IEEE Internet Comput 3:4–6
12. Rouse M (2015) 3Vs (volume, velocity & variety). http://whatis.techtarget.com/definition/3Vs. Accessed Apr 2015
13. Hitachi Data Systems (2015) Capitalize on big data. http://www.hds.com/assets/pdf/hitachi-webtech-educational-series-capitalize-on-big-data.pdf. Accessed 20 Mar 2015

14. Hurwitz J, Nugent A, Halper F, Kaufman M (2015) Structured data in a big data environment. www.dummies.com/howto/ content/structured-data-in-a-big-data-environment.html. Accessed 2 Apr 2015
15. Shaikh FK, Zeadally S, Siddiqui F (2013) Energy efficient routing in wireless sensor networks. In: Next-generation wireless technologies. Springer, London, pp 131–157
16. Rouse M (2006) Wireless sensor networks. http://searchdatacenter.techtarget.com/definition/sensor-network. Accessed 20 Feb 2015
17. Akyildiz IF, Vuran MC (2010) Wireless sensor networks, 4th edn. Wiley, New York
18. Rios LG, Diguez JEAI (2014) Big data infrastructure for analyzing data generated by wireless sensor networks. In: IEEE international congress on big data (BigData Congress), 2014. IEEE, pp 816–823
19. Jardak C, Riihijärvi J, Oldewurtel F, Mähönen P (2010) Parallel processing of data from very large-scale wireless sensor networks. In: Proceedings of the 19th ACM international symposium on high performance distributed computing. ACM, pp 787–794
20. Fan T, Zhang X, Gao F (2013) Cloud storage solution for WSN in internet innovation union. Int J Database Theory Appl 6(3):49–58
21. Ahmed K, Gregory M (2011) Integrating wireless sensor networks with cloud computing. In: Seventh international conference on Mobile Ad-hoc and Sensor Networks (MSN), 2011. IEEE, pp 364–366
22. Kwak H, Lee C, Park H, Moon S (2010) What is Twitter, a social network or a news media?. In: Proceedings of the 19th international conference on world wide web. ACM, pp 591–600
23. Ellison NB, Steinfield C, Lampe C (2007) The benefits of Facebook "friends:" social capital and college students' use of online social network sites. J Comput-Mediat Commun 12 (4):1143–1168
24. Gilbert E, Bakhshi S, Chang S, Terveen L (2013) I need to try this?: a statistical overview of pinterest. In: Proceedings of the SIGCHI conference on human factors in computing systems. ACM, pp 2427–2436
25. Shervington M (2015) What is google Plus? A complete user guide. http://www.martinshervington.com/what-is-google-plus/. Accessed 20 Apr 2015
26. Hochman N, Schwartz R (2012) Visualizing instagram: tracing cultural visual rhythms. In: Proceedings of the workshop on Social Media Visualization (SocMedVis) in conjunction with the sixth international AAAI conference on Weblogs and Social Media (ICWSM–12), pp 6–9
27. Watson I, Mullen J, Smith-Spark L (2015) CNN. Nepal earthquake: death toll passes 4,800 as rescuers face challenges. http://edition.cnn.com/2015/04/28/asia/nepal-earthquake/. Accessed on 05 May 2015
28. Ravilious K (2015) Nepal quake 'followed historic pattern'. http://www.bbc.com/news/science-environment-32472310. Accessed on 28 Apr 2015
29. Garg Y, Chatterjee N (2014) Sentiment analysis of Twitter feeds. In: Big data analytics. Springer International Publishing Switzerland, pp 33–52
30. Felemban E, Sheikh AA, Shaikh FK (2014) MMaPFlow: a crowd-sourcing based approach for mapping mass pedestrian flow. In: Proceedings of the 11th international conference on Mobile and Ubiquitous Systems: Computing, Networking and Services (MOBIQUITOUS '14)
31. Yousefi S, Mousavi MS, Fathy M (2006) Vehicular ad hoc networks (VANETs): challenges and perspectives. In: 6th international conference on ITS telecommunications proceedings, 2006. IEEE, pp 761–766
32. Ali F, Shaikh FK, Ansari AQ, Mahoto NA, Felemban E (2015) Comparative analysis of VANET routing protocols- on placement of road side units. Int J Wirel Pers Commun, Springer, pp 1–14, 2015. doi:10.1007/s11277-015-2745-z
33. Zhang Y, Zhao J, Cao G (2010) Roadcast: a popularity aware content sharing scheme in vanets. ACM SIGMOBILE Mobile Comput Commun Rev 13(4):1–14
34. Sutariya D, Pradhan SN (2010) Data dissemination techniques in vehicular ad hoc network. Int J Comput Appl 8(10):35–39

35. Dubey BB, Chauhan N, Kumar P (2010) A survey on data dissemination techniques used in VANETs. Int J Comput Appl 10(7):5–10
36. Talpur A, Baloch N, Bohra N, Shaikh FK, Felemban E (2014) Analyzing the impact of body postures and power on communication in WBAN. Procedia Comput Sci 32:894–899
37. Khelil A, Shaikh FK, Sheikh AA, Felemban E, Bojan H (2014) DigiAID: a wearable health platform for automated self-tagging in emergency cases, In: 4th international conference on wireless Mobile Communication and Healthcare (Mobihealth), 2014 EAI, pp 296,299
38. Aziz Z, Qureshi UM, Shaikh FK, Bohra N, Khelil A, Felemban E (2015) Revisiting routing in wireless body area networks. In: Emerging communication technologies based on wireless sensor networks: current research and future applications. CRC Press (to appear)
39. TDWI Best Practices Report (2015) Managing big data. http://tdwi.org/research/2013/10/tdwi-best-practices-report-managing-big-data.aspx?tc=page0. Accessed 01 Mar 2015
40. Dinh HT, Lee C, Niyato D, Wang P (2013) A survey of mobile cloud computing: architecture, applications, and approaches. Wirel Commun Mob Comput 13(18):1587–1611
41. Agrawal D, Das S, El Abbadi A (2010) Big data and cloud computing: new wine or just new bottles? Proc VLDB Endowment 3(1–2):1647–1648
42. Elazhary H (2014) Cloud computing for big data. MAGNT Res Rep 2(4):135–144
43. Gartner IT glossary (2013) Cloud computing. http://www.gartner.com/it-glossary/cloud-computing. Accessed 01 Apr 2015
44. Rodero-Merino L, Vaquero LM, Gil V, Galán F, Fontán J, Montero RS, Llorente IM (2010) From infrastructure delivery to service management in clouds. Futur Gener Comput Syst 26 (8):1226–1240
45. Patidar S, Rane D, Jain P (2012) A survey paper on cloud computing. In: Second international conference on Advanced Computing & Communication Technologies (ACCT), 2012. IEEE, pp 394–398
46. Khajeh-Hosseini A, Sommerville I, Bogaerts J, Teregowda P (2011) Decision support tools for cloud migration in the enterprise. In: IEEE international conference on Cloud Computing (CLOUD), 2011. IEEE, pp 541–548
47. Hashem IAT, Yaqoob I, Anuar NB, Mokhtar S, Gani A, Khan SU (2015) The rise of "big data" on cloud computing: review and open research issues. Inf Syst 47:98–115
48. Abadi DJ (2009) Data management in the cloud: limitations and opportunities. IEEE Data Eng Bull 32(1):3–12
49. Das S, Agrawal D, El Abbadi A (2009) Elastras: an elastic transactional data store in the cloud. USENIX HotCloud 2:7
50. Valduriez P (2009) Shared-memory architecture. In: Encyclopedia of database systems. Springer US, New York, pp 2638–2638
51. Jill Dyche (2015) Data as a service explained and defined. http://searchdatamanagement.techtarget.com/answer/Data-as-a-service-explained-and-defined Accessed on 20 Mar 2015
52. Demirkan H, Delen D (2013) Leveraging the capabilities of service-oriented decision support systems: putting analytics and big data in cloud. Decis Support Syst 55(1):412–421
53. Mathiprakasam M (2015) The road to analytics as a service. http://www.forbes.com/sites/oracle/2014/09/26/the-road-to-analytics-as-a-service/. Accessed on 20 Mar 2015
54. Poovendran R (2010) Cyber–physical systems: close encounters between two parallel worlds [point of view]. Proc IEEE 98(8):1363–1366
55. Shaikh FK, Zeadally S, Exposito E (2015) Enabling technologies for green internet of things. IEEE Syst J 99:1–12
56. Rajhans A, Cheng SW, Schmerl B, Garlan D, Krogh BH, Agbi C, Bhave A (2009) An architectural approach to the design and analysis of cyber-physical systems. Electronic Communications of the EASST, 21:1–10
57. CPS Steering Group (2008) Cyber-physical systems executive summary. CPS Summit
58. Simmon E, Kim KS, Subrahmanian E, Lee R, de Vaulx F, Murakami Y, Zettsu K, Sriram RD (2013) A vision of cyber-physical cloud computing for smart networked systems. NIST, Gaithersburg

59. Mokashi M, Alvi AS (2013) Data management in wireless sensor network: a survey. Int J Adv Res Comput Commun Eng 2:1380–1383
60. Padgavankar MH, Gupta SR (2014) Big data storage and challenges. Int J Comput Sci Inf Technol 5:2
61. Sathe S, Papaioannou TG, Jeung H, Aberer K (2013) A survey of model-based sensor data acquisition and management. In: Managing and mining sensor data. Springer US, New York, pp 9–50
62. Aggarwal CC (2013) Managing and mining sensor data. Springer Science & Business Media, New York
63. Chapman AD (2005) Principles and methods of data cleaning. GBIF, Copenhagen
64. Jeffery SR, Alonso G, Franklin MJ, Hong W, Widom J (2006) A pipelined framework for online cleaning of sensor data streams. IEEE, p 140
65. Elmenreich W (2002) Sensor fusion in time-triggered systems, Ph.D. thesis, Faculty of Informatics at the Vienna University of Technology, Austria. http://www.vmars.tuwien.ac.at/~wilfried/papers/elmenreich_Dissertation_sensorFusionInTimeTriggeredSystems.pdf
66. Hall David L, Llinas J (1997) An introduction to multisensor data fusion. Proc IEEE 85 (1):6–23
67. Castanedo F (2013) A review of data fusion techniques. Sci World J 2013:1–19
68. Kimura N, Latifi S (2005) A survey on data compression in wireless sensor networks. In: International conference on Information Technology: Coding and Computing (ITCC), 2005, vol. 2. IEEE, pp 8–13
69. Marcelloni F, Vecchio M (2008) A simple algorithm for data compression in wireless sensor networks. Commun Lett IEEE 12(6):411–413
70. Agrawal R, Faloutsos C, Swami A (1993) Efficient similarity search in sequence databases. Springer, Berlin/Heidelberg, pp 69–84
71. Gandhi S, Nath S, Suri S, Liu J (2009) Gamps: compressing multi sensor data by grouping and amplitude scaling. In: Proceedings of the 2009 ACM SIGMOD international conference on management of data. ACM, pp 771–784
72. Wang L, Deshpande A (2008) Predictive modeling-based data collection in wireless sensor networks. In: Wireless sensor networks. Springer, Berlin/Heidelberg, pp 34–51
73. Arion A, Jeung H, Aberer K (2011) Efficiently maintaining distributed model-based views on real-time data streams. In: Global Telecommunications Conference (GLOBECOM 2011). IEEE, pp 1–6
74. Xing K, Cheng X, Li J (2005) Location-centric storage for sensor networks. In: IEEE international conference on mobile adhoc and sensor systems conference. IEEE, p 10
75. Petit L, Nafaa A, Jurdak R (2009) Historical data storage for large scale sensor networks. In: Proceedings of the 5th French-speaking conference on mobility and ubiquity computing. ACM, pp 45–52
76. Diao Y, Ganesan D, Mathur G, Shenoy PJ (2007) Rethinking data management for storage-centric sensor networks. In: CIDR, vol. 7, pp 22–31
77. Dutta P, Culler DE, Shenker S (2007) Procrastination might lead to a longer and more useful life. In: The sixth workshop on Hot Topics in Networks (HotNets-VI) pp 1–7
78. Deshpande A, Madden S (2006) MauveDB: supporting model-based user views in database systems. In: Proceedings of the 2006 ACM SIGMOD international conference on management of data. ACM, pp 73–84
79. Kanagal B, Deshpande A (2008) Online filtering, smoothing and probabilistic modeling of streaming data. In: IEEE 24th international conference on Data Engineering, ICDE 2008. IEEE, pp 1160–1169
80. Bhattacharya A, Meka A, Singh AK (2007) Mist: distributed indexing and querying in sensor networks using statistical models. In: Proceedings of the 33rd international conference on very large data bases. VLDB Endowment, pp 854–865
81. Cugola G, Margara A (2012) Processing flows of information: from data stream to complex event processing. ACM Comput Surv (CSUR) 44(3):15

82. Babcock B, Babu S, Datar M, Motwani R, Widom J (2002) Models and issues in data stream systems. In: Proceedings of the twenty-first ACM SIGMOD-SIGACT-SIGART symposium on principles of database systems. ACM, pp 1–16
83. Luckham D (2002) The power of events, vol 204. Addison-Wesley, Reading
84. Golab L, Özsu MT (2003) Issues in data stream management. ACM Sigmod Rec 32(2):5–14
85. Dunkel J (2009) On complex event processing for sensor networks. In: International symposium on autonomous decentralized systems, 2009. ISADS'09. IEEE, pp 1–6
86. Miller S (2013) Big data analytics. Podcasts at Singapore Management University, Available at: http://ink.library.smu.edu.sg/podcasts/8
87. Big Data in the Cloud Converging Technologies-Intel (2014) http://www.intel.com/content/www/us/en/big-data/big-data-cloud-technologies-brief.html. Accessed on Apr 2015
88. Tang LA, Yu X, Kim S, Han J, Peng WC, Sun Y, Gonzalez H, Seith S (2012) Multidimensional analysis of atypical events in cyber-physical data. In: IEEE 28th international conference on Data Engineering (ICDE), 2012. IEEE, pp 1025–1036
89. Tang LA, Yu X, Kim S, Han J, Peng WC, Sun Y, Leung A, La Porta T (2012) Multidimensional sensor data analysis in cyber-physical system: an atypical cube approach. Int J Distrib Sens Netw 2012:1–19
90. Yu X, Tang LA, Han J (2009) Filtering and refinement: a two-stage approach for efficient and effective anomaly detection. In: Ninth IEEE international conference on Data Mining, 2009. ICDM'09. IEEE, pp 617–626
91. Tang LA, Yu X, Kim S, Han J, Hung CC, Peng WC (2010) Tru-alarm: trustworthiness analysis of sensor networks in cyber-physical systems. In: IEEE 10th international conference on Data Mining (ICDM), 2010. IEEE, pp 1079–1084
92. Xia F, Kong X, Xu Z (2011) Cyber-physical control over wireless sensor and actuator networks with packet loss. In: Wireless networking based control. Springer, New York, pp 85–102
93. Thouin F, Thommes R, Coates MJ (2006) Optimal actuation strategies for sensor/actuator networks. In: 3rd annual international conference on mobile and ubiquitous systems: networking & services, 2006. IEEE, pp 1–8
94. Conti M, Das SK, Bisdikian C, Kumar M, Ni LM, Passarella A, Roussos G, Tröster G, Tsudik G, Zambonelli F (2012) Looking ahead in pervasive computing: challenges and opportunities in the era of cyber–physical convergence. Pervasive Mob Comput 8(1):2–21
95. Guturu P, Bhargava B (2011) Cyber-physical systems: a confluence of cutting edge technological streams. International conference on advances in computing and communication

Chapter 7
Distributed Platforms and Cloud Services: Enabling Machine Learning for Big Data

Daniel Pop, Gabriel Iuhasz, and Dana Petcu

Abstract Applying popular machine learning algorithms to large amounts of data has raised new challenges for machine learning practitioners. Traditional libraries do not support properly the processing of huge data sets, so the new approaches are needed. Using modern distributed computing paradigms, such as MapReduce or in-memory processing, novel machine learning libraries have been developed. At the same time, the advance of cloud computing in the past 10 years could not be ignored by the machine learning community. Thus, a rise of cloud-based platforms has been of significance. This chapter aims at presenting an overview of novel platforms, libraries, and cloud services that can be used by data scientists to extract knowledge from unstructured and semi-structured, large data sets. The overview covers several popular packages to enable distributed computing in popular machine learning environments, distributed platforms for machine learning, and cloud services for machine learning, known as machine-learning-as-a-service approach. We also provide a number of recommendations for data scientists when considering machine learning approach for their problem.

Keywords Machine learning • Data mining • Cloud computing • Big data • Data scientist • Distributed computing • Distributed platforms

7.1 Introduction

Analyzing large amounts of data collected by companies, industries, and scientific domains is becoming increasingly important for all impacted domains. The data to be analyzed is no longer restricted to sensor data and classical databases, but it often includes textual documents and Web pages (text mining, Web mining), spatial data, multimedia data, or graph-like data (e.g., molecule configuration and social networks).

D. Pop (✉) • G. Iuhasz • D. Petcu
Institute e-Austria Timisoara, West University of Timisoara, Blvd. Vasile Parvan, nr. 4, 300223 Timişoara, Romania
e-mail: daniel.pop@e-uvt.ro

Z. Mahmood (ed.), *Data Science and Big Data Computing*,
DOI 10.1007/978-3-319-31861-5_7

Although, for more than two decades, parallel database products such as Teradata, Oracle, and Netezza have provided means to realize a parallel implementation of machine learning algorithms, expressing these algorithms in SQL code is a complex and difficult-to-maintain task. On the other side, large-scale installations of these products are expensive. Another reason for moving away from relational databases is the exponential growth of the unstructured data (e.g., audio and video) and semi-structured data (e.g., Web traffic data, social media content, sensor-generated data) in recent years. The needs of data science practitioners with respect to data analysis tools vary greatly across different domains, from medical statistics and bioinformatics to social network analysis or even in physics. This diversity is equally important for the advancement of machine learning tools and platforms. Consequently, in the past decade, researchers moved from the parallelization of machine learning algorithms and support in relational databases toward the design and implementation on top of novel distributed storage (e.g., NoSQL data stores, distributed file systems) and processing paradigms (e.g., MapReduce). From the business perspective, Software-as-a-Service (SaaS) model opened up new opportunities for machine learning providers, who moved the stand-alone tools toward cloud-based machine learning services.

In this chapter, we survey how distributed storage and processing platforms help data scientists to process large, heterogeneous sets of data. The tools, frameworks, and services included in this chapter share a common characteristic: all run on top of distributed platforms. Thus, parallelization of machine learning algorithms, either using multiple-core CPU or GPU, was not included here. The reader is referred to [42], a recent comprehensive study covering that topic. We also avoided commercial solution providers, small or big players, since their offerings are either based on distributed open-source packages or they do not disclose the implementation details.

In the first section, we briefly introduce the reader to the machine learning field, describing and classifying the types of problems and overviewing the challenges of applying traditional algorithms to large, unstructured data sets. The first category of tools considered in this survey covers tools, packages, and libraries that enable data scientists to use traditional environments for data analysis such as R Systems, Python, or statistics applications, in order to deal with large data sets. We survey, next, the distributed platforms for big data processing, either based on Apache Hadoop or Spark, as well as platforms specifically designed for distributed machine learning. We also include a section on scalable machine learning services delivered using Software-as-a-Service business model since they offer easy-to-use, user-friendly graphical interfaces supporting users in quickly getting and deploying models. The last section of the chapter summarizes our findings and provides readers with a collection of best practices in applying machine learning algorithms.

7.2 Machine Learning for Data Science

The broadest and simplest definition of machine learning is that *it is a collection of computational methods that use experience*, i.e., information available to the system to improve performance or to make predictions [28]. This information usually takes the form of electronic records collected and made available for analytical purposes. These records can take the form of pre-labeled training sets (usually by a human operator although this is not always the case). Another important source of data is that resulting from direct interactions with a given environment, either virtual, such as software interactions, network data, etc., or relying on real-world natural scenarios, such as weather phenomena, water level, etc. Data quality and quantity are extremely important in order to obtain an acceptable learned model. Machine learning relies on data-driven methods that combine fundamental concepts in the field of computer science with optimization, probability, and statistics [28].

There is a wide array of applications to which machine learning can and is being applied, such as taming (text mining and document classification), spam detection, keyword extraction, emotion extraction, natural language processing (NPL), unstructured text understanding, morphological analysis, speech synthesis and recognition, optical character recognition (OCR), computational biology, face detection, image segmentation, image recognition, fraud detection, network intrusion detection, board and video games, navigation in self-driving vehicles, planning, medical diagnosis, recommendation systems, or search engines. In all these applications, we can identify several types of learning-related issues, which are:

- *Classification* – to assign each item from a data set to a specific category, e.g., given a document, assign a domain (history, biology, mathematics) to which it belongs.
- *Regression and time series analysis* – to predict a real value for each item, e.g., future stock market values, rainfall runoff, etc.
- *Ranking* – to return an ordered set of features based on some user-defined criterion (e.g., Web search).
- *Dimensionality reduction (feature selection)* – to use for transforming initial large feature spaces into a lower-dimensional representation so that it preserves the properties of the initial representation.
- *Clustering* – to group items based on some predefined distance measure. It is usually used on very large data sets. In sociology, it can be used to group individuals into communities [30].
- *Anomaly detection* – to conduct observation or series of observations which do not resemble any pattern or data item in a data set [6, 37].

The most common classification techniques are called *linear classifiers*. In this case, classification is expressed in the form of a linear function. This function assigns scores to each possible category. Among the linear classifiers, we have linear regression, perceptron, and support vector machines (SVM) [28]. Another form of classification is based on kernel estimation in the form of the k-nearest neighbor (k-NN) algorithms. Decision trees such as C4.5 [28] are also used for this type of problem and are based on information theory (difference in entropy), which is used as the splitting criterion. Ensemble meta-algorithm-based techniques such as AdaBoost [28] are also used although they have questionable performance on noisy data sets. Some methods such as Classification and Regression Tree (CART) algorithm can be used for both regression and classification problems. Naïve Bayes, for example, can also be used for both types of problems.

Clustering algorithms are largely split according to their particular definition of what cluster model they use. Connectivity models (hierarchical clustering) are based on distance connectivity. Centroid models, such as k-means (k-M), represent each cluster with a single mean vector. Density models consider clusters as connected dense regions from the data space. DBSCAN and OPTICS [28] are two algorithms using this model. Statistical distribution-based models are also used.

Anomaly detection is a special case of either classification or clustering; thus, it uses mostly the same algorithms and methodologies. Feature selections' main goal is the reduction of the amount of recourses required to analyze big data set. They are extremely useful when no domain expert is available that could help in the reduction of the dimensionality of the available data. There are a number of general dimensionality reduction techniques such as principal component analysis (PCA), kernel PCA, multilinear PCA, and wrapping methods [25].

In machine learning, there are different types of training scenarios [28]. Arguably the most widely used type of training is called *supervised* learning. In this scenario, the learner receives a set of labeled data for training and validation. The learned prediction model can be then applied to a larger data set and identify all unseen data points. This type of learning is used for classification and regression (time series analysis). Supervised methods rely on the availability and accuracy of labeled data sets. In *unsupervised* learning, the learner receives unlabeled data that it has to group based on a distance measurement. In some scenarios, labeled data is extremely hard to come by; thus, training a classification model is often unfeasible. This type of learning is used for clustering, anomaly detections (a type of clustering), and dimensionality reduction.

In some cases, labeled data is only a small fraction of the overall training data set. This is called *semi-supervised* learning. The idea is that the distribution of unlabeled data can help the learner achieve a much better performance [11].

In *reinforcement* learning, the training is done using an evaluation function. This means that training and testing are much more interlaced than in other learning scenarios. The performance of an algorithm in a problem environment is continuously evaluated through the monitoring and evaluation of its performance. Favorable outcomes are rewarded, while unfavorable ones are punished. Reinforcement learning is used in genetic algorithm, neural networks, etc. *Online* learning is used when data is available in a sequential way. This means that the mapping between data sets and labels is established each time a new data point is received.

Due to the popularity of data analytics, machine learning techniques are being pursued by teams with complementary skills across very different businesses (finance, telecommunications, life sciences, etc.). This section aims to classify the diversity of groups of interests with respect to machine for big data. We must state upfront that there is no clear line between these perspectives, as competencies and expectations blur the edges and multidisciplinary teams are put in place to tackle complex scenarios. Some of the groups are:

- *Data scientists and machine learning practitioners*: One way of approaching the problem is from the data scientist's perspective. Statisticians and data scientists are now facing data set size explosion; thus, coping with large-size data sets is a must. These are users with strong mathematical background, proficient in statistics and mathematical software applications, such as R, Octave, MATLAB, Mathematica, Python, SAS Studio, or IBM's SPSS, but less experienced in coping with data sets of large dimensions, distributed computing, or software development. Their expectation is to easily reuse the algorithms already available in their preferred language and be able to run them against large data sets on distributed architectures (on-premise or cloud based). A later section in this chapter entitled "Distributed and Cloud-Based Execution Support in Popular Machine Learning Tools" overviews packages and tools available for this purpose.
- *Software engineers and developers*: Teams of software engineers often face client requirements asking for the transition from available (large) data warehouse to actionable knowledge. These are users with a vast experience in software development, skilled programmers in general-purpose programming languages, and they "speak" parallel and distributed computing. Deep mathematics and statistics are not necessarily their preferred playground, as they expect tools and libraries to enable them to integrate advanced ML algorithms in their systems and thus quickly get actionable results. They need fast, easy-to-customize (less number of parameters), and easy-to-integrate algorithms that run on distributed architectures and are able to fetch data from large data repositories. Tools addressing these requirements are discussed in a later section in this chapter entitled "Distributed Machine Learning Platforms."
- *Domain experts*: Domain experts (financial, telecommunications, physics, astronomy, biotechnologies, etc.) know their data best, but they are less experienced in ML algorithms and software tools. Ideally, they need off-the-shelf software applications, easy to install and use, or cloud-based Software-as-a-Service solutions allowing them to get insights on their data and produce reports and executable models for further usage. A later section on "Machine Learning as a Service" presents several machine learning services providers.

The dynamic of natural, social, and economic systems raises new challenges for data scientists, such as:

- *Massive data sets*. Data sets are growing faster, being common now to reach numbers of 100 TB or more. The Sloan Digital Sky Survey occupies 5 TB of storage, the Common Crawl Web corpus is 81 TB in size, and the 1000 Genomes Project requires 200 TB of space, just to name a few.
- *Large models*. Massive data sets need large models to be learned. Some deep neural networks are comprised of more than ten layers with more than a billion parameters [24, 25], collaborative filtering for video recommendation on Netflix comprises 1–10 billion parameters, and multitask regression model for simplest whole-genome analysis may reach 1 billion parameters as well.
- *Inadequate ML tools and libraries*. Traditional ML algorithms used for decades (k-means, logistic regression, decision trees, Naïve Bayes) were not designed for handling large data sets and huge models; they were not developed for parallel/distributed environments.
- *"Operationalization" of predictive models*. "Operationalize" refers to integrate predictive models into automated decision-making systems and processes on a large scale in order to deliver predictions to end users, who will ultimately benefit from them. Integrating these models into multiple platforms (Web, stand-alone, mobile) across different business units requires a high degree of customization, which slows deployment, drives up costs, and limits scalability.
- *Lack of clear contracts*. More recently, terms such as Analytics as a Service (AaaS) and Big Data as a Service (BDaaS) are becoming popular. They comprise services for data analysis similarly as IaaS offers computing resources. Unfortunately, the analytics services still lack well-defined service-level agreements available for IaaS because it is difficult to measure quality and reliability of results and input data, to provide promises on execution times and guarantees on methods for analyzing the data. Therefore, there are fundamental gaps on tools to assist service providers and clients to perform these tasks and facilitate the definition of contracts for both parties [2].
- *Inadequate staffing*. Market research shows that inadequate staffing and skills, lack of business support, and problems with analytics software are some of the barriers faced by corporations when performing analytics [36].

In the next three sections, we discuss various machine learning tools.

7.3 Distributed and Cloud-Based Execution Support in Popular Machine Learning Tools

Annual Nuggets survey [23] shows that R, Python, SQL, and SAS have been rated the preferred languages of choice for the past 3 years. One of the early trends matching cloud computing and data analysis was, around 2010s, the provision of virtual machine images (VMI) for these popular systems (R, Octave, or Maple) integrated within public cloud service providers, such as Amazon Web Services or Rackspace. After several proofs of concept were successfully built, such as

Table 7.1 Distributed processing and storage

Environment	Package	Distributed processing support	Distributed file system access
R	RHadoop	Hadoop	HDFS
	RHIPE	Hadoop	HDFS
	Segue for R	Amazon Elastic MapReduce	–
	RHive	HIVE	HIVE
	Snow	Socket-based, MPI, PVM	–
	H5	–	HDF5
	Pbd*	MPI	NetCDF
Python	pyDoop	Hadoop	HDFS
	Anaconda	Distributed and GPU	HDFS, HDF5
	IPython. parallel	Distributed and parallel	–

Cloudnumbers,[1] CloudStat,[2] Opani,[3] and Revolution R Enterprise,[4] the practice today is to provide VMI through the public cloud providers' marketplaces, such as Amazon Marketplace. One can find Amazon Machine Images (AMI), via the marketplace, for all the popular mathematical and statistics environments. Examples include Predictive Analytics Framework and Data Science Toolbox[5] that support both Python and R, BF Accelerated Scientific Compute for R with accelerated math libraries for boosted performance, or SAS University Edition for SAS Studio.

Much more effort has been invested in the development of plug-ins for the most popular machine learning platforms to allow data scientists to easily create and run time-consuming jobs over clusters of computers. This approach allows ML practitioners to reuse their existing code and adapt it for large data set processing, into the same environment they used for prototyping. It also leverages existing infrastructure (grids, clusters) for large-scale distributed computation and data storage. Table 7.1 synthesizes available plug-ins for distributed storage and processing for the most popular languages of big data: R and Python.

Since R is the preferred option among machine learning practitioners, several packages were developed in order to enable big data processing within R, most of them being available under CRAN[6] package Web page. These R extensions make possible to distribute the computational workload on different types of clusters, while accessing data from distributed file systems. First example is the RHadoop [33], a collection of five R packages, that enables R users to run MapReduce jobs on

[1] http://cloudnumbers.com

[2] http://cs.croakun.com

[3] http://opani.com

[4] www.revolutionanalytics.com

[5] http://datasciencetoolbox.org

[6] http://cran.r-project.org/web/packages/available_packages_by_name.html

Hadoop by writing R functions for mapping and reducing. Similarly, RHIPE[7] is another R package that brings MapReduce framework to R practitioners, providing seamless access to Hadoop cluster from within R environment. Using specific R functions, programmers are able to launch MapReduce jobs on the Hadoop cluster, with results being easily retrieved from HDFS. Segue[8] for R project makes it easier to execute MapReduce jobs from within the R environment on elastic clusters at Amazon Elastic MapReduce,[9] but lacks support for handling large data sets. RHive is an extension enabling distributed computing via HIVE in R, by a seamless integration between HQL (Hive Query Language) and R objects and functions. Snow (Simple Network of Workstations) [41] and its variants (snowfall, snowFT, doSnow) implement a framework that is able to express an important class of parallel computations and is easy to use within an interactive environment like R. It supports three types of clusters: socket based, MPI, and PVM. Support for manipulating large data sets in R is available in H5 plug-in, which provides an interface to the HDF5 API through S4 objects, supporting fast storage and retrieval of R objects to/from binary files in a language-independent format. The pbd* (pbdBASE, pbbMPI, pbdNCDF4, pbdSLAP, etc.) series is a collection of R packages for programming with big data, enabling MPI distributed execution, NetCDF file system access, or tools for scalable linear algebra.

As far as Python is concerned, we should start by mentioning pyDoop,[10] a Python MapReduce and HDFS API for Hadoop [26]. Anaconda[11] is a free, scalable Python distribution for large-scale data analytics and scientific computing. It is a collection of Python packages (NumPy, SciPy, Pandas, IPython, Matplotlib, Numba, Blaze, Bokeh) that enables fast large data set access, GPU computation, access to distributed implementations of ML algorithms, and more. IPython.parallel[12] provides a sophisticated and powerful architecture for parallel and distributed computing [14] that enables IPython to support many different styles of parallelism including single program multiple data (SPMD), multiple program multiple data (MPMD), message passing using MPI, data parallel, and others. In a tutorial at PyCon 2013, Grisel [15] presented how scikit-learn [32], a popular open-source library for machine learning in Python, can be used to perform distributed machine learning algorithms on a cheap Amazon EC2 cluster using IPython.parallel and StarCluster.[13] We should note as well that most of the libraries and frameworks considered in the next sections offer Python language bindings, but we choose not to include them in this section.

---

[7] http://www.stat.purdue.edu/~sguha/rhipe/doc/html/index.html

[8] http://code.google.com/p/segue

[9] http://aws.amazon.com/elasticmapreduce

[10] https://github.com/crs4/pydoop

[11] https://store.continuum.io/cshop/anaconda

[12] http://ipython.org/ipython-doc/dev/parallel/

[13] http://star.mit.edu/cluster/

Other mathematical and statistics environments have seen similar interest in embracing big data processing. For example, HadoopLink[14] is a package that allows MapReduce programs being implemented in Mathematica and to run them on a Hadoop cluster. It looks more like a proof of concept (PoC), being stalled since 2013. MATLAB has its Parallel Computing Toolbox which extends the capabilities of MATLAB MapReduce and Datastore[15] in order to run big data application. MATLAB Distributed Computing Server also supports running parallel MapReduce programs on Hadoop clusters.[16]

There are extensions to traditional machine learning libraries that enable execution on top of Hadoop or Spark clusters. Weka [16], one of the most popular libraries for data mining, supports both Hadoop and Spark execution through Weka Hadoop integration [17]. There is also a commercial distribution, Pentaho [34], that offers a complete solution for big data analytics, supporting all phases of an analytics process – from preprocessing to advanced data exploration and visualization, which uses distributed Weka execution for analytics. Another example is the KNIME's [4] big data extension,[17] which enables the access to Hadoop via Hive. RapidMiner [20] has Radoop[18] that enables the deployment of workflows on Hadoop.

7.4 Distributed Machine Learning Platforms

After distributed processing and storage environments (Hadoop, Dryad, MPI) reached an acceptable level of maturity, they became an increasingly appealing foundation for the design and implementation of new platforms for machine learning algorithms. These provide users out-of-the-box algorithms, which are run in parallel mode over a cluster of (commodity) computers. These solutions do not use statistics, or mathematics software packages, rather they offer self-contained, optimized implementations in general-purpose programming languages (C/C++, Java) of state-of-the-art ML methods and algorithms. This section focuses on ML platforms specifically designed for distributed and scalable computing. Table 7.2 summarizes some recent platforms.

The IBM Research Lab has been one of the pioneers who invested in distributed machine learning frameworks. Nimble [12] and SystemML [13] are two high-level conceptual frameworks supporting the definition of ML algorithms and their execution on Hadoop clusters. Nimble, a sequel to IBM's Parallel Machine Learning Toolbox [31], features a multilayered framework enabling developers to express

[14] https://github.com/shadanan/HadoopLink

[15] http://www.mathworks.com/help/matlab/large-files-and-big-data.html

[16] http://www.mathworks.com/help/distcomp/big-data.html

[17] https://www.knime.org/knime-big-data-extension

[18] https://rapidminer.com/products/radoop/

Table 7.2 Distributed ML frameworks

Name	License	ML Problem	Distr. Env.	Comm.	Lang.
Petuum	Open source (Sailing Lab)	DL, CLS, CLU, RGR, MET, TOP	Clusters or Amazon EC2, Google CE	Medium	C++
Jubatus	LGPL v2.1	CLS, RGR, ANO, CLU, REC, Graph	Zookeeper	Medium	C++
MLlib (MLBase)	Apache 2.0	RGR, CLS, REC, CLU	Spark	Large	Scala, Java
Mahout	Apache 2.0	Collaborative filtering, CLS, CLU, DR, TOP	Hadoop, Spark, H2O	Medium	Java
Oryx	Apache 2.0	REC, CLS, RGR, CLU	Hadoop, Spark	Low	Java
Trident-ML	Apache 2.0	CLS, RGR, CLU, DR	Storm	Low	Java
H2O	Apache 2.0	DL, RGR, CLS, CLU, DR	Hadoop	Medium	Java
GraphLab Create	Apache 2.0	CLU, CLS, RGR, DL, REC	Hadoop, Spark, MPI	High	C++
Vowpal Wabbit	Ms-PL	CLS, RGR, CLU	Hadoop	Medium	C++
Deeplearning4J	Apache 2.0	DL	Hadoop, Spark, AWS, Akka	Medium	Java, Scala
Julia's MLBase	MIT License	CLS	Julia		Julia
Flink-ML	Apache 2.0	CLS, RGR, CLU, REC	Flink Hadoop	Low	Scala
DryadLINQ			Dryad, Hadoop YARN	None	C#, LINQ
Nimble	NA	CLU, FRQ, ANO,	Hadoop	None	Java
SystemML	NA	RGR, PageRank	Hadoop	None	DML

ANO anomaly detection, *CLS* classification, *CLU* clustering, *DL* deep learning, *DR* dimensionality reduction, *FRQ* frequent pattern, *MET* metrics learning, *REC* recommendation, *RGR* regression, *TOP* topic modeling

their ML algorithms as tasks, which are then passed to the next layer, an architecture-independent layer, composed of one queue of DAGs of tasks, plus worker thread pool that unfolds this queue. The bottom layer is an architecture-dependent layer that translates the generic entities from the upper layer into various runtimes, the only distributed environment supported within the proof of concept being Hadoop alone. The layered architecture of the system hides the low-level control and choreography details of most of the distributed and parallel programming paradigms (MR, MPI, etc.), it allows developers to compose parallel ML algorithms using reusable (serial and parallel) building blocks, but also it enables portability and scalability. SystemML proposes an R-like language (declarative machine learning language) that includes linear algebra primitives and shows how it can be optimized and compiled down to MapReduce. Authors report an extensive performance evaluation on three ML algorithms (group nonnegative matrix factorization, linear regression, PageRank) on varying data and Hadoop cluster sizes. These two systems are purely research endeavors, and they are not available to the community.

Most of the frameworks rely on Hadoop's MapReduce paradigm and the underlying distributed file storage system (HDFS) because it simplifies the design and implementation of large-scale data processing systems. Only a few frameworks (e.g., Jubatus, Petuum, GraphLab Create) have tried to propose novel distributed paradigms, customized to machine learning for big data, in order to optimize the complex, time-consuming ML algorithms.

Recognizing the limitations and difficulties of adapting general-purpose distributed frameworks (Hadoop, MPI, Dryad, etc.) to ML problems, a team at CMU under E. P. Xing lead designed a new framework for distributed machine learning able to handle massive data sets and cope with big models. Petuum[19] (from Perpetuum Mobile) [8, 43] takes advantage of data correlation, staleness, and other statistical properties to maximize the performance for ML algorithms, realized through core features such as a distributed Parameter Server and a distributed Scheduler (STRADS). It may run either on-premise clusters or on cloud computing resources like Amazon EC2 or Google Compute Engine (GCE).

Jubatus[20] is a distributed computing framework specifically designed for online machine learning on big data. A loose model sharing architecture allows it to efficiently train and share machine learning models by defining three fundamental operations, viz., update, mix, and analyze [19]. Comparing to Apache Mahout, Jubatus offers stream processing and online learning, which means that the model is continuously updated with each data sample that is coming in, by fast, not memory-intensive algorithms. It requires no data storage nor sharing, only model mixing. In order to efficiently support online learning, Jubatus operates updates on local models and then each server transmits its model difference that are merged and distributed back to all servers. The mixed model improves gradually thanks to all servers' work.

GraphLab Create,[21] formerly GraphLab project [27], is a framework for machine learning that expresses asynchronous, dynamic, graph-parallel computation while ensuring data consistency and achieving a high degree of parallel performance, in both shared-memory and distributed settings. It is an end-to-end platform enabling data scientists to easily create intelligent apps at scale, from cleaning the data, developing features, training a model, and creating and maintaining a predictive service. It runs on distributed Hadoop/YARN clusters, as

[19] http://petuum.org

[20] http://jubat.us

[21] https://dato.com/products/create

well on local machine or on EC2, and it exposes a Python interface for an easy accessibility.

Apache Mahout [29] is a scalable machine learning framework built on top of Hadoop that features a rich collection of distributed implementations of machine learning and data mining algorithms. Although initially created on top of Hadoop, starting with version 0.10, it supports additional execution engines such as Spark and H20, while Flink[22] is a project in progress. The same release introduces Mahout-Samsara, a new math environment created to enable users to develop their own extensions, using Scala language, based on general linear algebra and statistical operations. Mahout-Samsara comes with an interactive shell that runs distributed operations on a Spark cluster. This makes prototyping or task submission much easier and allows users to customize algorithms with a whole new degree of freedom.

H2O[23] and MLlib [10] are two of the most actively developed projects. Both feature distributed, in-memory computations and are certified for Apache Spark (MLlib being part of Spark), as well as for Hadoop platforms. This in-memory capability means that in some instances these frameworks outperform Hadoop-based frameworks [44]. MLlib has been shown to be more scalable than Vowpal Wabbit. One important distinction when comparing H2O with other MapReduce applications is that each H2O node (which is a single JVM process) runs as a mapper in Hadoop. There are no combiners nor reducers. Also, H2O has more built-in analytical features and a more mature REST API for R, Python, and JavaScript than MLlib.

Vowpal Wabbit[24] [38] is an open-source, fast, out-of-core learning system, currently sponsored by Microsoft research. It has an efficient implementation of online machine learning, using the so-called hash trick [35] as the core data representation, which results in significant storage compression for parameter vectors. VW reduces regression, multiclass, multi-label, or structured prediction problems to a weighted binary classification problem. A Hadoop-compatible computational model called AllReduce [1] has been implemented in order to eliminate MPI and MapReduce drawbacks, which relate to machine learning. Using this model, a 1000-node cluster was able to learn a terafeature data set in one hour [1].

Julia[25] is high-level, high-performance, and dynamic programming language. It is designed for computing and provides a sophisticated compiler, distributed parallel execution, and numerical accuracy and has an extensive mathematical function library. Comparing to traditional MPI, Julia's implementation of message passing is "one sided," thus simplifying the process management. Furthermore, these operations typically do not look like "message send" and "message receive" but rather resemble higher-level operations like calls to user functions. It also provides a

[22] https://flink.apache.org/

[23] http://0xdata.com/product/

[24] https://github.com/JohnLangford/vowpal_wabbit/

[25] http://julialang.org/

powerful browser-based notebook using IPython. It also possesses a built-in package manager and it is able to call C functions directly. It is specially designed for parallelism and distributed computation. It also provides a variety of classification, clustering, and regression analysis packages[26] implemented in Julia.

Another framework focusing on real-time online machine learning is Trident-ML [22], built on top of Apache Storm, a distributed stream-processing framework. It processes batches of tuples in a distributed way, which means that it can scale horizontally. However, Storm does not allow state updates to append simultaneously, a shortage that hinders distributed model learning.

The Apache Oryx 2[27] framework is a realization of the lambda architecture built on top of Spark and Apache Kafka. It is a specialized framework that provides real-time, large-scale machine learning. It consists of three tiers: lambda, machine learning, and application. The lambda tier is further split up into batch, speed, and serving tier, respectively. Currently, it has only three end-to-end implementations for the batch, speed, and serving layers (collaborative filtering, k-means clustering, classification, and regression based on random forest). Although it has only these three complete implementations, its main design goal is not that of a traditional machine learning library but more of a lambda architecture-based platform for MLlib and Mahout. At this point, it is important to note several key differences between Oryx 1[28] and Oryx 2. Firstly, Oryx 1 has a monolithic tier for lambda architecture, while Oryx 2 has three as mentioned in the previous paragraph. The streaming-based batch layer in Oryx 2 is based in Spark, while in the first version, it was a custom MapReduce implementation in the computational layer. Two of the most important differences relate to the deployment of these frameworks. Oryx 2 is faster yet more memory hungry than the previous version because of its reliance on Spark. Second, the first version supported local (non-Hadoop) deployment, while the second version does not.

DryadLINQ[29] [5] is LINQ[30] (Language Integrated Query) subsystem developed at Microsoft Research on top of Dryad [21], a general-purpose architecture for execution of data-parallel applications. A DryadLINQ program is a sequential program composed of LINQ expressions performing arbitrary side effect-free transformations on data sets and can be written and debugged using standard . NET development tools. The system transparently translates the data-parallel portions of the program into a distributed execution plan, which is passed to the Dryad execution platform that ensures efficient and reliable execution of this plan. Following Microsoft's decision to focus on bringing Apache Hadoop to Windows systems, this platform has been abandoned, and Daytona project took off, which has recently became Windows Azure Machine Learning platform.

---

[26] http://mlbasejl.readthedocs.org/en/latest/

[27] http://oryxproject.github.io/oryx/

[28] https://github.com/cloudera/oryx

[29] http://research.microsoft.com/en-us/projects/dryad/

[30] http://msdn.microsoft.com/netframework/future/linq/

Deeplearning4J[31] is an open-source distributed deep-learning library written in Java and Scala. It is largely based on ND4J library for scientific computation that enables GPU, as well as native code integration. It is also deployable on Hadoop, Spark, and Mesos. The main difference between this library and the others mentioned above is that it is mainly focused on business use cases, not on research. This means that some features, such as parallelism, is automatic, meaning that worker nodes are set up automatically.

Apache SAMOA[32] is a distributed streaming machine learning framework. It contains abstractions for distributed streaming machine learning algorithms. This means that users can focus on implementing distributed algorithms and not worry about the underlying complexities of the stream processing engines it supports (Storm, S4, Samza, etc.).

7.5 Machine Learning as a Service (MLaaS)

This section focuses on Software-as-a-Service providers' provision of machine learning services as MLaaS. These services are accessible via RESTful interfaces, and in some cases, the solution may also be installed on-premise (e.g., ersatz). The favorite class of machine learning problems addressed by these services is predictive modeling (BigML, Google Prediction API, EigenDog), while clustering and anomaly detection receive far less attention. We did not include in this category the fair number of SQL over Hadoop processing solutions (e.g., Cloudera Impala, Hadapt, Hive), because their main target is not machine learning problems, rather fast, elastic, and scalable SQL processing of relational data using the distributed architecture of Hadoop.

Table 7.3 presents current MLaaS solutions as well as some of their key characteristics. We have identified four characteristics: machine learning problem support, data sources, model exporting, and model deployment. It is easily observable that most MLaaS are designed to deal with common problems such as classification, regression, and clustering. When it comes to data acquisition facilities, all platforms support data upload (various formats csv, arff, etc.); some even feature integration with different storage solutions (S3, HDFS, etc.). Predictive model training, verification, and visualization are supported by all the solutions listed in Table 7.3; however, not all support predictive model exporting via PMML.[33] Last but not least, some platforms support local Web services as well as cloud deployment. In the next few paragraphs, we detail some of the more important services from Table 7.3.

[31] http://deeplearning4j.org

[32] https://samoa.incubator.apache.org/

[33] http://www.dmg.org/v4-1/GeneralStructure.html

Table 7.3 Machine learning as a service

Name	ML problems	Data source	Model export	Deployment
Azure ML	CLS, RGR, CLU, ANO	Upload, Azure	None	Cloud
PredictionIO	RGR, CLS, REC, CLU	Upload, Hbase	None	Local, cloud
Ersatz Labs	DL	Upload	None	Cloud, local
ScienceOps[a] (ScienceBox)	RGR, CLS, REC, CLU	S3, Upload	PMML	Cloud, local
Skymind	DL	Upload	None	Cloud, Local
BigML[b]	CLS, RGR, CLU	Upload, S3, Azure, OData	PMML	Cloud
Amazon ML	CLS, RGR, CLU	S3, Redshift Upload	None	Cloud
BitYota[c]	CLS, RGR, CLU	S3, Azure	None	Cloud
Google Prediction API	CLS, RGR, CLU, ANO	Upload, Google Cloud Storage	PMML	Cloud
EigenDog[d]	CLU, RGR	Upload, S3	None	Local, cloud
Metamarkets[e]	CLU, ANO	Upload, HDFS	None	Local (Druid), cloud
Zementis ADAPA[f]	CLS, RGR, CLU	S3, Azure, Upload, SAP HANA	PMML	Local, cloud

[a]https://yhathq.com/products/scienceops
[b]http://bigml.com
[c]http://bityota.com
[d]https://eigendog.com/\#home
[e]http://metamarkets.com/
[f]http://zementis.com/products/adapa/amazon-cloud/

Windows Azure Machine Learning, formerly project Daytona, was officially launched in February 2015 as a cloud-based platform for big data processing. This comes with a rich set of predefined templates for data mining workflows, as well with a visual workflow designer that allows end users to compose complex machine learning workflows. In addition, it supports the integration of R and Python scripts within workflows and is able to run the jobs on Hadoop and Spark platforms. The built models are deployed in a highly scalable cloud environment and can easily be accessed via Web services [39].

PredictionIO[34] is based on an open-source software such as Spark. This means that the solution can be deployed and hosted on any infrastructure. This is in sharp contrast with Azure, which requires the data to be uploaded into Azure. Also, it is possible to write custom distributed data processing tasks in Scala while on Azure custom scripts can only be run on a single node. There are no restrictions on the size of the training data not on the number of concurrent request. It can be deployed on Amazon WS, Vagrant, Docker, or even starting from source code.

[34] https://prediction.io/

The recent popularity of deep learning has resulted in the creation of various services which bundle deep-learning libraries (Theano, pylearn2, Deeplearning4j, etc.) into a MLaaS format. Some good examples are Ersatz Labs[35] and Skymind.[36] These provide similar services and support distributed as well as GPU deployment.

Amazon Machine Learning service[37] allows users to train predictive models in the cloud. It targets a similar use case as Azure Machine Learning from Microsoft and Google's Predictive API. It has similar features to many large-scale learning applications including visualization and basic data statistics. The exact learning algorithm it uses is not known; however, it is similar to Vowpal Wabbit. There are some limitations such as the inability to export the learned model or to access data which is not stored inside Amazon (Amazon S3 or Redshift).

Google Prediction API[38] is Google's cloud-based machine learning tools that can help to analyze data. It is closely connected to Google Cloud Storage[39] where training data is stored and offers its services using a RESTful interface, client libraries allowing programmers to connect from Java, JavaScript, .NET, Ruby, Python, etc. In the first step, the model needs to be trained on data, supported models being classification and regression for now. After the model is built, one can query this model to obtain predictions on new instances. Adding new data to a trained model is called Streaming Training and it is also nicely supported. Recently, PMML preprocessing feature has been added, i.e., Prediction API supports preprocessing your data against a PMML transform specified using PMML 4.0 syntax and does not support importing of a complete PMML model that includes data. Created models can be shared as hosted models in the marketplace.

7.6 Related Studies

Since 1995, when Thearling [40] presented a massively parallel architecture and the algorithms for analyzing time series data, allegedly, one of the first approaches to parallelization of ML algorithms, many implementations were proposed for ML algorithm parallelization for both shared and distributed systems. Consequently, many studies tried to summarize, classify, and compare these approaches. We will address in this section only the most recent ones.

Upadhyaya [42] presents an overview of machine learning efforts since 1995 onward grouping the approaches based on prominent underlying technologies: those employed on GPUs (2000–2005 and beyond), those using MapReduce technique (2005 onward), the ones that did not consider neither MapReduce nor GPUs

[35] http://www.ersatzlabs.com/

[36] http://www.skymind.io/about/

[37] http://docs.aws.amazon.com/machine-learning/latest/mlconcepts/mlconcepts.html

[38] https://developers.google.com/prediction/

[39] https://developers.google.com/storage/

(1999–2000 and beyond), and, finally, few efforts discussing the MapReduce technique on GPU. Contrasting to this extensive overview, we focus on more recent distributed and cloud-based solutions, regardless if they are coming from academia or industry.

The book *Scaling Up Machine Learning: Parallel and Distributed Approaches* by Bekkerman et al. [3] presents an integrated collection of representative approaches, emerged in both academic (Berkeley, NYU, University of California, etc.) and industrial (Google, HP, IBM, Microsoft) environments, for scaling up machine learning and data mining methods on parallel and distributed computing platforms. It covers general frameworks for highly scalable ML implementations, such as DryadLINQ and IBM PMLT, as well as specific implementations of ML techniques on these platforms, like ensemble decision trees, SVM, or k-means. The book is a good starting point, but it does not aim at providing a structured view on how to scale out machine learning for big data applications, which is central to our study.

A broader study is conducted by Assunção et al. [2], who discuss approaches and environments for carrying out analytics on clouds for big data applications. Model development and scoring, i.e., machine learning, is one of the areas they considered, alongside other three: data management and supporting architectures, visualization and user interaction, and business models. Through a detailed survey, they identify possible gaps in technology and provide recommendations for the research community on future directions on cloud-supported big data computing and analytics solutions. With respect to this study, our work goes into deeper details on the specific topic of distributed machine learning approaches, synthesizing and classifying existing solutions to give data scientists a comprehensive view of the field.

Interesting analyses have been made available through online press and blogs [7, 9, 18]; they have reviewed open-source or commercial players for big data analytics and predictions.

7.7 Conclusion and Guidelines

Analyzing big data sets gives users the power to identify new revenue sources, develop loyal and profitable customer relationships, and run the organization more efficiently and cost-effectively – overall, giving them competitive advantage over other competitions. Big data analytics is still a challenging and time-demanding task that requires important resources, in terms of large e-infrastructure, complex software, skilled people, and significant effort, without any guarantee on ROI. After reviewing more than 40 solutions, our key findings are summarized below.

Both, research and industry, have invested efforts in developing "as-a-Service" solutions for big data problems (Analytics-as-a-Service, Data-as-a-Service, Machine-Learning-as-a-Service) in order to benefit of the advantages cloud computing provides such as resources on demand (with costs proportional to the actual usage), scalability, and reliability.

Existing programming paradigms for expressing large-scale parallelism (MapReduce, MPI) are the de facto choices for implementing distributed machine learning algorithms. The initial enthusiastic interest devoted to MapReduce has been balanced in recent years by novel distributed architectures specifically designed for machine learning problems. Nevertheless, Hadoop remains the state-of-the-art platform for processing large data sets stored on HDFS, either in MapReduce jobs or using higher-level languages and abstractions.

Although state-of-the-art tools and platforms provide intuitive graphical user interfaces, current environments lack an interactive process, and techniques should be developed to facilitate interactivity in order to include analysts in the loop, by providing means to reduce time to insight. Systems and techniques that iteratively refine answers to queries and give users more control of processing are desired.

From the perspective of resource management and data processing, new frameworks able to combine applications from multiple programming models (e.g., MPI, MapReduce, workflows) on a single solution need to be further investigated. Optimization of resource usage and energy consumption, while executing data-intensive applications, is another challenging research direction in the next decade.

Developing software that meets the high-quality standards expected for business-critical applications remains a challenge, and quality-driven development methodology and new tools need to be created. Building on the principles of model-driven development and on popular standards, e.g., UML or MARTE, such an approach will guide the simulation, verification, and quality evolution of big data applications.

Further exploiting the scalability, availability, and elasticity of cloud computing for model building and exposing of prediction and analytics as hosted services is opening a competitive and challenging market. Tools and frameworks to support the integration of mobile and sensor data into cloud platforms need to be further developed.

At the end of this chapter, we review some of the best practices that the recently published literature recommends, namely:

- Understand the business problem. Having a well-defined problem, knowing specific constraints available for the problem under investigation, can greatly improve performance of the ML algorithms.
- Understand the ML task. Is it supervised or unsupervised? What activities are required to get the data labeled? The same features (attributes, domains, labels) need to be available at both times, training and testing. Pick a machine learning method appropriate to the problem and the data set. This is the most difficult task, and here are some questions you should consider: Do human users need to understand the model? Is the training time a constraint for your problem? What is an acceptable trade-off between having an accurate answer and having the answer quickly? Keep in mind that there is no single best algorithm; experiment with several algorithms and see which one gives better results for your problem.
- In case of predictive modeling, carefully select and partition the data at hand in training and validation set, which will be used to build the model, versus the test

set that you will use to test the performance of your model. More data for training the model results in better predictive performance. Better data always beats a better algorithm, no matter how advanced it is. Visualize the data with at least univariate histograms. Examine correlations between variables.

- Well prepare your data. Deal with missing and invalid values (misspelled words, values out of range, outliers). Take enough time, because no matter how robust a model is, poor data will yield poor results.
- Evaluate your model using confusion matrix, ROC (receiver operating characteristic) curve, precision, recall, or F1 score. Do not overfit your model, because the power lies in good prediction of unseen examples.
- Use proper tools for your problems. For low-level programming environments you might find difficult to use, try first the machine learning services offered by cloud service providers, which are easy to use and powered by state-of-the-art algorithms.

Acknowledgments This work was supported by the European Commission H2020 co-funded with project DICE (GA 644869).

References

1. Agarwal A, Chapelle O, Dudik M, Langford J (2014) A reliable effective terascale linear learning system. J Mach Learn Res 15:1111–1133
2. Assunção MD, Calheiros RN, Bianchi S, Netto MAS, Buyya R (2014) Big data computing and clouds: trends and future directions. J. Parallel Distrib Comput. http://dx.doi.org/10.10.16/j.jpdc.2014.08.003
3. Bekkerman R, Bilenko M, Langford J (eds) (2012) Scaling up machine learning: parallel and distributed approaches. Cambridge University Press, Cambridge
4. Berthold MR, Cebron N, Dill F, Gabriel TR, Kötter T, Meinl T, Ohl P, Sieb C, Thiel K, Wiswedel B (2008) KNIME: The Konstanz Information Miner. In: Preisach C, Burkhardt H, Schmidt-Thieme L, Decker R (eds) Studies in classification, data analysis, and knowledge organization. Springer, Berlin/Heidelberg
5. Budiu M, Fetterly D, Isard M, McSherry F, Yu Y (2012) Large-scale machine learning using DryadLINQ. In: Bekkerman R, Bilenko M, Langford J (eds) Scaling up machine learning. Cambridge University Press, Cambridge
6. Chandola V, Banerjee A, Kumar V (2009) Anomaly detection: a survey. ACM Comput Surv 41(3):15:1–15:58
7. Charrington S (2012) Three new tools bring machine learning insights to the masses, February, Read Write Web. http://www.readwriteweb.com/hack/2012/02/three-new-tools-bring-machine.php
8. Dai W et al (2015) High-performance distributed ML at scale through parameter server consistency models, AAAI
9. Eckerson W (2012) New technologies for big data. http://www.b-eye-network.com/blogs/eckerson/archives/2012/11/new_technologie.php
10. Franklin M et al (2015) MLlib: Machine Learning in apache Spark
11. Gander M et al (2013) Anomaly detection in the cloud: detecting security incidents via machine learning, trustworthy eternal systems via evolving software, data and knowledge, vol 379. Springer, Berlin/Heidelberg, pp 103–116

12. Ghoting A, Kambadur P, Pednault E, Kannan R (2011) NIMBLE: a toolkit for the implementation of parallel data mining and machine learning algorithms on MapReduce. In: Proceedings of the 17th ACM SIGKDD international conference on Knowledge Discovery and Data mining KDD'11, ACM, New York, NY, USA, pp 334–342
13. Ghoting A, Krishnamurthy R, Pednault E, Reinwald B, Sindhwani V, Tatikonda S, Tian Y, Vaithyanathan S (2011) SystemML: declarative machine learning on MapReduce. In: Proceedings of the 2011 I.E. 27th International Conference on Data Engineering (ICDE '11). IEEE Computer Society, Washington, DC, USA, pp 231–242
14. Granger B, Perez F, Ragan-Kelley M (2011) Using IPython for parallel computing. http://minrk.github.com/scipy-tutorial-2011. Accessed 13 May 2015
15. Grisel O (2013) Advanced machine learning with scikit-learn, PYCON tutorial. https://us.pycon.org/2013/schedule/presentation/23/
16. Hall M et al (2009) The WEKA data mining software: an update. ACM SIGKDD Explor Newsl 11(1):10–18
17. Hall M (2013) Weka and Spark – http://markahall.blogspot.co.nz/. Accessed 13 May 2015
18. Harris D (2015) 5 low-profile startups that could change the face of big data. http://gigaom.com/cloud/5-low-profile-startups-that-could-change-the-face-of-big-data/. Accessed 15 July 2015
19. Hido S, Tokui S, Oda S (2013) Jubauts: an open source platform for distributed online machine learning, NIPS workshop on Big Learning, Lake Taho
20. Hofmann M, Klinkenberg R (2013) RapidMiner: data mining use cases and business analytics applications. Chapman &Hall/CRC, Boca Raton
21. Isard M et al. (2007) Dryad: distributed data-parallel programs from sequential building blocks. SIGOPS Oper Syst Rev 41:59–72. doi:10.1145/1272998.1273005
22. Jain A, Nalya A (2014) Learning storm. Packt Publishing, Birmingham
23. Nuggets KD (2014) http://www.kdnuggets.com/polls/2014/languages-analytics-data-mining-data-science.html. Accessed 15 May 2015
24. Krizhevsky A, Sutskever I, Hinton GE ImageNet (2012) Classification with deep convolutional neural networks. NIPS 2012: neural information processing systems, Lake Tahoe, Nevada
25. Le Q, Ranzato MA, Monga R, Devin M, Chen K, Corrado G, Dean J, Ng A (2012) Building high-level features using large scale unsupervised learning, international conference in machine learning, Edinburgh, UK
26. Leo S, Zanetti G (2010) Pydoop: a Python MapReduce and HDFS API for Hadoop. In: Proceedings of the 19th ACM international symposium on high performance distributed computing, Chicago, IL, USA, pp 819–825
27. Low Y et al. (2012) Distributed GraphLab: a framework for machine learning and data mining in the cloud. In: Proceedings of the VLDB endowment, vol 5, no 8, August 2012, Istanbul, Turkey
28. Mohri M, Rostamizadeh A, Talwalkar A (2012) A foundations of machine learning. The MIT Press, Cambridge, MA
29. Owen S, Anil R, Dunning T, Friedman E (2011) Mahout in action. Manning Publications Co., Shelter Island
30. Patcha A, Park JM (2007) An overview of anomaly detection techniques: existing solutions and latest technological trends. Comput Netw Elsevier, North-Holland, Inc., 51:3448–3470
31. Pednault E, Yom-Tov E, Ghoting A (2012) IBM parallel machine learning toolbox. In: Bekkerman R, Bilenko M, Langford J (eds) Scaling up machine learning. Cambridge University Press, New York
32. Pedregosa F et al (2011) Scikit-learn: machine learning in Python. J Mach Learn Res 12:2825–2830
33. Piccolboni A (2015) RHadoop. https://github.com/RevolutionAnalytics/RHadoop/wiki. Accessed 13 May 2015
34. Roldn MC (2013) Pentaho data integration beginner's guide. Packt Publishing, Birmingham

35. Rosen J et al (2013) Iterative MapReduce for large scale machine learning, CoRR, abs/1303.3517
36. Russom P Big data Analytics (2011) TDWI best practices report, The Data Warehousing Institute (TDWI) Research
37. Sagha H, Bayati H, Millán JDR, Chavarriaga R (2013) On-line anomaly detection and resilience in classifier ensembles. Pattern Recogn Lett, Elsevier Science Inc., 34:1916–1927
38. Shi Q et al (2009) Hash kernels for structured data. J Mach Learn Res JMLR.org, 10:2615–2637
39. Elston SF (2015) Data science in the cloud with Microsoft Azure Machine Learning and R, O'Reilly
40. Thearling KK (1995) Massively parallel architectures and algorithms for time series analysis. In: Nadel L, Stien D (eds) Lectures in complex systems. Addison-Wesley, Reading
41. Tierney L, Rossini AJ, Snow NL (2009) A parallel computing framework for the R system. Int J Parallel Prog 37:78–90. doi:10.1007/s10766-008-0077-2
42. Upadhyaya SR (2013) Parallel approaches to machine learning – a comprehensive survey. J Parallel Distrib Comput 73(3):284–292. ISSN 0743–7315. http://dx.doi.org/10.1016/j.jpdc.2012.11.001
43. Wei D, Wei J, Zheng X, Kim JK, Lee S, Yin J, Ho Q, Xing EP (2013) Petuum: a framework for iterative-convergent distributed ML. arxiv.org/abs/1312:7651
44. Zaharia M et al. Spark: cluster computing with working sets. In: Proceedings of the 2nd USENIX conference on hot topics in Cloud, Computing, USENIX Association, pp 10–10

Chapter 8
An Analytics-Driven Approach to Identify Duplicate Bug Records in Large Data Repositories

Anjaneyulu Pasala, Sarbendu Guha, Gopichand Agnihotram, Satya Prateek B, and Srinivas Padmanabhuni

Abstract Typically, the identification and analysis of duplicate bug records of a software application are mundane activities, carried out by software maintenance engineers. As the bug repository grows in size for a large software application, this manual process becomes erroneous and a time-consuming activity. Automatic detection of these duplicate bug records will reduce the manual effort spent by the maintenance engineers. It also results in the reduction of costs of software maintenance. There are two types of duplicate bug records: (1) the records that describe the same problem using similar vocabulary, and (2) the records that describe different problems using dissimilar vocabulary but share the same underlying root cause. Each of these types of records needs a different set of techniques to identify the duplicate bug records. In this chapter, we explain the various machine learning techniques that are used to detect both types of duplicate bug records. Some of these duplicate bug records reappear, that is, they show up continuously over a long period of time. Here, we present a framework that can be used to automate the entire process of detection of both types of duplicates and recurring bug records. Using the framework, we conducted empirical studies on the open-source Chrome bug data records that are accessible online and the results are reported.

Keywords Software maintenance • Mining software repositories • Duplicate data records • Vector space model • Clustering techniques • Co-occurrence models • Bug fix

A. Pasala (✉) • S. Guha • G. Agnihotram • S. Prateek B • S. Padmanabhuni
Infosys Labs, Infosys Ltd., Electronics City, Hosur Road, 560 100 Bangalore, India
e-mail: Anjaneyulu_Pasala@infosys.com; Sarbendu_Guha@infosys.com; Gopichand_A@infosys.com; srinivas_p@infosys.com

Z. Mahmood (ed.), *Data Science and Big Data Computing*,
DOI 10.1007/978-3-319-31861-5_8

8.1 Introduction

In recent years, the software industry has been growing in leaps and bounds. It was estimated that in 2014, the global software market was valued at $299 billion [1]. The maintenance activities alone account for 60–80 % of the overall software budget [2]. In the USA alone, the cost of maintenance was calculated to be more than $70 billion for ten billion lines of existing code [3]. It has been reported that more than 50 % of software personnel are engaged in various activities [4] of software maintenance. That shows the enormity of software maintenance activities in organizations.

Therefore, organizations are increasingly focusing on cost optimization techniques for software maintenance. One such technique is an automation of mundane but highly important activities relating to software maintenance. One of the areas where automation is used is in bug tracking and handling systems. Bug tracking and handling systems are being extensively used by organizations to allow software users to report on issues faced or bugs detected while using the software application. These are called bug reports or bug records. The set of bug records for a particular software application is termed as a bug repository. For large software applications, the number of bugs reported grows exponentially in size, resulting in a large data repository. Sometimes, these repositories grow to the levels of millions and millions of records in number. Hence, there arises the challenge of managing and analyzing these large bug data repositories.

One major challenge of these systems is managing and analyzing duplicates in such massive data repositories. As several records come in concurrently, a large number of these records tend to be duplicates. Studies have shown that duplicate bug records can amount to as much as 20 % of total bugs reported [5]. As a result of this, maintenance engineers spend a lot of time and effort, dealing with issues that are being handled elsewhere or have already been dealt with. Thus, automatic detection of duplicate data records helps the maintenance engineers to dispose of issues faster and in a timely fashion. Additionally, it enables them to gain the confidence of software users.

Typically, issue management is carried at multiple levels. In software maintenance parlance, service level 1 is where the bugs or incidents are initially received immediately after they are reported. The maintenance engineers are assigned bugs which are subsequently resolved. However the amount of analytics that gets carried out at level 1 is very minimal. The analytics are usually carried at the level 2 where the bugs reported over a period of time are analyzed for bug patterns and recurring bugs.

For instance, a well-known software organization develops and maintains 200 software applications. On average 900 bugs or incidents are reported every month on each of these applications by the users. It results in 2.16 millions of bug records in a year for the organization to analyze. Among 900 reported bugs, close to 250 of them are duplicates. It shows the enormity of duplicate bugs reported by the users. Hence, there arises a need for a system that automatically identifies and

Table 8.1 Examples of apple-to-apple pairs

1.	Gmail app should allow associating notifications/ ringtones with individual labels	Gmail should allow per-label notification settings
2.	No master password option	Master password option is missing

Table 8.2 Examples of apple-to-orange pairs

1.	Home if force closing	ClassCastException in the Launcher with certain widgets on Donut/master
2.	Version check for browser type causes failure – emulation mode to mimic IE ver. response for such queries	Can't watch Netflix

groups all these duplicate bugs present in a massive data repository together. Such a system would enable the maintenance engineer to identify frequently occurring bugs and forward them to level 3 for further analysis and closure. At level 3, the root cause for these duplicates or recurring bugs is identified and resolved. In this way the organization ensures that the bug is resolved and it does not recur in the future. This is how the prevention of recurring bugs is handled manually in the organizations. This manual process becomes a highly error-prone and time-consuming activity in the case of massive bug repositories. And the cost grows exponentially with the size of the repository. Therefore, automatic detection of these duplicate bug records will reduce the manual effort spent by maintenance engineers. Hence, it will result in the reduction of costs spent on software maintenance.

Magnus et al. [6] describe two types of bug data records: (1) records that describe the same problem using similar vocabulary are called *apple-to-apple* pairs, and (2) records that describe different problems textually (using dissimilar vocabulary) but share the same underlying cause are called *apple-to-orange* pairs. Some examples of reported *apple-to-apple* and *apple-to-orange* pairs are shown in Tables 8.1 and 8.2, respectively. In our analysis, we find that around 11 % of the validated duplicates in the Chrome bug repository are *apple-to-orange* pairs.

Currently, the *apple-to-apple* pairs can be detected using natural language processing (NLP) and information retrieval (IR) techniques as they are textually similar. However, the *apple-to-orange* pairs cannot be detected using these techniques as they generally use different vocabulary to describe the problem they narrate.

Given a dataset of validated duplicate data records, our first task is to identify all the *apple-to-apple* pairs in it. A couple of well-known techniques are described to detect such *apple-to-apple* pairs. The idea being that any duplicate pair that cannot be detected using simple NLP techniques is likely to be an *apple-to-orange* pair. Given the list of *apple-to-orange* pairs, we build a matrix of the words found in the training data that keeps track of the number of times a word, in one bug of an *apple-*

to-orange pair, occurs along with another word in the second bug of that same pair. After that, when the user searches for duplicate bug data records, the word matrix is used to retrieve the keywords that are most common among the query and appended with the keywords. This expanded query is then used to search for the duplicates. This is the process we aim to use to detect the *apple-to-orange* pairs.

The organization of this chapter is as follows. Section 8.2 briefly presents the different techniques proposed and used to identify the similar records in the literature. Section 8.3 discusses our proposed framework to identify both types of duplicate bug records. The next two sections, 8.4 and 8.5, describe in detail the techniques presented to identify the duplicate records of *apple-to-apple* pairs and *apple-to-orange* pairs, respectively. These are explained using example data repository available online. Section 8.6 presents the case study conducted on Chrome bug data and its results. Section 8.7 briefly explains the proposed recurring duplicate bug's prevention framework. Finally, Section 8.8 provides conclusions and future challenges of bug's detection in big data repositories and the challenges of analytics to be carried in the future.

8.2 Literature Survey

The most well-known information retrieval (IR) technique used in identifying the duplicate records is vector space model (VSM). It was initially used by Magnus et al. [6]. In their approach a defect report is represented as a weighted vector of words, and the similarity between two records is calculated as the distance between their vectors. Using this approach, they were able to filter out 40 % of the marked duplicate bug records. Nicholas et al. [7] build on Magnus' model by using a linear model classifier using textual similarity and a clustering algorithm combined. The clustering is done on a graph built with nodes as defect records and edges link records with similar text. They analyzed bug records from the Mozilla data and were able to filter out 8 % of the duplicate bug records.

Further machine learning approaches such as Latent Dirichlet Allocation (LDA), a topic modeling method, along with clustering have also been successfully used. These approaches use cluster–topic distribution and topic–word distribution for extracting keywords from the corpus [8]. Support Vector Machine (SVM) technique has also been used with various features of keywords like frequency, part of speech, distance from the beginning of document, etc. to detect textually similar data records.

Statistical approaches such as the term frequency, which is the number of times a word appears in the corpus, are also used extensively along with the above approaches. These approaches improve the detection of duplicates in the text documents. Further improvements in the results are achieved using a term frequency–inverse document frequency (TF-IDF), mainly to obtain the frequency of the word in the corpus [9].

Xiaoyin et al. [10] present an approach that involves execution information along with NLP techniques. In their approach, when a new bug record arrives, its language information and execution information are compared with the existing bug record information. Based on this comparison, a small number of existing bug records are suggested to the triager as the similar bug records to the new bug. Finally, the triager determines whether the new bug record is a duplicate of the existing bug record.

Chengnian et al. [11] leveraged discriminative models for information retrieval to detect duplicate bug records more accurately. They claim that their technique could result in 17–31 %, 22–26 %, and 35–43 % relative improvement over state-of-the-art techniques. Furthermore, retrieval function (REP) has been used to measure the similarity between two bug records [12]. It fully utilizes the information available in a bug record including not only the similarity of textual content in summary and description fields but also similarity of non-textual fields such as product, component, version, etc. For more accurate measurement of textual similarity, they extended BM25F weighting technique, especially for duplicate record retrieval. They also used a two-round stochastic gradient descent to automatically optimize REP for specific bug repositories in a supervised learning manner.

Yuan et al. [13] extended Nicholas' work by improving the accuracy of automated duplicate bug record identification. They used a support vector classifier to identify a duplicate bug. They conducted empirical studies on bug records from Mozilla bug tracking system and found that they could improve the accuracy of finding duplicate records.

Anahita et al. [14] extended the state of the art by using contextual information, relying on prior knowledge of software quality, software architecture, and system development topics, to improve bug deduplication detection. However, Ashish et al. [15] used a character N-gram-based model for duplicate bug detection. They use a feature extraction module that extracts all the character N-grams from the title and description of each bug record and then calculate their similarity score as a function of shared N-gram characters. It is more promising in detecting duplicate bug data records submitted by developers.

Techniques also have been proposed to detect certain types of bug records with different vocabulary such as synonym replacement, semantic matching using WordNet, etc. However, the existing techniques can only detect duplicate bug records with similar text and cannot detect dissimilar duplicate bug records as they do not share the common words. Also, synonym replacement techniques do reasonably well only when two bug records describe the same problem using different words but totally fail in case of dissimilar duplicate bug records. This is because while the underlying cause for the two bugs may be the same, they are describing separate problems so the vocabulary for the two will be completely different. There is no system, currently, where both types of duplicates can be detected simultaneously in a real-time scenario.

Hence, there is a need for a method and system for detection of duplicate bug records by expanding the queries using a word matrix. The word matrix models the underlying relationships between the words present in the two dissimilar bug records and by extension models the relation between the bug records themselves. Further, it can also be extended such that it can be used in online scenario for detection of all types of duplicates.

8.3 The Proposed System to Identify Duplicate Records

In this section, we propose and present a system to detect both types of duplicate bug data records in an issue management system. A high-level depiction of the system is shown in Fig. 8.1.

The system concurrently searches for both types of duplicate records using different techniques. The resulting bug records from these searches are aggregated and presented as a single list of duplicates. When the user submits a query (or a bug arrives at the system through online process), the search for duplicate bug records is carried out both with the normal query and with the expanded query. The details of the expanded query is discussed in the subsequent sections. The normal query is used to get the duplicate records that are similar in vocabulary of the query itself. Hence, these records are termed as *apple-to-apple* pairs of duplicates. The expanded query is used to get the duplicates that are dissimilar in vocabulary of this query but represent a common cause to the problem of this bug occurrence. These duplicate records are termed as *apple-to-orange* pairs. The records thus fetched in both searches are then aggregated and displayed as a list of duplicates to the user. In the following sections, we discuss the different techniques that are used to detect the *apple-to-apple* and *apple-to-orange* pairs.

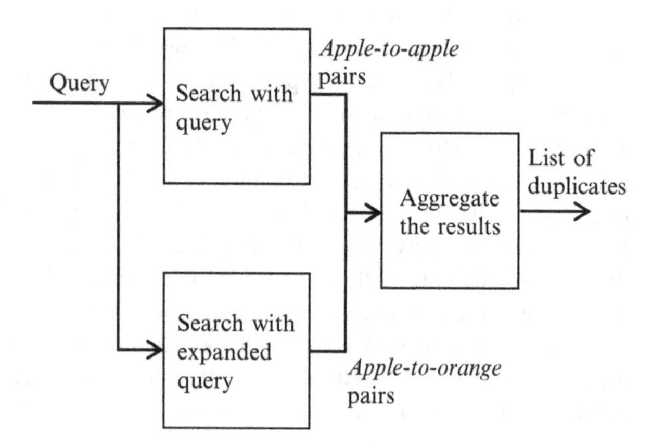

Fig. 8.1 A system to detect all duplicate bugs

8.4 Detecting Apple-to-Apple Pairs

In this section, we describe two well-known machine learning techniques to detect *apple-to-apple* records. The techniques are vector space model (VSM) and clustering approaches.

8.4.1 Vector Space Model

In VSM [16, 17], both the collection of records and the query are represented as vectors of unique words present in these records. These vectors usually result in a high-dimensional Cartesian space depending on the number of unique words present in these records. Each vector component corresponds to a unique word in the vocabulary of these records. Given a query vector and a set of record vectors, it ranks the records by computing a similarity measure between the query vector and each record vector. This ranking is based on comparing the angle between these vectors. The smaller the angle, the more similar are the vectors representing the respective record and the query. The similarity in vectors means the record and query representing it are similar in terms of the words present in the vocabulary of the respective record and the query. Hence, these two are duplicate in terms of the vocabulary they represent. For example, consider that records d_1, d_2, and d_3 and a query q are having the following vocabulary:

d_1 = <Gossip, Gossip, Gossip, Gossip, Gossip, Jealous>
d_2 = <Gossip, Gossip, Gossip, Gossip, Gossip, Gossip, Gossip, Jealous, Jealous, Jealous, Jealous, Jealous, Jealous, Jealous>
d_3 = <Jealous, Jealous, Jealous, Jealous, Jealous, Gossip>
q = <Gossip, Gossip, Gossip, Jealous, Jealous, Jealous>

These records can then be represented in a two-dimensional vector as shown in Fig. 8.2.

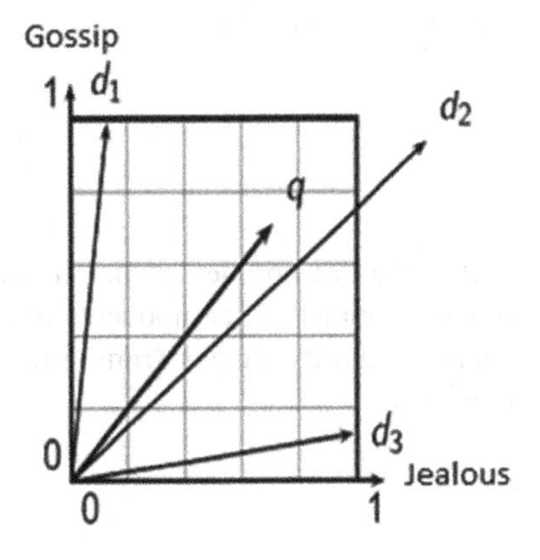

Fig. 8.2 Vector space representation of records and query

The records $d_1, d_2,$ and d_3, and the query q are distributed in a vector space defined by the terms "Gossip" and "Jealous." Here the record d_2 is most similar to query q because the angle between d_2 and q is the smallest. A mathematical notion to the VSM technique has been defined as follows.

In general, two n -dimensional vectors, say, vectors x and y, are represented as

$$\vec{x} = < x_1, x_2, \ldots, x_n >$$
$$\vec{y} = < y_1, y_2, \ldots, y_n >$$

We have

$$\vec{x}.\vec{y} = |\vec{x}||\vec{y}|\cos\theta$$

where $\vec{x}.\vec{y}$ represents the dot product between the vectors and $|\vec{x}|$ and $|\vec{y}|$ represent the magnitude of these vectors, respectively. The dot product is also defined as

$$\vec{x}.\vec{y} = \sum_{i=1}^{n} x_i y_i$$

The length of a vector is computed from the Euclidean distance formula as

$$|\vec{x}| = \sqrt{\sum_{i=1}^{n} x_i^2}$$

Substituting and rearranging these three equations, we get

$$\cos\theta = \frac{\vec{x}}{|\vec{x}|} * \frac{\vec{y}}{|\vec{y}|} = \frac{\sum_{i=1}^{n} x_i y_i}{\sqrt{\sum_{i=1}^{n} x_i^2}\sqrt{\sum_{i=1}^{n} y_i^2}}$$

If $\theta = 0$, then cos $\theta = 1$ so the two vectors are collinear and as similar as possible. If $\theta = \pi/2$, then $cos\,\theta = 0$ and vectors are orthogonal and as dissimilar as possible.

In VSM, given a record vector $\vec{d}$ and a query vector $\vec{q}$, the cosine similarity $\text{sim}(\vec{d}, \vec{q})$ is given as

$$\text{sim}\left(\vec{d}, \vec{q}\right)\frac{\vec{d}}{\left|\vec{d}\right|} * \frac{\vec{q}}{|\vec{q}|}$$

It is the dot product of the record vector and query vector normalized to unit length. Provided all components of the vectors are nonnegative, the value of this cosine similarity ranges from 0 to 1, with its value increasing with increasing similarity.

Naturally, for a collection of even modest size of data records, this vector space model produces vectors having millions of dimensions. Though this high dimensionality may appear inefficient, in many circumstances the query vector is sparse with all but a few components being zero. For example, consider the entire Shakespeare's *Romeo and Juliet* novel represented as a set of records by considering each sentence as a record. Also consider the vector corresponding to a query *<William Shakespeare marriage>*. The vector will have only three nonzero components. To compute the length of the vector or its dot product with these record vectors, we need to consider the components corresponding to these three terms only. On the other hand, a record vector typically has a nonzero component for each unique term contained in the record. However, the length of a record vector is independent of the query. It may be precomputed and stored in a frequency or positional index along with other record-specific information. Otherwise, it may also be applied to normalize the record vector in advance, with the components of the normalized vector taking the place of term frequencies in the posting lists. Based on these concepts, the weightage techniques are formulated.

8.4.1.1 Weightage Techniques

As discussed above, in VSM, the records are represented as a set of vectors. Such vectors are represented by their projections on each axis, which represent the terms in the entire corpus. These projections of the vector along each axis (term) have magnitude values which can be treated as weight of that term. Therefore, we propose to use the weights of the terms.

Term frequency–inverse document frequency (TF-IDF) is the most popular weighting scheme. In TF-IDF method [17], weights are assigned to terms in a vector that represents a query or a record. In representing a record or a query as a vector, a weight is assigned to each term that represents the value of corresponding component of the vector. The assignment of weights to text tokens in a TF-IDF scheme potentially changes the unstructured text data into a set of numeric data, and thus it becomes easier to perform mathematical and statistical operations on the data. The weight of a term denotes the magnitude of the component of the record/query vector along the dimension of the vector represented by that term. When assigning a weight in a record vector, the TF-IDF weights are computed by taking the product of a function of the term frequency ($f_{t,d}$) and a function of the inverse document frequency ($1/N_t$), where N_t is the number of documents in which the term t occurs. When assigning weight to a query vector, the within-query term frequency q_t may be substituted for $f_{t,d}$ in essence treating the query as a tiny document. A TF-IDF weight is a product of functions of the term frequency and inverse document frequency. A common error is the use of the raw term frequency or $f_{t,d}$. This may lead to poor performance as very commonly occurring words will get assigned higher weightage compared to distinctive words which may occur infrequently.

The IDF function typically relates the document frequency (N_t) to the total number of records in the collection (*N*). The basic intuition behind the IDF is that a term appearing in many records *should be assigned a lower weight or importance than a term appearing* in only a few records. Of the two functions, IDF comes closer to having a "standard form":

$$\mathrm{IDF} = \log\left(N/N_t\right)$$

The basic intuition behind the TF function is that a term appearing many times in a record should be assigned a higher weight for that record than for a record in which it appears less number of times. Another important consideration for the TF function is that the value should not necessarily increase linearly with $f_{t,d}$. Although two occurrences of a term should be given more weight than single occurrence, they should not necessarily be given twice the weight. The following definition of TF meets these requirements:

$$\mathrm{TF} = \begin{cases} \log\left(f_{t,d}\right) + 1 & \text{if} \quad f_{t,d} > 0 \\ 0 & \text{otherwise} \end{cases}$$

When the above equation is used with a query vector, $f_{t,d}$ is replaced with the term q_t, the query term frequency of t in q.

8.4.1.2 Example

Consider the text fragments from Shakespeare's *Romeo and Juliet*, Act I, Scene I, as listed in Table 8.3.

In Table 8.3, there are five records in the collection and the "sir" word appears in four of them. Therefore, the IDF value for "sir" is

$$\log\left(N/f_{\mathrm{sir}}\right) = \log\left(5/4\right) \approx 0.32$$

Also, "sir" word appears twice in record 2. Therefore, the TF-IDF value for the corresponding component of its vector is

Table 8.3 Text fragment collection

Record ID	Record content
1	Do you quarrel, sir?
2	Quarrel sir! No, sir!
3	If you do sir, I am for you. I serve as good a man as you
4	No better
5	Well sir!

$$(\log (f_{\text{sir}}, 2) + 1)*\log (N / f_{\text{sir}}) = (\log 2 + 1)*\log(5/4) \approx 0.64$$

Computing the TF-IDF values for the remaining components and the remaining records gives the following set of vectors:

$$\vec{d_1} \approx < 0.00, 0.00, 0.00, 0.00, 1.32, 0.00, 0.00, 0.00, 0.00, 0.00, 0.00, 1.32, 0.00, 0.32, 0.00, 1.32 >$$
$$\vec{d_2} \approx < 0.00, 0.00, 0.00, 0.00, 0.00, 0.00, 0.00, 0.00, 0.00, 0.00, 1.32, 1.32, 0.00, 0.64, 0.00, 0.00 >$$
$$\vec{d_3} \approx < 2.32, 2.32, 4.64, 0.00, 1.32, 2.32, 2.32, 4.64, 2.32, 2.32, 0.00, 0.00, 2.32, 0.32, 0.00, 3.42 >$$
$$\vec{d_4} \approx < 0.00, 0.00, 0.00, 2.32, 0.00, 0.00, 0.00, 0.00, 0.00, 0.00, 1.32, 0.00, 0.00, 0.00, 0.00, 0.00 >$$
$$\vec{d_5} \approx < 0.00, 0.00, 0.00, 0.00, 0.00, 0.00, 0.00, 0.00, 0.00, 0.00, 0.00, 0.00, 0.00, 0.32, 2.32, 0.00 >$$

where the components are sorted alphabetically according to their corresponding terms. Normalizing these vectors by dividing with their lengths produces unit vectors, as follows:

$$\vec{d_1}/|\vec{d_1}| \approx < 0.00, 0.00, 0.00, 0.00, 0.57, 0.00, 0.00, 0.00, 0.00, 0.00, 0.00, 0.57, 0.00, 0.14, 0.00, 0.57 >$$
$$\vec{d_2}/|\vec{d_2}| \approx < 0.00, 0.00, 0.00, 0.00, 0.00, 0.00, 0.00, 0.00, 0.00, 0.00, 0.67, 0.67, 0.00, 0.33, 0.00, 0.00 >$$
$$\vec{d_3}/|\vec{d_3}| \approx < 0.24, 0.24, 0.48, 0.00, 0.14, 0.24, 0.24, 0.48, 0.24, 0.24, 0.00, 0.00, 0.24, 0.03, 0.00, 0.35 >$$
$$\vec{d_4}/|\vec{d_4}| \approx < 0.00, 0.00, 0.00, 0.87, 0.00, 0.00, 0.00, 0.00, 0.00, 0.00, 0.49, 0.00, 0.00, 0.00, 0.00, 0.00 >$$
$$\vec{d_5}/|\vec{d_5}| \approx < 0.00, 0.00, 0.00, 0.00, 0.00, 0.00, 0.00, 0.00, 0.00, 0.00, 0.00, 0.00, 0.00, 0.14, 0.99, 0.00 >$$

To rank these five records with respect to the query < "quarrel", "sir" >, we first construct the unit query vector as stated below:

$$\frac{\vec{q}}{|\vec{q}|} \approx < 0.00, 0.00, 0.00, 0.00, 0.00, 0.00, 0.00, 0.00, 0.00, 0.00, 0.00, 0.97, 0.00, 0.24, 0.00, 0.00 >$$

The dot product between these unit query vector and each unit record vector gives the cosine similarity values. The cosine similarity values for all these records are computed and tabulated in Table 8.4. Based on these cosine similarity values, the final record rankings are 2,1,5,3,4.

8.4.2 *Clustering Approaches*

Clustering approaches [18] are used in unsupervised methods to group the similar text records together which will help the user in identifying the *apple-to-apple*

Table 8.4 Cosine similarity for text fragment collection

Record ID	Cosine similarity
1	0.59
2	0.73
3	0.01
4	0.00
5	0.03

pairs. These approaches group the similar records together by identifying the same words in the records. There are different clustering approaches available, e.g., K-means clustering [8, 19] and supervised methods such as the nearest neighbor (NN) method [20, 21].

We use tokenization technique and TF-IDF scores to identify similar words in the records so that they can be grouped together in one group. Tokenization is the process of chopping up a given stream of text or character sequence into words, phrases, or some meaningful words. This information is used as input for further processing such as obtaining the similar phrases, *apple-to-apple* and *apple-to-orange* phrases. In the following sub-section, we describe the K-means clustering approach to identify similar records.

8.4.2.1 K-Mean Clustering

It is one of the most efficient methods of clustering approaches. From the given set of data records, words or phrases are obtained using tokenization and form the k different clusters. Each of these clusters characterized with a unique centroid (e.g., mean of the words in that cluster) is partitioned using the K -means algorithm. The elements belonging to one cluster are close to the centroid of that particular cluster and dissimilar to the elements belonging to the other clusters.

The "k" in the K -means algorithm refers to the number of groups to be assigned for a given dataset. If "n" objects have to be grouped into "k" clusters, cluster centers have to be initialized. These objects are referred to words or phrases. Further, each object is assigned to its closest cluster center. And the center of the cluster is updated until there is no change in each cluster center. From these centers, we can define a clustering by grouping objects according to the center the object is assigned to. The data is partitioned into k clusters where each cluster's center is represented by the mean value of the objects in the cluster. Each of these clusters forms the *apple-to-apple* pairs. The algorithm is explained as follows.

8.4.2.2 Algorithm: Cluster Algorithm to Group Similar Bug Data Records

Input: Data records and k – the number of clusters
Output: A set of k clusters:

Step 1: Choose k number of clusters to be determined.
Step 2: Choose C_k centroids randomly as the initial centers of the clusters.
Step 3: Repeat:

3.1: Assign each object to its closest cluster center using Euclidean distance.
3.2: Compute new cluster center by calculating mean points.

Step 4: Until:

4.1: No change in cluster center OR.
4.2: No object changes its clusters.

The flow of different activities of the K -means algorithm is shown in Fig. 8.3.

8.4.2.3 Example

From the given documents or records, derive the keywords using TF-IDF approach. After that, cluster the similar keywords together form the *apple-to-apple* pair in each of the cluster.

Consider the given four clusters as shown in Table 8.5. In cluster 1 we are having similar words (Ab–Ab). In cluster 2, we are having (Ghi–Ghi) similar words. Similarly, in cluster 3, we are having similar keywords (Dep–Dep). Likewise, in cluster 4, we are having similar keywords (Rff–Rff). This algorithm helps in identifying the duplicates of *apple-to-apple* pairs.

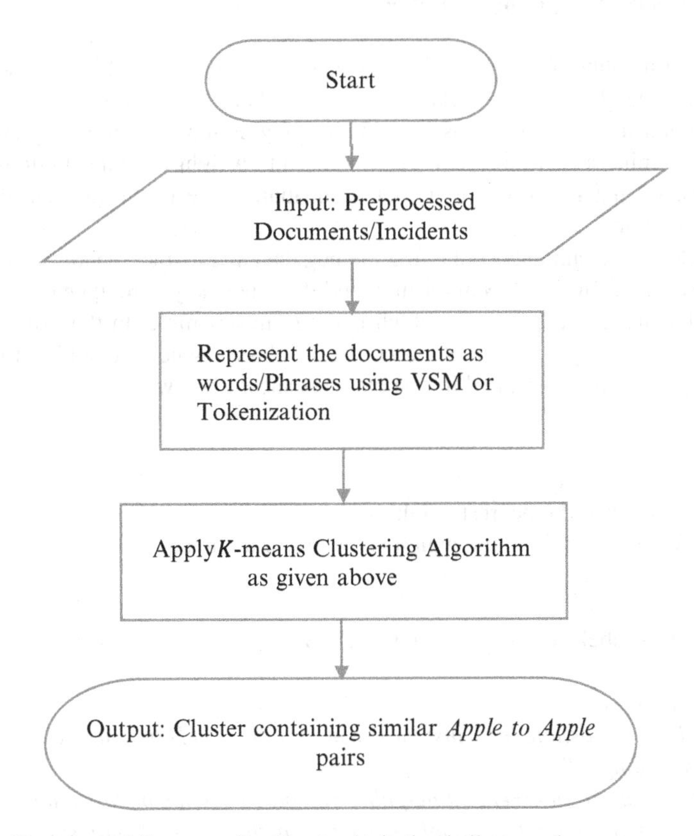

Fig. 8.3 Flow chart of K -means algorithm to obtain the similar records

Table 8.5 Identifying the similar words together using K-means clustering

Cluster1	Cluster2	Cluster3	Cluster4
Ab	Ghi	Dep	Rff
Ab	Ghi	Dep	Rff
Ab	Ghi	Dep	Rff
Ab	Ghi	Dep	Rff
Ab	Ghi	Dep	Rff
Ab	Ghi	Dep	Rff
Ab	Ghi	Dep	Rff
Ab	Ghi	Dep	Rff
Ab	Ghi	Dep	Rff
Ab	Ghi	Dep	Rff
Ab	Ghi	Dep	Rff
Ab	Ghi	Dep	Rff
Ab	Ghi	Dep	Rff
Ab	Ghi	Dep	Rff

8.4.2.4 Nearest Neighbor Classifier

The nearest neighbor (NN) classifier is a supervised approach which is also called as K-nearest neighbor (K -NN) classifier [15, 21]. This approach is a nonparametric approach and used for both classification and regression. It is the simplest of all machine learning algorithms, and it can be used to weight the contributions of the neighbors, so that the nearer neighbors contribute more to the average than the distant neighbor.

Consider the documents containing the bug data and extract the keywords using TF-IDF scheme. Run the NN algorithm, and if the new keyword appears, it will be classified as its similar keyword which is the nearest sample. In this way we will group all similar keywords. In other words, the classes which are similar to *apple-to-apple* pairs will be assigned in one class, as shown below:

Input

- D - > Training data
- K -> Number of the nearest neighbors
- S -> New unknown sample which is a word or a phrase

Output

- C -> Class label assigned to new sample S

The Algorithm Is

- *Let $I_1, I_2, \ldots, I_k$ denote the k instances from training set D that are nearest to new unknown sample S.*
- *C = The class from k-nearest neighbor samples with maximum count.*
- *k = 1; then the output is simply assigned to the class of that single nearest neighbor.*

8.5 An Approach to Detect Apple-to-Orange Pairs

In this section, we describe an approach to detect the *apple-to-orange pairs* of duplicates. The approach is based on building word co-occurrence model. The idea is to model the underlying relations between two dissimilar duplicates and use the model thus built for future detection of duplicates. To accurately learn these relations and build the model, we need to analyze the *apple-to-orange pairs* that have been previously identified and stored during the maintenance of the software manually by the maintenance engineers. Typically, this process is called learning or training phase of the technique. After building the model, we use the model thus built in our system to provide real-time detection of both *apple-to-apple* and *apple-to-orange pairs*. This process is carried automatically by the system as and when a new bug record arrives. The process is called an online phase of the technique. Therefore, the process of identifying duplicates in dissimilar textual records consists of two phases as shown in Fig. 8.4. These phases are further described in detail in the following subsections, respectively.

8.5.1 Training Phase

The process of building a model using the previously known data records is known as the training phase. In this phase, we train the model that will be used to expand

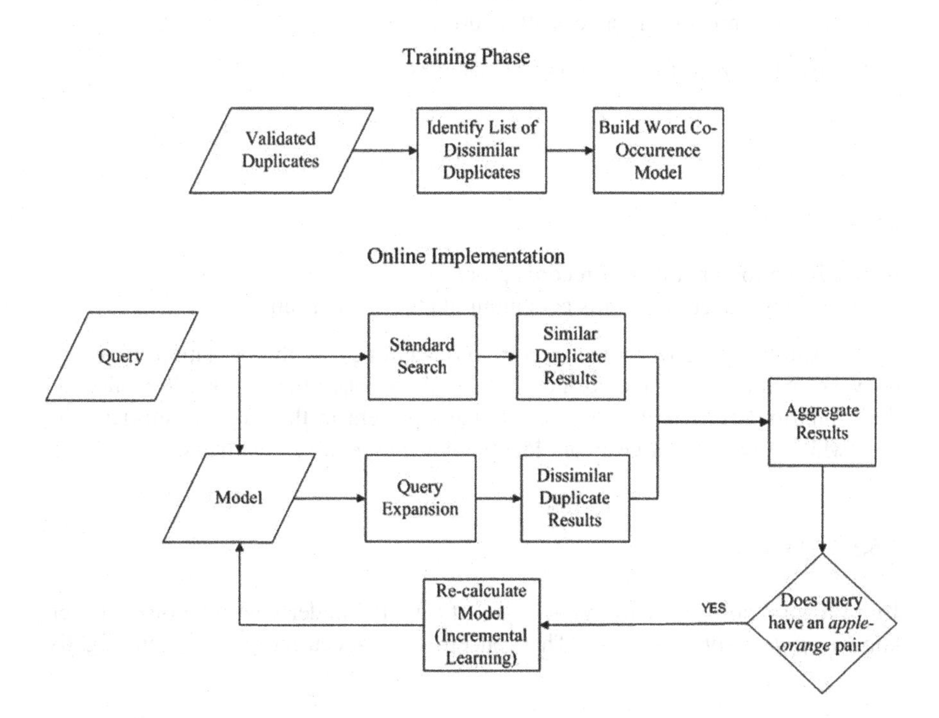

Fig. 8.4 Approach to detect apple-to-orange pairs

the query that is submitted by the user of the duplicate bug record detection system. For training the model, we need the previous history of duplicate bug records that has been manually verified by the maintenance engineers and marked as duplicates. The training phase can be divided into two steps, namely, identify the *apple-to-orange* pairs and build a word co-occurrence model. These steps are elaborated here.

8.5.1.1 Step 1

Identify the *apple-to-orange pairs* present in bug record history. Initially to train our model, we have to identify the list of *apple-to-orange* pairs present in the validated duplicate dataset. This is done by using vector space model. Typically each record is represented by a vector with one component in the vector for every term in the entire vocabulary present in the bug records processed till that time. These components are usually calculated using the TF-IDF weighting scheme:

$$w_i = tf_i \cdot idf_i$$

where

- w_i is the weight assigned to each term in the vector.
- tf_i represents the term frequency, i.e., the number of occurrences of a term in that record.
- idf_i represents the inverse-document frequency and is a measure of whether a term is common or rare across all records.

  Inverse-document frequency is calculated as

$$idf_i = \log \frac{D_{\text{tot}}}{D_{t_i}}$$

where

- D_{tot} is the total number of records processed.
- D_{t_i} is the number of records containing that particular term.

The similarity between two records is calculated as the deviation of angles between each record or the cosine of the angle between the vectors. We calculate the similarity between all the duplicate pairs present in the bug repositories. The ones which have no similarity are identified as *apple-to-orange* pairs.

8.5.1.2 Step 2

Build a word co-occurrence model by capturing the underlying relations between known *apple-to-orange* pairs. The concept of co-occurrence has been slightly modified in our approach than in the normal use. We do not consider the frequency of co-occurrence between two words belonging to the same pair. Rather we consider only the number of times a word in one bug record occurs along with a word in that record's validated duplicate. The model is represented in a word matrix which is of size NxN (N being the size of the vocabulary), and the value of the cell a_{ij} will be the co-occurrence score between the word with index i and the word with index j. This score is representative of the relationship between the two words; the greater the score, the more related the two words are. As there will be huge number of words in our vocabulary, representing the model as a simple two-dimensional array will not be feasible. As the majority of the words do not co-occur, there is a high level of sparsity in the matrix which allows us to use simpler sparse representations. In this way, we plan to capture the relationships between *apple-to-orange* pairs. For example, if "server failure" and "login issue" are two validated dissimilar duplicates, then we map server with login in the matrix. The next time any server issue is reported, then we can use this model to predict that there might have been some sort of login issue also. We build the co-occurrence model for all the known *apple-to-orange* pairs in the available bug record history.

8.5.2 Online Phase

In a real-time scenario with new bug records constantly being reported by the users of the software application, our system performs two searches for each of these bug records (or queries) as shown in Fig. 8.1. These are (1) the standard text similarity algorithm that is used to find all the similar duplicates for that bug record and (2) the technique to find the *apple-to-orange* pairs present in the repository with extended query. The procedure has the following three steps:

1. *Compare and augment* the query with the word co-occurrence model built during the training phase. It has the following two subphases:
   a. *Extract* the keywords relevant to the query from the co-occurrence model. This is done by picking up those related keywords that are most common to all the words present in the query. These are the words that have the highest co-occurrence score in the co-occurrence model.
   b. *Expand* the query with the words extracted from the model. The above word matrix is then used to expand the query submitted by the user. To do this, the query is first processed and all the stop words in the query are removed. Then for each word in the query, we extract the top 5 most commonly co-occurring words from the matrix. All these related words are added into a collection, sorted according to their count. The word with the greatest count would be the word that is most common among the words of the original query. Thus from the collection, we choose the top 5 words based on their count. These words

are added to the original query which is used to search for duplicates. Thus the resulted query is called expanded query.

2. *Search* the repository with the above expanded query. This will return a list of records that include duplicates that are dissimilar in text to the original query. More details on the procedure of search are given below.
3. *Retrain* the model with the new data if maintenance engineer validates it to be part of an *apple-to-orange* pair. The process of learning wherein model parameters are changed or tweaked whenever new data points emerge is known as incremental learning.

Finally, we merge the results obtained via the two searches to provide the user with an aggregated list of duplicates.

The final search is performed on the indexed data using three separate retrieval algorithms implemented in Lucene. These are Okapi BM25 and language modeling using Jelinek–Mercer and Dirichlet smoothing.

8.5.3 *Okapi BM25*

The BM25 weighting scheme also known as Okapi weighting scheme is one of the state-of-the-art information retrieval (IR) techniques and has been in widespread use since its introduction [22]. It is a probabilistic model, which uses three key document attributes, namely, term frequency, document frequency, and average document length. The scoring function is implemented as

$$S(q,d) = \sum_{t \in q} IDF(t) \cdot \frac{tf_t^d}{k_1\left((1-b) + b\frac{l_d}{avl_d}\right) + tf_t^d}$$

where

- tf_t^d is the frequency of term t in document d.
- l_d is the length of document d.
- avl_d is the average document length along the collection.
- k_1 is a parameter used to tune the saturation of the term frequency and is usually between 1.2 and 2.
- b is a length tuning parameter whose range is [0,1].

The IDF for term t is computed as

$$IDF(t) = \log \frac{D_{\text{tot}} - D_t + 0.5}{D_t + 0.5}$$

where

- D_{tot} is the total number of documents.
- D_t is the number of documents where that particular term occurs.

8.5.4 Language Modeling with Smoothing

The central idea behind language modeling is to estimate a model for each document and then rank the documents by the query likelihood according to the estimated language model. A big problem while calculating the maximum likelihood is that if a particular term has not been encountered before, then the maximum likelihood would become 0 as the probability of the individual term would be 0. To avoid this, a technique known as *smoothing* is used which adjusts the likelihood estimator to account for that type of data sparseness. Here, we use two types of smoothing methods Jelinek–Mercer and Dirichlet priors. The Jelinek–Mercer method involves a simple mixture of a document-specific distribution and a distribution estimated from the entire collection:

$$p_\lambda(t|d) = (1-\lambda)p_{ml}(t|d) + \lambda p_{ml}(t|c)$$

where $p_{ml}(t|d)$ represents the conditional probability of the term belonging to that document and $p_{ml}(t|c)$ is the conditional probability of the term belonging to that particular language.

In the Dirichlet smoothing method, the language model is built from the whole collection as the prior distribution of a Bayesian process. The conjugate priors for this form a Dirichlet distribution hence the name:

$$p_\mu(t|d) = \frac{tf_t^d + \mu p(t|d)}{l_d + \mu}$$

Here, $p(t|d)$ is the conditional probability of the term belonging to the document. According to Zhai et al. [9], the optimal value for λ in Jelinek–Mercer is 0.7, and μ in Dirichlet prior is 2000.

8.6 Implementation and Case Study

The prototype of duplicate bug identifier is built in Java using the open-source API Lucene (Apache 2.0 licence). Lucene allows easy implementation of indexing and searching using the VSM algorithm described above. It calculates the TF-IDF values for each word and implements its custom scoring technique to find the similarity between bug records. We modify this custom scoring technique to include some of the more recent and powerful algorithms such as Okapi BM25 and language modeling as discussed earlier.

As there are often huge number of words in bug record vocabulary, representing the model as a standard matrix becomes highly memory intensive, and it will not be possible to run any program in memory. However, as a majority of the words do not co-occur, there is a high level of sparsity (the presence of zeros) in the matrix which allows us to use simpler sparse representations such as compressed row storage (CRS) or compressed column storage (CCS). These representations save memory by storing subsequent nonzero values in contiguous memory locations and ignoring the zeros. This sparse matrix multiplication was done using another open-source API called l4aj which has functions for CRS as well as CCS matrix representations. We used CRS for sparse matrix representation.

8.6.1 Datasets

Typically a bug record contains several fields like summary, description, platform, priority, etc. which are in a natural language format. There are also other fields with non-textual information as attachments. The status of the bug is set by the developer or triager who checks if the bug record is valid or not. And if the bug record has been raised before or caused by the same issue as another bug record, then it is marked as a duplicate. A typical example of a bug record from Chrome dataset, taken from Chrome issue tracker, is shown in Fig. 8.5. We have used the Chrome bug record dataset available on the Google code website. The Chrome dataset consists of 142,286 issues out of which 25,756 issues (18.1 %) were marked by the maintenance engineers as duplicates. These numbers on this bug data are from the date which we have accessed. Figure 8.6 shows a sample of the validated duplicate bug record dataset.

We calculate the similarity between all the duplicate pairs, and the ones which have no similarity are identified as *apple-to-orange* pairs. From Fig. 8.7, it is noticed that 2769 out of 25,756 duplicate pairs in chrome were *apple-to-orange* pairs.

ID ▾	Pri ▾	Mstone ▾	ReleaseBlock ▾	Area ▾	Feature ▾	Status ▾	Owner ▾	Summary + Labels ▾
129413	1	21	Dev	UI	—	Assigned	ygorshe...@chromium.org	Login screen on test image did not detect USB to Ethernet dongle until disconnected and reconnected several times LabBreak
130679	1	21	Dev	UI	Ash-WM	Assigned	sky@chromium.org	Pressing ctrl+t from incognito window opens the new tab in normal window
130881	0	—	Dev	—	Extensions	Assigned	asargent@chromium.org	ExtensionDownloader::FetchUpdatedExtension crashes browser on every launch ext
130759	0	21	Beta	UI	—	Untriaged	—	chrome://chrome/settings/cookies completely broken
124881	1	21	Beta	WebKit	Ash, GPU, GPU-Compositing	Assigned	r...@chromium.org	Calendar rasters very slowly through tiled SkPicture playback path

Fig. 8.5 Bug records in Chrome issue tracker

ID	Summary	Description	Dup ID	Dup Summary	Dup Description
46273	Flash crasl	Developer Build 494	46286	Flash plugin cra	Platform
45250	Browser C	Official Build 48356	46283	notifier::Media	notifierM
45188	Bad Korea	official build 47796 U	46151	Chrome in Linu:	Chrome V
46113	Zygote bu	49136 UbuntuWhen	46114	Zygote bug?	49136 Ub
46138	ThreadSar	E.g. see for the repor	46086	possible data ra	Chromium
46041	views::Tal	EnvironmentVmware	46042	Browser Crash (	Official B
46256	Access Vic	devOS Windows XP S	46042	Browser Crash (	Official B
46027	Google Ch	Go to www.google.co	46013	Replace Google	1. Navigat
45916	[Mac - Fir	49004 Chromium cra:	46003	Crash : Dictiona	Platform
45985	Devnagari	Script Hindi fonts are	45984	Devnagari Script	Hindi fon
45937	flash does	Chromium Developer	45938	abc.go.com say:	Chromium
45936	WebM	6.0.422 youtube. 10.6	45935	WebM is slow c	6.0.422 yc
45868	Tracking b	This is necessary to sy	45869	Tracking bug fo	This is nec
45842	Wrench m	Google Chrome Offici	45830	Wrench menu i	Classic-st

Fig. 8.6 Validated duplicate pairs in Chrome dataset

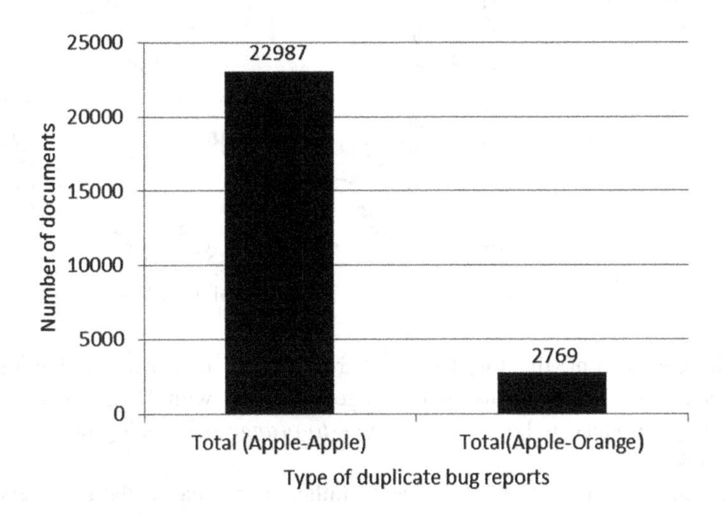

Fig. 8.7 No. of apple-to-orange pairs found in Chrome dataset

The *apple-to-orange* dataset was split into 60 % training and 40 % testing datasets. This split was done randomly and repeated five times for the purpose of generalization.

Among the 1108 *apple-to-orange* pairs used for testing, our tool detected around 205 pairs. However the number of bug records detected in the top 10 is quite low with the best result being only 14 records coming among the top 10. The average recall rates at 10 for all three techniques are shown in Table 8.6, where *recall rate at k* is defined as the fraction of duplicate pairs that were detected in the top k results of the search. A "recall rate at 10" of 1 means that all the duplicate bug report pairs

Table 8.6 Recall rate of apple-to-orange pairs after using query expansion

Rank size	Okapi BM25	Jelinek–Mercer	Dirichlet priors
10	0.003	0.002	0.015
20	0.010	0.008	0.029
30	0.017	0.014	0.039
50	0.034	0.029	0.059
100	0.065	0.068	0.081

Table 8.7 Performance statistics of our technique versus standard search

Number of bug records	10,000	20,000	50,000	100,000	150,000
Number of *apple-to-apple* pairs	112	348	1116	2147	2796
Number of *apple-to-orange* pairs	1181	2907	8317	18,685	25,755
Time taken for standard search (milliseconds)	63	62	62	63	75
Time taken for search with query expansion (seconds)	1.773	5.431	22.56	65.874	98.5

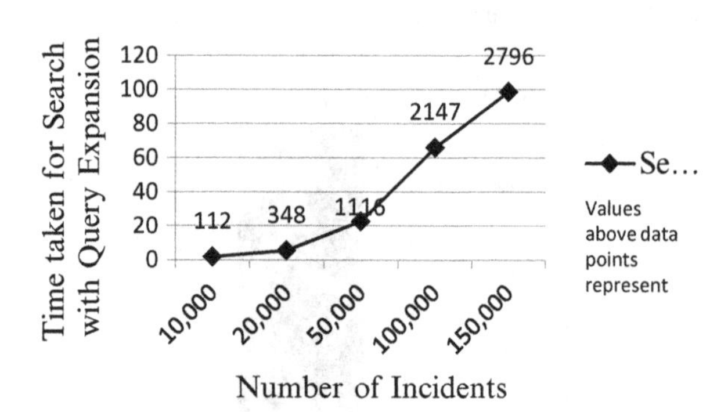

Fig. 8.8 Performance statistics of the technique

were detected among the top 10 search results. We can observe that the best performance was obtained using language modeling with Dirichlet smoothing. Still it only amounts to 1.5 % of the *apple-to-orange* pairs being detected in the top 10 ranks.

The performance of the technique is rather low because there is very little training dataset. We can see that a significant number of duplicates that were not being detected at all before are becoming visible now and with more data that is spread over a longer period of time. It will be possible to increase the rank of those results and improve the recall rates.

The technique as such does not add much time costs to the standard search algorithms. Table 8.7 describes the time overhead in detecting the duplicates using VSM and our technique. The program was run on a machine with a 32 bit single core processor and 4GBs of RAM. We can observe that there is an increase in time taken to display the results as the number of bug's increases, but at the maximum for 150,000 bugs present, it can return the results for a new query in less than 2 min on even a slower machine.

Figure 8.8 is a graphical representation of the time taken to get the results for a bug given the total number of bugs already present in the repository. The values displayed over each data point correspond to the number of *apple-to-orange* pairs that have been used to train the model.

8.7 Recurring Bug Prevention Framework

It is observed that some of the bugs reappear now and then. Such bugs are close to 20 % of the total bugs reported by the end users for that particular software application. Preventing these bugs will save the effort and time of maintenance engineers. These types of bugs need to be monitored over a period of time and a root cause will be identified. Once the root cause is identified, a solution to the problem can be designed and implemented. Based on this approach, we propose the following technical solution that automates the process of identifying the recurring bugs and how well it prevents them to recur in the future. The framework consists of four components as shown in Fig. 8.9. Brief detail of the components appears below.

8.7.1 Knowledge-Enriched Bug Repository

Typically, this component is a bug repository system that stores all the reported bugs by the users for that particular software from the start. This repository is maintained using any issue tracking system. The term knowledge enriched is used because in addition to storing the bug details as reported by the user, it also stores specific information retrieval-related information. This additional information consists of whether it is duplicate of some existing bug, whether it belongs to a particular cluster, when that particular bug group had been fixed, what the solution is, when it is applied, etc. Also if it has reappeared, it marks and links all of the

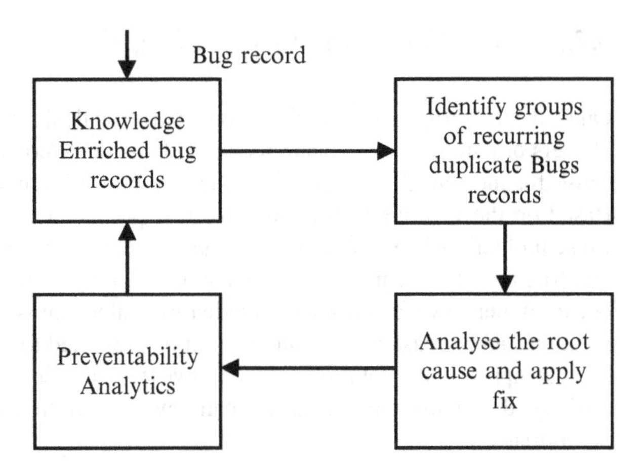

Fig. 8.9 Recurring bug's prevention framework

duplicates. We can use this information to speed up the bug clustering process and also to aid in the calculation of number of bugs prevented.

8.7.2 *Identify Groups of Recurring Duplicate Bugs*

At a given point of time, say once a month (or as decided by the organization based on its policies toward bug analytics), using IR techniques the bugs in the knowledge-enriched bug repository can be clustered into groups of similar bugs. It is useful for identification of recurring bug patterns. IR technique is executed on the above repository of reported bugs. One way is to iterate through all the bugs in the repository and identify those duplicates with score greater than a threshold. Like in ticket clustering tool, we can give the user the ability to vary the number of search results and the threshold for the score. Two possible algorithms to identify this bug patterns are (1) VSM and (2) clustering techniques.

Initial analysis on the Chrome data using VSM showed that out of 40,000 there were around 2500 groups with cluster size greater than 10 and with a moderate score (0.75) threshold. This grouping took 6 min which is not a huge overhead considering the number of bugs reported every month is considerably lesser. Further, through tooling we can also reduce the number of bugs to be grouped each time increasing the performance.

Secondly, we used K-means algorithm to group the bugs into categories based on similarity and provide the user the ability to choose number of clusters or give keywords to aid and fine-tune the clustering. The problem is that the grouping would be broader in nature and would not be feasible for large data. From the previous example, we could see that there were more than 2500 clusters for the Chrome incident data. Therefore, clustering technique could be helpful for small datasets, which is a limitation.

8.7.3 *Root Cause Analysis and Bug Fix*

Once the patterns are identified, root cause analysis can be performed on the clusters or patterns of duplicate/recurring bugs detected. The identification of root cause for the repeatedly occurring bug is more of technical and manual process. Based on the identified root cause for that particular bug, a solution to this root cause problem is identified. The required solution is implemented. As the solutions are stored for this recurring bug in the issue tracking system as and when these bugs occurred, hence we can use machine learning algorithms to analyze all the applied solutions, and a reasonable summary can be presented to the maintenance engineer so that appropriate solution can be invented and implemented. Based on the analysis, the maintenance engineer can devise a solution that is more suitable and appropriate.

8.7.4 Preventability Analytics

In this step, we store the information regarding the bug fixes in the knowledge-enriched bug repository to aid in the calculation of preventability. For example, storing the time at which a particular pattern of reoccurring bugs are fixed can be used to track further occurrences of the same. That is, if no further incidents of that type are encountered, then we can use that to demonstrate the effectiveness of preventive maintenance.

Let us assume that at some point of time $t_{1,}$ bug B_1 has occurred the first time. For the duration of "n" days, it was observed that there are "m" times the bug that has been reported. That is, the B_m has occurred at time t_2. At t_2, the root cause analysis is carried and the bug fixed with likely permanent solution at time t_3. When we carry the duplicate analysis at time t_4, the duplicate bug doesn't appear at all. Based on this analysis, we can conclude that the bug is not recurring. Therefore, we will assume that all such bugs have been disappeared. Hence, using statistical techniques the percentage of bugs detected and closed and prevented can be derived.

8.8 Conclusions

In this chapter, the practical challenges of identifying duplicate bugs in large bug data repositories are explained in detail with examples. To address these challenges, a set of machine learning techniques are proposed and discussed. Based on the proposed techniques, a tool has been built. Using the tool a case study has been conducted on a practical open-source dataset which is available online. The results have been presented in terms of the number of duplicate bugs detected in this dataset.

The chapter proposed VSM and cluster algorithms to identify duplicate bugs which are similar in vocabulary. It also proposed a method to detect the textually dissimilar duplicate records by capturing the underlying root cause relations between the two bug records. It used an expanded query by aggregating the words present in it and also words captured through the co-occurrence relationships between words present in the existing duplicate records of that repository. Both types of duplicates are captured concurrently using two searches, one with the original query and the other with the expanded query. This is a novel way of identifying duplicate data records. It expands further by identifying patterns in the history of previously validated duplicates. One of the limitations of the method is that it requires a priory properly triaged bug data records which are used to build the model. The process of triaging though is often neglected in organizations and there can be a shortage of available data records.

Though we have conducted a case study on practical data, the results are not quite encouraging on identification of apple-*to-orange* types of duplicates especially on the top 10 lists of duplicates. Further, empirical analysis is required to validate these proposed methods of duplicate identification and results. Then only one can analyze the efficiency of the proposed methods. Further, there was a limitation on the dataset size we have used. We have to experiment on a huge datasets and conclude the results. Certainly the recurring bugs' prevention framework takes a longer time to implement in an organization and also to validate the framework. Therefore a lot of empirical analysis is warranted.

Further, the bug record not only contains textual information but also contains the images of the screens. It is observed that few bugs have a simple screenshots attached to the bug information for detailed explanation. However, these records are out of our scope, as we perform analytics on text only. These images contain more accurate information about the nature of the failure that occurred during the access of the software application. To handle this additional information, given a bug report with a screenshot of the error attached along with it, we can use optical character recognition (OCR) techniques to extract any textual information that it contains, and the text can be added to the description of the bug. This enhanced description can now be used by the duplicate bug detection algorithms to provide improved duplicate bug results. This could further increase the processing time for duplicate search and building co-occurrence matrix. This shows the need for further work in this direction to improve both the results fetched and as well as for improving the processing time. Hence, there exists a wider scope for expanding our work in applying analytics to large bug repositories.

References

1. Datamonitor (2011) Software: Global Industry Guide 2010. MarketLine
2. Canfora ACG (1995), Software maintenance. In: 7th international conference on software engineering and knowledge engineering, pp 478–486
3. Sutherland J (1995) Business objects in corporate information systems. ACM Comput Surv 27 (2):274–276
4. Jones C (2006) http://www.compaid.com/caiinternet/ezine/capersjones-maintenance.pdf
5. Anvik J, Hiew L, Murphy GC (2005) Coping with an open bug repository. In: Proceedings of the OOPSLA workshop on eclipse technology exchange, pp 35–39
6. Alexandersson M, Runeson ONP (2007) Detection of duplicate defect reports using natural language. In: Processing 29th international conference on software engineering, pp 499–510
7. Jalbert N, Weimer W (2008) Automated duplicate detection for bug tracking systems. In: IEEE international conference on dependable systems and networks with FTCS and DCC, pp 52–61
8. Pudi V, Krishna PR (2009) Data mining, 1/e. Oxford University Press, India. 2009.x
9. Zhai C, Lafferty J (2004) A study of smoothing methods for language models applied to information retrieval. ACM Trans Inf Syst 22(2):179–214
10. Wang X, Zhang L, Xie T, Anvik J, Sun J (2008) An approach to detecting duplicate bug reports using natural language and execution Information. In: Proceedings of ICSE 2008, Leipzig, Germany
11. Sun C, Lo D, Jiang XWJ, Khoo S-C (2010) A discriminative model approach for accurate duplicate bug report retrieval. In: The Proceedings of ICSE 2010, Cape Town, South Africa

12. Sun C, Lo D, Khoo S-C, Jiang J (2011) Towards more accurate retrieval of duplicate bug records. In: The Proceedings of IEEE/ACM international conference of automated software engineering, IEEE Computer Society, Washington, DC, pp 253–262
13. Tian Y, Sun C, Lo D (2012) Improved duplicate bug report identification. In: Proceedings of 16th European conference on software maintenance and reengineering
14. Alipour A, Hindle A, Stroulia E (2013) A contextual approach towards more accurate duplicate bug detection. In: Mining Software Repositories (MSR), IEEE Press Piscataway, pp 183–192
15. Sureka A, Jalote P (2010) Detecting duplicate bug report using character N-Gram-based features. In: Proceedings of the Asia Pacific software engineering conference, pp 366–374
16. Manning C, Raghavan P, Schütze H (2008) Introduction to information retrieval. Cambridge University Press, New York
17. Salton G, Wong A, Yang CS (1975) A vector space model for automatic indexing. Commun ACM 18(11):613–620
18. Han J, Kamber M, Pei J (2011) Data mining: concepts and techniques: concepts and techniques. Elsevier, Amsterdam
19. Forgy EW (1965) Cluster analysis of multivariate data: efficiency versus interpretability of classifications. Biometrics 21:768–769
20. Dasarathy BV (1990) Nearest neighbor (NN), norms: NN pattern classification techniques ISBN: 0-8186-8930-7
21. Shakhnarovish G, Darrell T, Indyk P (2005) Nearest-neighbour methods in learning and vision. MIT Press, Cambridge. ISBN 0-262-19547-X
22. Robertson SE, Walker S, Jones S (1995) Okapi at TREC-3. In: Proceedings of the third text retrieval conference Gaithersburg, USA, pp 109–126

Part III
Big Data Tools and Analytics

Chapter 9
Large-Scale Data Analytics Tools: Apache Hive, Pig, and HBase

N. Maheswari and M. Sivagami

Abstract The Apache Hadoop is an open-source project which allows for the distributed processing of huge data sets across clusters of computers using simple programming models. It is designed to handle massive amounts of data and has the ability to store, analyze, and access large amounts of data quickly, across clusters of commodity hardware. Hadoop has several large-scale data processing tools and each has its own purpose. The Hadoop ecosystem has emerged as a cost-effective way of working with large data sets. It imposes a particular programming model, called MapReduce, for breaking up computation tasks into units that can be distributed around a cluster of commodity and server class hardware and thereby providing cost-effective horizontal scalability. This chapter provides the introductory material about the various Hadoop ecosystem tools and describes their usage with data analytics. Each tool has its own significance in its functions in data analytics environment.

Keywords Hadoop • Pig • Hive • HBase • HDFS • MapReduce • Data analytics • Big data

9.1 Introduction

Big data refers to large or complex poly-structured data, including video, text, sensor logs, and transactional records, that flows continuously through and around organizations. Traditional data processing applications are not enough to handle such big volumes of data. So, there are various challenges like data capture, data sharing, data analysis, storage, security, and visualization. Organizations across public and private sectors have made a strategic decision to turn big data into competitive advantage. Extracting the required information from big data is similar to refining the business intelligence from transactional data. The big data requires "extract, transform, and load" (ETL) infrastructure [1] to handle it efficiently.

N. Maheswari (✉) • M. Sivagami
School of Computing Science and Engineering, VIT University, Vandalur-Kelambakkam Road, 600 127 Chennai, Tamil Nadu, India
e-mail: maheswari.n@vit.ac.in; msivagami@vit.ac.in

Z. Mahmood (ed.), *Data Science and Big Data Computing*,
DOI 10.1007/978-3-319-31861-5_9

When the source data sets are large, fast, and unstructured, the traditional ETL is not suitable, because it is too complex to develop and too expensive to operate and takes too long to execute.

Yahoo, Google, Facebook, and other companies have extended their services to web scale; the amount of data they collect routinely from user interactions online would have besieged the capabilities of traditional IT architecture. In the interest of advancing the development of core infrastructure components rapidly, they released relevant code for many of the components into open source. Apache Hadoop has emerged from these components, as the standard for managing big data.

Apache Hadoop is an open-source distributed software platform for storing and processing data. Written in Java, it runs on a cluster of industry-standard servers configured with direct-attached storage. Using Hadoop, petabytes of data can be stored reliably on tens of thousands of servers while scaling performance cost-effectively by merely adding inexpensive nodes to the cluster. Apache Hadoop has the distributed processing framework known as MapReduce (Fig. 9.1) that helps programmers solve data-parallel problems by subdividing large data sets into small parts and processing independently. The system splits the input data set into multiple chunks, each of which is assigned a map task that can process the data in parallel. Each map task reads the input as a set of key-value pairs and produces a transformed set of key-value pairs as the output. The framework shuffles and sorts

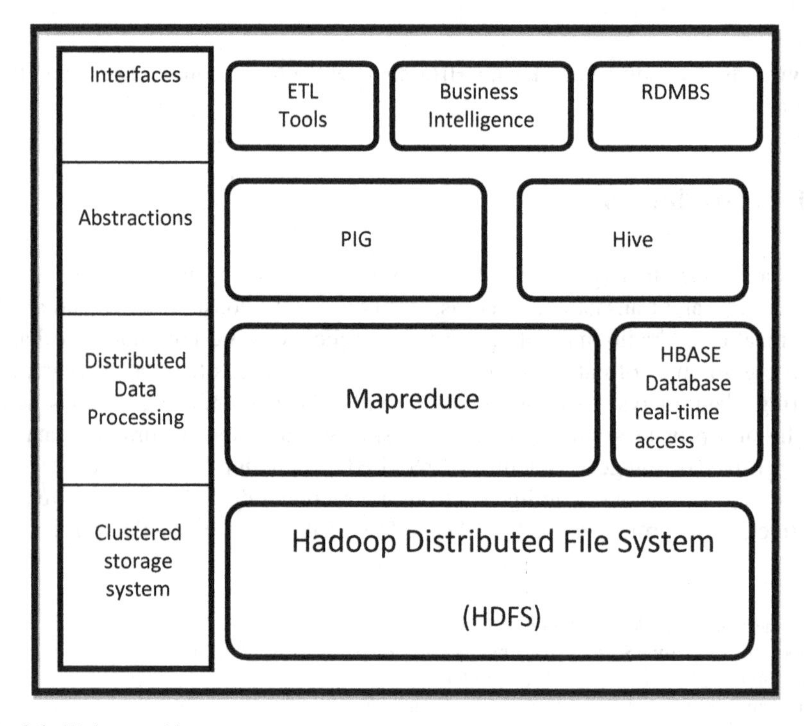

Fig. 9.1 Hadoop architecture

outputs of the *map* tasks, sending the intermediate key-value pairs to the *reduce* tasks, which group them into final results. MapReduce uses JobTracker and TaskTracker mechanisms to schedule tasks, monitor them, and restart any that fail.

The Apache Hadoop platform also includes the Hadoop Distributed File System (HDFS), which is designed for scalability and fault tolerance. HDFS stores large files by dividing them into blocks (usually 64 or 128 MB each) and replicating the blocks on three or more servers. HDFS provides APIs for MapReduce applications to read and write data in parallel. Capacity and performance can be scaled by adding Data Nodes, and a single NameNode mechanism manages data placement and monitors server availability. HDFS clusters may have thousands of nodes and each can hold petabytes of data.

Apache Hadoop includes many other components that are useful for ETL along with MapReduce and HDFS.

In the following sections, more details about the large-scale data analytics tools have been provided. Section 9.2 describes the Hive compilation stages, Hive commands, partitioning and bucketing in Hive, and Hive performance and challenges. Section 9.3 describes the Pig compilation stages, Pig commands, Pig user-defined function, Pig scripts and issues, and challenges in Pig. Section 9.4 highlights the HBase architecture, HBase commands, and HBase performance. Conclusions are given in Sect. 9.5.

9.2 Apache Hive

The size of data sets being collected and analyzed in the industry for business intelligence is growing rapidly, making traditional warehousing solutions prohibitively expensive. Apart from ad hoc analysis [2] and business intelligence applications used by analysts across the company, a number of Facebook products are also based on analytics.

The entire data processing infrastructure in Facebook was developed around a data warehouse built using a commercial RDBMS. The data generated was growing very fast – for example, it grew from a 15 TB data to a 700 TB data set. The infrastructure was so inadequate to process the data and the situation was getting worse. So there was an urgent need for infrastructure that could scale along with large data. As a result, the Hadoop technology has been explored to address the scaling needs. The same jobs that had taken more than a day to complete could now be completed within a few hours using Hadoop. But Hadoop was not easy for end users, especially for those users who were not familiar with MapReduce. End users had to write MapReduce programs for simple tasks like getting row counts or averages. Hadoop lacked the expressiveness of popular query languages like SQL, and as a result, users ended up spending hours to write programs for even simple analysis. To analyze the data more productively, there must be a need to improve the query capabilities of Hadoop. As a result, Hive was introduced in

January 2007. It aimed to bring the familiar concepts of tables, columns, partitions, and a subset of SQL to the unstructured world of Hadoop, along with maintaining the extensibility and flexibility of the Hadoop. Hive and Hadoop are used extensively now for different kinds of data processing. Hive was open-source and since then has been used and explored by a number of Hadoop users for their data processing needs.

Apache Hive is a Hadoop ecosystem tool. It acts as a data warehouse infrastructure built on Hadoop for providing data aggregation, querying, and analysis. Data warehouse applications maintain large data sets and can be mined for analytics. Most of the data warehouse applications use SQL-based relational databases. In order to do data analytics using Hadoop, data has to be transferred from warehouse to Hadoop. Hive simplifies the process of moving the data from data warehouse applications to Hadoop.

Hive QL is an SQL-like language provided by Hive to interact with it. It facilitates a mechanism to project the structure on the data and query the data. The traditional MapReduce pattern can be connected in Hive as a plug-in component. Hive does not create Java MapReduce programs. It uses the built-in generic mapper and reducer modules which are driven by an XML file containing the job plan. The generic modules serve like small language interpreters, and the "language" to force the computation is encoded using XML.

9.2.1 Hive Compilation and Execution Stages

The CLI (command line interface) is the most popular way to use Hive. The CLI allows statements to be typed one at a time. It also facilitates to run "Hive scripts" which are collection of Hive statements.

Hive components are *shell*, *thrift server*, *metastore*, *driver*, *query compiler*, and *execution engine*. The shell allows the interactive queries to execute in CLI [3]. A thrift service provides remote access from other processes. Access is provided by using *JDBC* and *ODBC*. They are implemented on top of the thrift service. All Hive installations require a metastore service, which Hive uses to store table schemas and other metadata. It is typically implemented using tables in a relational database. The driver manages the life cycle of Hive QL statement during compilation, optimization, and execution. Query compiler is invoked by the driver when Hive QL statement is given. It translates the query statement into directed acyclic graph of MapReduce jobs.

The driver submits the DAG-based MapReduce jobs to the execution engine of the Hive and is available in Hadoop. By default, Hive uses a built-in *Derby SQL* server, which provides limited single process storage. For example, when using Derby, two simultaneous instances of the Hive CLI cannot run. A simple web interface known as Hive Web Interface provides remote access to Hive.

9.2.2 *Hive Commands*

Apache Hive works in three modes [4]: local, distributed, and pseudo distributed modes. Local mode refers to the files from local file system and the jobs run all the tasks in a single JVM instance. Distributed mode accesses the files from HDFS and the jobs are run as several instances in the cluster. By default, Hive stores the table data in `file:///user/hive/warehouse` for local mode and in `hdfs://nnamenode_server/user/hive/ warehouse` for the distributed mode. Pseudo distributed mode is almost identical to distributed, but it is a one-node cluster.

This section explores the data definition and data manipulation language parts of Hive QL which is used to create and store data into the Hive tables, extract data into the file system, and manipulate data with queries.

The syntax and examples have been provided for the Hive commands. These commands can be executed in the `hive` > command line interface (CLI).

9.2.2.1 Databases

The database in Hive is essentially a *catalog* or *namespace* of tables [4]. Databases are very useful for larger clusters with multiple teams and users and help in avoiding table name collisions. It is also common to use databases to organize production tables into logical groups. In the absence of database name, the default database is used. The simplest syntax for creating a database is shown in the following example:

```
CREATE (DATABASE|SCHEMA) [IF NOT EXISTS]
database_name
[COMMENT database_comment]
[LOCATION hdfs_path]
[WITH DBPROPERTIES
(property_name=property_value, ...)];
create database mydb;
```

The uses of SCHEMA and DATABASE are interchangeable – they mean the same thing. CREATE DATABASE was added in Hive 0.6. The WITH DBPROPERTIES clause was added in Hive 0.7.

The USE database command informs the Hive of the database where the data is to be stored or accessed from. It helps to group the logical related tables to avoid the name collision:

```
use database_name;
use mydb;
```

The SHOW command displays the list of databases available in Hive:

```
show databases;
```

9.2.2.2 Tables

The CREATE TABLE command creates a table with the given name. An error is thrown if a table or view with the same name already exists. One can use IF NOT EXISTS statement to skip the error. It should be noted that:

- Table names and column names are case insensitive but SerDe and property names are case sensitive.
- Table and column comments are string literals (single quoted).
- The TBLPROPERTIES clause tags the table definition with the user's own metadata key-value pairs. Some predefined table properties also exist, such as last_modified_user and last_modified_time which are automatically added and managed by Hive. Other predefined table properties include: TBLPROPERTIES ("comment" = "*table_comment*") and TBLPROPERTIES ("hbase.table. name" = "*table_name*") – see HBase Integration.

To specify a database for the table, either issue the USE database_name statement prior to the CREATE TABLE statement (in Hive 0.6 and later) or qualify the table name with a database name ("database_name.table.name" in Hive 0.7 and later). The keyword "default" can be used for the default database:

```
CREATE [TEMPORARY] [EXTERNAL] TABLE [IF NOT EXISTS]
[db_name.]table_name -- (Note: TEMPORARY available in Hive
0.14.0and later)
 [(col_namedata_type [COMMENT col_comment], ...)]
 [COMMENT table_comment]
 [PARTITIONED BY (col_namedata_type [COMMENT
col_comment], ...)]
 [CLUSTERED BY (col_name, col_name, ...) [SORTED BY
(col_name [ASC|DESC], ...)] INTO num_buckets BUCKETS]
 [ROW FORMAT row_format]
 [STORED AS file_format]
 [LOCATION hdfs_path]
 [TBLPROPERTIES (property_name=property_value,
...)] -- (Note: Available in Hive 0.6.0 and
later)
[AS select_statement]; -- (Note: Available in
Hive 0.5.0and later; not supported for external
tables)
```

```
CREATE [TEMPORARY] [EXTERNAL] TABLE [IF NOT EXISTS]
[db_name.]table_name
 LIKE existing_table_or_view_name
 [LOCATION hdfs_path];
[ROW FORMAT row_format]
 [STORED AS file_format]
```

Use STORED AS TEXTFILE [4] if the data needs to be stored as plain text files. TEXTFILE is the default file format. Use STORED AS SEQUENCEFILE if the data needs to be compressed. Use STORED AS ORC if the data needs to be stored in ORC file format. The Optimized Row Columnar (ORC) file format provides an efficient way to store Hive data. Use ROW FORMAT SERDE for the serialization:

```
CREATE TABLE IF NOT EXISTS mydb.employees (
name STRING COMMENT 'Employee name',
salary FLOAT COMMENT 'Employee salary',
subordinates ARRAY<STRING>COMMENT 'Names of
subordinates',
deductionsMAP<STRING, FLOAT>
COMMENT 'Keys are deductions names, values are
percentages',
address STRUCT<street:STRING, city:STRING,
state:STRING, zip:INT>
COMMENT 'Home address')
COMMENT 'Description of the table'
TBLPROPERTIES ('creator'='me', 'created_at'='2012-
01-02 10:00:00', ...)
LOCATION '/user/hive/warehouse/mydb.db/employees';
```

The SHOW TABLES command lists all the tables and views in the current working database:

```
show tables;
```

The DESCRIBE table command displays the structure of the table:

```
describe table table_name
describe table employees.
```

9.2.2.3 Loading Data into Table

The LOAD command loads the data either from local file system or from HDFS to the Hive warehouse:

```
Load data local /path of a file/ overwrite into
table
table_ name.
load data local inpath '/home/ponny/mar.txt'
overwrite into table dt;
```

9.2.2.4 Retrieving Data from Table

The SELECT..CASE command helps to retrieve data based on the various conditions specified in each case:

```
select fieldnames from tablename where condition
select count(*), avg(salary) FROM employees;
select * FROM employees WHERE country = 'US' AND
state = 'CA';

select fieldnames case condition then action .... end
from table name
select *, case when c>60 then 'first'
when c<60 then 'second'
else 'third'
end from ma;
```

9.2.2.5 Drop Command for Database and Tables

The DROP command deletes the database/table from Hive:

```
DROP (DATABASE|SCHEMA) [IF EXISTS] database_name
[RESTRICT|CASCADE];
drop database mydp;

drop table tablename
drop table employee;
```

9.2.3 Partitioning and Bucketing

A simple query in Hive reads the entire data set even with the WHERE clause filter. This becomes a bottleneck for running MapReduce jobs over a large table [5, 6]. This issue can be overcome by implementing partitions in Hive. Hive makes it very easy to implement partitions by using the automatic partition scheme when the table is created. It organizes tables into partitions, a way of dividing a table into related parts based on the values of partitioned columns such as date, city, and department. Using partition, it is easy to query a portion of the data.

In Hive's implementation of partitioning, data within a table is split across multiple partitions. Each partition corresponds to a particular value(s) of partition column(s) and is stored as a sub-directory within the table's directory on HDFS. Though the selection of partition key is always a sensitive decision, it should always be a low cardinal attribute, e.g., if the data is associated with time dimension, then date could be a good partition key. Similarly, if data has an association with location, e.g., a country or state, then it is a good idea to have hierarchical partitions like country/state. When the table is queried and if it has partitions, only the required partitions of the table are queried, thereby reducing the I/O and time required by the query. Partitioning can carry out using static, dynamic, or hybrid mode. Data type of the partitioning column has to be able to be converted to a string in order to be saved as a directory name in HDFS.

9.2.3.1 Dynamic Partitioning

Dynamic partitioning has been done based on the attribute/attributes and not the value. Table is dynamically partitioned based on the value of the corresponding attribute in the insertion phase. The following steps are required for the partition.

Non-partitioned Table Creation

```
create table dt(a int,b string,c int,,d int,e int)
row format delimited fields terminated by ','
stored as textfile;
```

Load Data into Non-partitioned Table

```
load data local inpath '/home/ponny/mar.txt'
overwrite into table dt;
```

Dynamic Partition Table Creation

```
create table dtp(b string, c int, d int, e int) partitioned by
(a int) row format delimited fields terminated by ',' stored
as text file;
```

Flag Setting

```
set hive.exec.dynamic.partition = true;
set hive.exec.dynamic.partition.mode = nonstrict;
```

Load Data into Partitioned Table

```
insert overwrite table dtp partition(a) select
b,c,d,e,a from dt;
```

Here, the table dtp is partitioned by the attribute "a." It will create n partitions based on the total number of unique values of the attribute "a."

9.2.3.2 Static Partitioning

Static partitioning is done based on the value/values of attribute/attributes. Table is partitioned based on the value of the partitioned attribute in the insertion phase. The following steps are required for the partition.

Non-partitioned Table Creation

```
create table dt(a int,b string,c int,,d int,e int)
row format delimited fields terminated by ','
stored as textfile;
```

Load Data into Non-partitioned Table

```
load data local inpath '/home/ponny/mar.txt'
overwrite into table dt;
```

Static Partition Table Creation

```
create table stp(a int,,b string,,d int,e int)
partitioned by (c int) row format delimited fields
terminated by ',' stored as textfile;
```

Flag Setting

```
set hive.exec.dynamic.partition.mode = strict;
```

Load Data into Partitioned Table

```
insert overwrite table stp partition(c = 90)
select a,b,d,e from dt where c = 90;
```

Here the table stp is partitioned by the value of the attribute "c." It will create a partition based on the given value of the attribute "c." It will create a directory with the name "c = 90," and the partitioned values are stored in the file inside this directory.

9.2.3.3 Hybrid Partitioning

Hybrid partitioning is a combination of dynamic and static partitioning. It should have a minimum of one attribute pertaining to static partition and it must be mentioned first in the order. If there are multiple partitioning columns, their order is significant, since that translates to the directory structure in HDFS. The following steps are required for the partition.

Non-partitioned Table Creation

```
create table dt(a int,b string,c int,,d int,e int)
row format delimited fields terminated by ','
stored as textfile;
```

Load Data into Non-partitioned Table

```
load data local inpath '/home/ponny/mar.txt'
overwrite into table dt;
```

Hybrid Partition Table Creation

```
create table stpy(a int,b string,e int) partitioned
by (c int,d int) row format delimited fields
terminated by ',' stored as textfile;
```

Flag Setting

```
set hive.exec.dynamic.partition.mode = strict;
```

Load Data into Partitioned Table

```
insert overwrite table stpy partition(c = 90,d)
select a,b,e,d from dt where c = 90;
```

Here the table `stpy` is partitioned based on the unique values of `d` for the records which have the value of `c` = 90. For example, a table named Emp1 contains employee data such as id, name, dept, and yoa (i.e., year of appointment). In order to retrieve the details of all employees who joined in 2011, a query searches the whole table for the required information. However, if the employee data is dynamically partitioned with the year and stored in a separate file, it reduces the query processing time. The following example shows how to partition a file and its data:

The following file contains employee data table *Empl*:

```
/tabl/employeedata/file1
id, name, dept, yoa
1, anbu, TP, 2011
2, jeeva, HR, 2011
3, kunal,SC, 2012
4, Praba, SC, 2012
```

The above data is partitioned into two files using year:

```
/tabl/employeedata/2011/file2
1, anbu, TP, 2011
2, jeeva, HR, 2011

/tabl/employeedata/2012/file3
3, kunal,SC, 2012
4, Praba, SC, 2012
```

In Hive, as data is written to disk, each partition of data will be automatically split into different folders. During a read operation, Hive will use the folder structure to quickly locate the write partitions and also return the partitioning columns as columns in the result set. This can dramatically improve query performance, but only if the partitioning scheme reflects common filtering. Partitioning feature is very useful in Hive; however, a design that creates too many partitions may optimize some queries but damages other important queries. The other drawback is too many partitions lead to the large number of unnecessary Hadoop files and directories. This creates overhead on the NameNode since it must keep all metadata for the file system in memory. If the number of partitions rises above the certain threshold, it can be run into "out of memory" errors when MapReduce jobs are being generated. In this condition, even simple select statement can fail. In order to solve this issue, Java heap size can be tuned/partitioning [7] and the scheme can be modified to result as fewer partitions.

9.2.3.4 Bucketing

Bucketing is another technique [8] for decomposing data sets into more manageable parts. For example, suppose a table using the date as the top-level partition and the employee id as the second-level partition leads to too many small partitions. To overcome this, Hive provides bucketing concept. Bucketing has several advantages. The number of buckets is fixed so it does not fluctuate with data. If two tables are bucketed by employee_id, Hive can create a logically correct sampling. Bucketing also aids in doing efficient map-side joins, etc. Bucketing happens by using a hash algorithm and then a modulo on the number of buckets. So, a row might get inserted into any of the buckets. If the employee table had the buckets and employee_id had used as the bucketing column, the value of this column will be hashed by a user-defined number into buckets. Records with the same employee_id will always be stored in the same bucket. Assuming the number of employee_id is much greater than the number of buckets, each bucket will have huge employee_ids. Creating table, like CLUSTERED BY (employee_id) INTO XX BUCKETS, can be specified, where XX is the number of buckets. Bucketing can be used for sampling of data, as well as for joining two data sets much more effectively. It reduces the I/O scans during the join process if the process is happening on the same keys (columns). It also reduces the scan cycles to find a particular key because bucketing ensures that the key is present in a certain bucket.

Example

Flag setting

```
Set hive.enforce.bucketing=true;
```

Creation of buckets

```
create table dtb(a int,b string, c int, d int, e int)
  clustered by (d) into 3 buckets row format delimited fields
  terminated by ',' stored as textfile;
```

Data insertion

```
insert overwrite table dtb select a,b,c,d,e from dt;
```

Basically, both partitioning and bucketing slice the data for executing the query much more efficiently than on the non-sliced data. The major difference is that the number of slices will keep on changing in the case of partitioning as data is modified, but with bucketing, the number of slices is fixed which are specified during table creation.

9.2.4 External Table

In the managed table [3, 4], data is loaded using LOAD or INSERT command. Hive will create appropriate directory and copy files into those directories. When dropping those tables, Hive removes directory and data files both because Hive owns data. So they are called Hive managed table. Here is an example:

```
create external table det(a int, c int, b string,d
int) row format delimited fields terminated by ','
location '/user/ms/extq';
```

It is very common that data analytics applications need to use data within Hive which is owned by some other. In that case, the data file is already there in HDFS directory but it is owned and used by some other tool like Pig. In the case of managed table, it is necessary to copy data into Hive warehouse. So it would be better if there is a facility to map that data into Hive, perform some queries, prepare output, remove mapping, and then leave data at the original place as it was. For this task, Hive provides external table concept. In order to create external table, the keyword external is to be mentioned in create table command. In external tables, Hive assumes that it does not own data or data files. It means that when an external table is dropped, Hive will remove metadata about external table but will leave table data as it was.

9.2.5 Hive Performance

Hive is most suited for *data warehouse* applications, where relatively static data is analyzed, fast response times are not required, and the data is not changing rapidly. Besides, Hive has its own challenges: it does not support transactions since it is not a full database. It does not provide record-level update, insert, or delete. New tables can be created from queries or output query results to files.

Input formats play a critical role in Hive performance. For example, JSON, the text type of input formats, is not a good choice for a large production system where data volume is really high. These types of readable formats actually take a lot of space and have some overhead of parsing (e.g., JSON parsing). To address these problems, Hive comes with columnar input formats like RCFile, ORC, etc. Columnar formats reduce the read operations in analytics queries by allowing each column to be accessed individually.

Hadoop can execute MapReduce jobs in parallel, and several queries executed on Hive automatically use this parallelism. However, single, complex Hive queries are normally translated to a number of MapReduce jobs that are executed by default in a sequence. Often though, some of a query's MapReduce stages are not interdependent and could be executed in parallel. They can take advantage of spare capacity on a cluster and improve cluster utilization while at the same time reducing the overall query execution time. Vectorization allows Hive to process a batch of rows together instead of processing one row at a time [9]. Each batch consists of a column vector which is usually an array of primitive types. Operations are performed on the entire column vector, which improves the instruction pipelines and cache usage.

Hive adds extensions to provide better performance in the context of Hadoop and to integrate with external programs. It provides a familiar programming model for people who know SQL and it also eliminates lots of hassle to code in Java. Hive makes it easier for developers to port SQL-based applications to Hadoop, compared with other Hadoop languages and tools. It focuses primarily on query part of SQL and better reflects the underlying MapReduce process. Pig is a simple scripting language and helps to write complex MapReduce transformations easily.

9.3 Apache Pig

Hadoop MapReduce has high coding complexity and it is highly favored with the programmers with high Java programming skills. Here, when the input file size increases, the execution time for MapReduce also increases proportionally. So there is a requirement for a platform which supports ease of programming as well as enhanced extensibility. In this context, Apache Pig reveals itself as a novel approach in data analytics. Pig was initially developed at Yahoo [10] and later it was open sourced. It allows people using Hadoop to focus more on analyzing large data sets and spend less time to write mapper and reducer programs. Apache Pig is a platform that supports substantial parallelization which enables the processing of huge data sets.

Pig is an open-source high-level data processing language. This high-level data processing language is known as *Pig Latin* which simplifies programming because of ease of expressing the code in Pig Latin. Pig Latin can work on schema-less or inconsistent environments and can operate on the available data as soon as it is loaded into the HDFS. The Pig compiler compiles the script in a choice of evaluation mechanisms and utilizes optimization opportunities in the script. This avoids tuning the program manually. The progress in Pig compiler makes the Pig Latin program get an automatic speedup. Pig Latin defines a set of transformations on a data set such as aggregate, join, and sort. The statement written in Pig script is translated into MapReduce so that it can be executed within Hadoop. Pig Latin describes a directed acyclic graph (DAG) where the edges are data flows and nodes are operators that process the data. Pig Latin stores the data after the execution in HDFS. It can be extended using user-defined functions (UDFs), in which the user can write the customized functions in Java, Python, etc., and those functions can be invoked directly from the Pig Latin.

9.3.1 Modes of User Interaction with Pig

Pig has two execution types:

- Local mode: To run Pig in local mode, users have access to a single machine. All files are installed and run using the local host and file system. Local mode can be specified using the -x flag (pig -x local). Local mode does not support parallel mapper execution with Hadoop.

- MapReduce mode: To run Pig in MapReduce mode, users need a setup of Hadoop cluster with HDFS installation. MapReduce mode is the default mode and users do not need to specify it using the -x flag (Pig or Pig -x MapReduce).

  Pig allows three modes of user interaction:

- Interactive mode: The user is presented with an interactive shell called Grunt, which accepts Pig commands.
- Batch mode: In this mode, a user submits a pre-written script called Pig script, containing a series of Pig commands, typically ending with STORE. The semantics are similar to interactive mode.
- Embedded mode: Pig Latin commands can be submitted through method invocations from a Java program. This option permits dynamic construction of Pig Latin programs, as well as dynamic control flow.

9.3.2 Pig Compilation and Execution Stages

A Pig script has different stages before being executed, as shown in Fig. 9.2. These stages are described below.

9.3.2.1 Parsing

The role of the parser [11] is to verify that the program is syntactically correct and whether all referenced variables are defined. Type checking and schema inference can also be done by parser. Other checks, such as verifying the ability to instantiate classes corresponding to user defined, also occur in this phase. A canonical logical plan with one-to-one correspondence between Pig Latin statements and logical

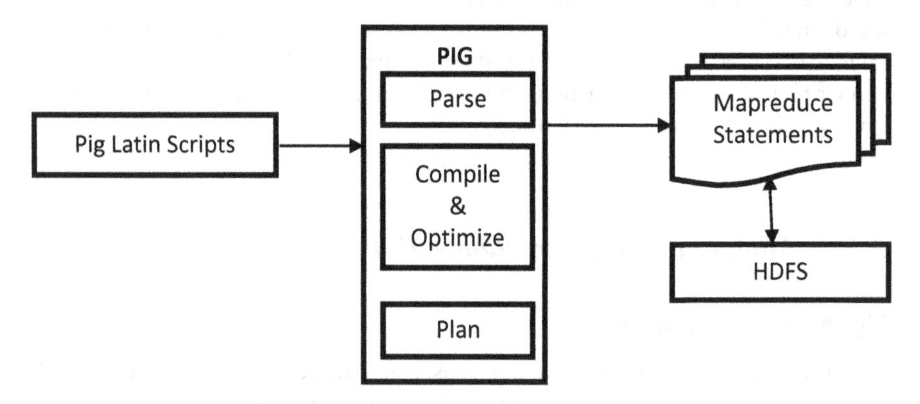

Fig. 9.2 Pig compilation and execution stages

operators which are arranged in directed acyclic graph (DAG) is the output of the parser. Logical optimizer optimizes the logical plan generated by the parser. In this stage, logical optimizations such as projection and pushdown are carried out.

9.3.2.2 Compile and Optimize

The optimized logical plan is compiled into a series of MapReduce jobs, which then pass through another optimization phase. An example of MapReduce-level optimization is utilizing the MapReduce combiner stage in order to perform early partial aggregation [2], in the case of distributive or algebraic aggregation functions.

9.3.2.3 Plan

Topologically, the DAG of optimized MapReduce jobs is sorted and then submitted to Hadoop HDFS for execution in that order. Pig (Hadoop job manager) usually monitors the Hadoop execution status, and the user periodically gets the reports on the progress of the overall program. Warnings or errors that arise during execution are logged and reported to the user.

9.3.3 *Pig Latin Commands*

Pig's interactive shell *Grunt* allows to execute Pig commands manually. It is typically used for ad hoc data analysis and the Grunt shell supports some basic utility commands. Typing *help* will print out a help screen of such utility commands. The command *qui*t can be used to exit the Grunt shell.

Pig Latin statements are the basic constructs to process data using Pig. A Pig Latin statement is an operator that takes a relation as input and produces another relation as output (except LOAD and STORE which read data from and write data to the file system).

A relation can be defined as follows:

- A relation is a bag (more specifically, an outer bag).
- A bag is a collection of tuples.
- A tuple is an ordered set of fields.
- A field is a piece of data.

The names (aliases) of relations, Pig Latin functions, and fields are case sensitive. The names of parameters and all other Pig Latin keywords are also case insensitive. Fields are referred to by positional notation or by name (alias). Positional notation is generated by the system. Positional notation is indicated with the dollar sign ($) and begins with zero (0), for example, $0, $1, and $2. Pig Latin statements may include expressions and schemas. It can be extended to multiple lines and must end with a semicolon. Basically, Pig Latin statements are processed using multi-query execution.

9.3.3.1 LOAD Command

This command places the data from a file into a relation. It uses the PigStorage load function as default with USING option unless specified. The data can be given a schema name using the AS option:

```
alias = LOAD 'file' [USING function] [AS schema];
```

9.3.3.2 DUMP Command

This helps to execute the statements mentioned above DUMP [12]. It displays the content of a relation. It helps in debugging. The relation should be small enough for printing on screen. The LIMIT operation can be applied on an alias to make sure it is small enough for display:

```
DUMP alias;
```

9.3.3.3 STORE Command

This helps to place the data from a relation into a directory. The directory must not exist when this command is executed. Pig will create the directory and store the relation in files named part-*nnnnn* in it. It uses the PigStorage store function as default with the USING option unless specified:

```
STORE alias INTO 'directory' [USING function];
```

In the following example, the names (aliases) of the relations are A and B:

```
A = load 'student' using PigStorage() AS
  (name:chararray, age:int,gpa:float);
B = limit A 4;
dump B;

(Ashu,18,5.0F)
(Maya,19,5.8F)
(Belly,20,7.9F)
(Joney,18,8.8F)
```

The limit command allows to specify how many tuples (rows) are to be returned back.

9.3.3.4 DESCRIBE Command

It displays the schema of a relation:

```
describe alias;
describe A;
```

9.3.3.5 ILLUSTRATE Command

It displays step-by-step process of how data is transformed, starting with a load command, to arrive at the resulting relation. To keep the display and processing manageable, only a (not completely random) sample of the input data is used to simulate the execution:

```
illustrate alias;
illustrate A;
```

9.3.3.6 Expressions

In Pig Latin, expressions are language constructs used with the FILTER, FOREACH, UNION, GROUP, and SPLIT operators. The most salient characteristic about Pig Latin as a language is its relational operators. These operators, as mentioned in the following subsections, define Pig Latin as a data processing language.

9.3.3.7 UNION Command

This combines multiple relations together, whereas SPLIT partitions a relation into multiple ones:

```
a = load 'A' using PigStorage(',') as (a1:int,
a2:int, a3:int);
b = load 'B' using PigStorage(',') as (b1:int,
b2:int, b3:int);
c = UNION a, b;
dump c;
```

9.3.3.8 SPLIT Command

The SPLIT operation on relation s sends a tuple to f1 if its third field ($2) is greater than 4000 and to f2 if it is not equal to 10,000. It is possible to write conditions such that some rows will go to both f1 and f2 or to neither:

```
split s into f1 if $2>4000,f2 if $2 ! = 10000;
dump f1;
dump f2;
```

9.3.3.9 FILTER Command

The FILTER operator trims a relation down to only tuples that pass a certain condition:

```
p = filter s by $2>=5000;
dump p;
```

SPLIT can be simulated by multiple FILTER operators.

9.3.3.10 GROUP Command

When grouping a relation, the result will be a new relation with two columns, "group" and the name of the original relation. The group column has the schema of what it is grouped by. The new relations contain the tuples with the same group key. Group key is the attribute on which the grouping has been performed:

```
l = group s by $1;
dump l;
```

The group operator in Pig is a blocking operator and forces the Hadoop MapReduce job. All the data is shuffled, so that rows in different partitions that have the same grouping key group together. So grouping has overhead.

9.3.3.11 FOREACH Command

This can be used to project specific columns of a relation into the output. Arbitrary expressions also can be applied on specific columns to produce the required output. Nested form of FOREACH facilitates to do more complex processing of tuples:

```
h = foreach s generate a*b;
dump h;
```

9.3.4 Pig Scripts

Pig scripts can run inside the Grunt shell and can be useful in debugging Pig scripts. Pig Latin scripts are about combining the Pig Latin statements together. Pig scripts can be executed in the Grunt shell using *exec* and *run* commands. The *exec* command executes a Pig script in a separate space from the Grunt shell. The command *run* executes a Pig script in the same space as Grunt (also known as *interactive mode* [13]). It has the same effect as manually typing in each line of the script into the Grunt shell. Pig Latin programs can be written in a sequence of steps where each step is a single high-level data transformation. Pig Latin scripts are generally structured as follows:

- A LOAD statement to read data from the file system.
- A series of "transformation" statements to process the data.
- A DUMP statement to view results or a STORE statement to save the results.
- DUMP or STORE statement is required to generate output.

9.3.4.1 ct.pig

```
u = load '/home/ponny/mar.txt' as (f1:chararray);
y = foreach u generate FLATTEN(TOKENIZE(f1));
r = group y all;
c = foreach r generate COUNT($1);
dump c;
```

9.3.4.2 comp.pig

```
u = load '/home/ponny/mar.txt' as (f1:chararray);
y1 = foreach u generate FLATTEN(TOKENIZE(f1)) as
word;
j = filter y1 by word=='abc';
k = group j all;
l = foreach k generate COUNT(j.$0);
dump l;
```

The left-hand side of equal sign in the Pig Latin commands refers to the relation but not a single row; and on the right-hand side, the required operations can be done for each row. The *ct.pig* script counts the total number of words in a file. The *comp. pig* counts the occurrence of the particular word "abc" in a file. The script always finds the count of the word "abc." The script can be made to work for any user-specific word by parameter substitution. The third line of the above *comp.pig* script can be replaced by the following statement for parameter substitution. For naming the parameters in Pig Latin, the parameter name should start with the dollar ($) symbol:

```
j = filter y1 by word=='$inp';
```

Here, $inp is the placeholder for the word entered by the user at the time of execution. The variable name "$inp" in the above command is enclosed with single quotes so that the word *abc* in the following execution command does not have the quotes:

```
exec -param inp = abc comp.pig
```

9.3.5 User-Defined Functions (UDFs) in Pig

Pig provides UDF as a piggy bank repository [14, 15]. These allow the users to define their own user-defined functions for their requirement. UDFs can be implemented using three languages, Java, Python, and Java script. Pig provides extensive support for Java functions and limited support for other two languages. In Java, UDFs can be developed using accumulator and algebraic interface.

9.3.5.1 Predefined UDF

Pig scripts can access the UDFs from the piggy bank repository. The sample script using the same has been given below:

```
register
/home/ponny/hadoop/pig/pig/contrib/piggybank/java/
piggybank.jar;
define
Upper.org.apache.pig.piggybank.evaluation.string.
UPPER();
u = load '/home/ponny/mar.txt' as (f1:chararray);
l = foreach u generate Upper($0);
```

9.3.5.2 Customized Java UDFs

The steps below describe the user-defined uppercase conversion function using Java UDF.

Step 1: Write a Java function, e.g.,

```
upper.java

package myudf;
import java.io.IOException;
import org.apache.pig.EvalFunc;
import org.apache.pig.data.Tuple;
public class upper extends EvalFunc<String>
{
public String exec(Tuple input) throws IOException
{
if (input == null || input.size() == 0)
```

```
return null;
try{
String str = (String) input.get(0);
return str.toUpperCase();
}
catch(Exception e){
throw new IOException("Caught exception processing
input row", e);
}}}
```

Step 2: Compile the Java function, e.g.,

```
cd myudf
javac -cp /home/ponny/hadoop/pig/pig/pig-0.11.1.jar
upper.java
```

Step 3: Make a jar file, e.g.,

```
cd
jar -cf myudfs.jar myudf
```

Step 4: Develop a Pig script (s.pig) which uses the above UDF, e.g.,

```
s.pig
register myudfs.jar;
a = load '/home/ponny/myudf/b.csv' using
PigStorage(',') as (l:chararray,c:int);
k = foreach a generate myudf.upper(l);
dump k;
```

Pig is a higher-level data processing layer on top of Hadoop. Its Pig Latin language provides programmers a more easy way to specify data flows. It supports schemas in processing structured data, yet it is flexible enough to work with unstructured text or semi-structured XML data. It is extensible with the use of UDFs. It simplifies two aspects of MapReduce programming – data joining and job chaining.

Pig gains popularity in different areas of computational and statistical analysis. It is widely used in processing weblogs and wiping off corrupted data from records. Pig helps to build the behavior prediction models based on the user interactions with a website. This feature can be used in displaying ads and news stories as per the users' choice. It can also be updated as the behavior prediction model with the users' data. This feature is largely used in social networking sites. When it comes to small data or scanning multiple records in random order, Pig cannot be considered as effective as MapReduce. It reduces the time consumed for working on the complex programming constructs for implementing a typical MapReduce paradigm using Java but still has the advantages of MapReduce task division method with the help of the layered architecture and inbuilt compilers of Apache Pig. Hive and Pig were designed for ad hoc batch processing of potentially large amount of data by leveraging MapReduce. To store enormous amount of data, HBase can be used as a key-value data store with low latency than HDFS.

9.4 Apache HBase

Hadoop works with various data formats such as arbitrary, semi-, or even unstructured and accesses them only in a sequential manner. Hadoop has to search the entire data set even for the simplest jobs. A huge data set when processed results in another huge data set, which should also be processed sequentially. So, a new technique is required to access any point of data in a single unit of time in Hadoop environment. HDFS in Hadoop is optimized for streaming access of large files. The files stored in HDFS are accessed through MapReduce to process them in batch mode. HDFS files are write-once files and read-many files. There is no concept of random writes in HDFS and it also doesn't perform random reads very well.

HBase as an open-source project [16] provides the solution for it. It leverages the fault tolerance provided by the Hadoop Distributed File System (HDFS). HBase is a part of the Hadoop ecosystem and it provides random real-time read/write access to data in the Hadoop File System.

It is a distributed column-oriented database, built on top of the Hadoop file system, and horizontally scalable. Other random access databases are Cassandra, CouchDB, Dynamo, MongoDB, etc. Column-oriented databases are those that store data tables as sections of columns of data, rather than as rows of data. Facebook, Twitter, Yahoo, Netflix, StumbleUpon, and Adobe use HBase internally to process data. Data model of HBase is similar to Google's big table [17] and it stores data as key-value pairs. HBase allows low latency access to small amounts of data from within a large data set. Single rows can be accessed quickly from a billion row table. HBase has flexible data model to work with and data is indexed by the row key. It scans faster across the tables. Scaling is in terms of writes as well as total volume of data.

The table schema defines only column families, which are the key-value pairs. A table has multiple column families and each column family can have any number of columns. Subsequent column values are stored contiguously on the disk. Each cell value of the table has a time stamp. It provides data replication across clusters. In Table 9.1, employee and occupation are shown as column families.

Table 9.1 Employee details

Row key	Employee		Occupation	
Eid	*Name*	*Department*	*Designation*	*Salary*
1	Anu	Sales	Manager	$12,000
2	Keen	Marketing	Associate manager	$10,000

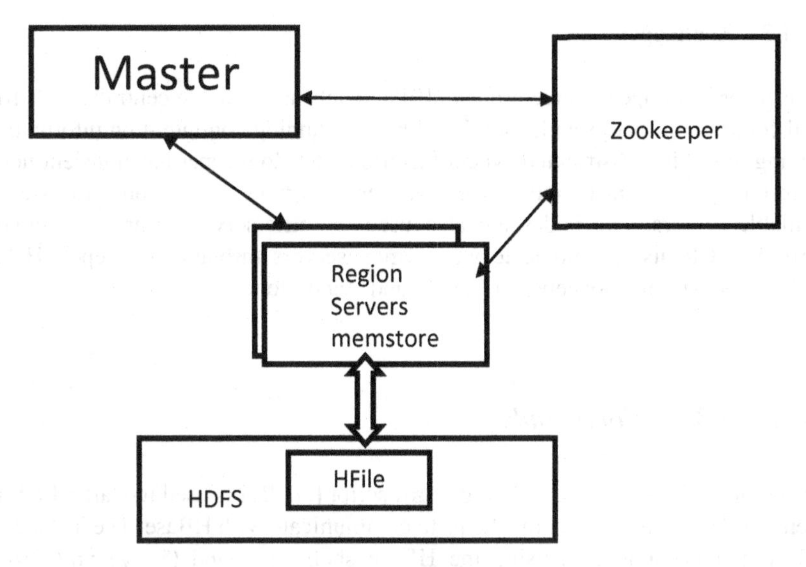

Fig. 9.3 HBase architecture

9.4.1 HBase Architecture

HBase architecture [18] as shown in Fig. 9.3 has master server and region servers as major components. Region servers can be added or removed as per requirement.

9.4.1.1 Region Server

Regions are tables. They are split up and spread across the region servers. Region servers have regions and stores. The regions in region servers handle data-oriented operations by communicating with the client. The region server also performs read and write requests for all the regions. It finalizes the size of the region, based on the region size thresholds.

The store in the region server has memory store and HFiles. Memstore in store is similar to a cache memory. Initially, the data is stored in memstore and transferred to HFiles as blocks. Later, the memstore is flushed.

9.4.1.2 Master Server

The master server takes help from Apache Zookeeper to handle the load balancing of region servers. It assigns regions to the region servers and unloads the busy servers by shifting the regions to less occupied servers. It maintains the state of the clusters. It takes the responsibility of schema changes and other metadata operations like creation of tables and column families.

9.4.1.3 Zookeeper

Zookeeper is an open-source project [19] and acts as a reliable centralized distributed coordinator that provides services like maintaining configuration information, naming, providing distributed synchronization, etc. Zookeeper has transient nodes representing different region servers. Master servers use these nodes to discover available servers. The nodes are also used to track server failures or network partitions. Clients communicate with region servers through Zookeeper. HBase itself takes care of Zookeeper in pseudo and stand-alone modes.

9.4.2 HBase Commands

The command `bin/start-hbase.sh` script [20, 21] is used to start all HBase daemons. HBase has a shell that helps to communicate with HBase. The instance of HBase can be connected using the HBase shell command ($ `./bin/hbase shell`). The HBase shell prompt ends with a > character. The HBase shell prompt will be `hbase(main):001:0>`.

Some of the commands and their output are discussed below.

9.4.2.1 Table Creation

Use the CREATE command to create a new table. Table and the column family names must be specified:

```
hbase(main):001:0>create 'test1', 'cf1' (test - table name
cf1- column family name)
0 row(s) in 0.4170 seconds
=>Hbase::Table - test1
```

9.4.2.2 List

Display the information about the created table "test1" using the LIST command:

```
hbase(main):002:0>list 'test1'
TABLE
test1
 1 row(s) in 0.0180seconds
 =>["test1"]
```

9.4.2.3 Put

Write data into the table using the PUT command:

```
hbase(main):002:0>put 'test1', 'row1', 'cf1:x', 'xy'
0 row(s) in 0.080seconds
hbase(main):003:0>put 'test1', 'row2', 'cf1:y', 'sb'
0 row(s) in 0.01seconds
```

Here, two values have been inserted. The first insert is at row1, column cf1:x, with a value of "xy." Columns in HBase are comprised of a column family prefix, cf1 in this example, followed by a colon and then a column qualifier suffix, x in this case.

9.4.2.4 Get

This command helps to get a single row of data at a time:

```
hbase(main):006:0>get 'test1', 'row1'
COLUMN CELL
cf:x timestamp=1421762485768, value=xy
 1 row(s) in 0.035seconds
```

9.4.2.5 Scan

Another way to get data from HBase is with the SCAN command. This command can be used to scan all the rows from the table. The scan can be limited for a number of records:

```
hbase(main):006:0>scan 'test1'
ROW COLUMN+CELL
row1 column=cf1:x, timestamp=1421762485768,
value=xy
row2 column=cf1:y, timestamp=1421762491785, value=sb
 2 row(s) in 0.0230seconds
```

9.4.2.6 Disable/Enable a Table

Table can be disabled or enabled, using the DISABLE or ENABLE command:

```
hbase(main):008:0>disable 'test1'
 0 row(s) in 1.1820seconds
hbase(main):009:0>enable 'test1'
 0 row(s) in 0.1770seconds
```

9.4.2.7 Drop

To delete a table, this command can be used:

```
hbase(main):011:0>drop 'test1'
0 row(s) in 0.137seconds
```

9.4.2.8 Exit/Quit

To quit the HBase shell and disconnect from the cluster, the QUIT command is available. But HBase will be running in the background.

9.4.2.9 Stop

In the same way that the `bin/start-hbase.sh` script is used to start all HBase daemons, the bin/stop-hbase.sh script stops them:

```
$ ./bin/stop-hbase.sh
stoppinghbase...................
$
```

After issuing this command, it can take several minutes for the processes to shut down. Use the jps to be sure that the HMaster and HRegionServer processes are shut down.

HBase is more suitable for the applications with large amounts of data which require frequent random read/write operations. HBase does not use Hadoop's MapReduce capabilities directly, though it can integrate with Hadoop to serve as a source or destination of MapReduce jobs. HBase is designed to support queries of massive data sets and is optimized for read performance. For writes, HBase seeks to maintain consistency. In contrast to "eventually consistent" [17], HBase does not offer various consistency level settings. Due to HBase's strong consistency, writes can be slower. It prevents data loss from cluster node failure via replication. Hadoop MapReduce is used together with HBase to process data in an efficient way. HBase is used in the organizations with multi-petabyte databases, running them as mission-critical data stores. It also helps for click stream data storage and time series analysis.

9.5 Conclusion

The amount of existing data for processing is constantly increasing and becomes more diverse. This chapter provides an insight into the various tools that can help in analyzing big data. Apache Hadoop is one of the widely used software libraries to perform large-scale data analytics tasks on clusters of computers in parallel.

Pig is a high-level data flow language and serves as an execution framework for parallel computation. Hive acts as a data warehouse infrastructure that provides data summarization and ad hoc querying. HBase is a NOSQL, scalable, distributed database that supports structured data storage for large tables. HBase is for key-value data store and retrieve, but queries cannot be performed on rows (key-value pairs). So Hive can be used as a query layer to an HBase data store – Pig also offers a way of loading and storing HBase data. These tools provide the major functionality to store both structured and unstructured data and perform refined processing and analysis. These tools also play pivotal role in designing scalable data analytics system. This chapter describes how the acquisition of large amounts of data is analyzed with the help of data analytics tools for further process.

References

1. Intel (2013) White paper: extract, transform and load Big data with Hadoop. Available at: hadoop.intel.com. Accessed 30 July 2015
2. Ashish et al (2010) Hive – a Petabyte scale data warehouse using hadoop. IEEE International Conference on Data Engineering, November 2010
3. Edward C et al (2012) Programming Hive. O'Reilly Media Inc, Sebastopol
4. Apache (2014) Language manual. Available at: https://cwiki.apache.org/confluence/display/Hive/LanguageManual+DDL#LanguageManualDDL-Overview. Accessed 10 May 2014
5. Tutorials point (2013) Hive partitioning. Available at: http://www.tutorialspoint.com/Hive/Hive_partitioning.html. Accessed 15 June 2014
6. Rohit R (2014) Introduction to Hive's partitioning. Available at: http://java.dzone.com/articles/introduction-Hives. Accessed 25 Jan 2015
7. Peschka J (2013) Introduction to Hive partitioning. Available at: http://www.brentozar.com/archive/2013/03/introduction-to-hive-partitioning/. Accessed 8 Aug 2015
8. Thrive school (2013) Available at: http://thriveschool.blogspot.in/2013/11/Hive-bucketed-tables-and-sampling.html. Accessed 10 Jan 2015
9. Philip N (2014) 10 best practices for Apache Hive.Available at: www.qubole.com/blog/big-data/hive-best-practices. Accessed 15 July 2015
10. Petit W (2014) Introduction to Pig. Available at: http://bigdatauniversity.com/bdu-wp/bdu-course/introduction-to-pig/#sthash.HUcw7EZe.dpuf. Accessed 20 June 2014
11. Apache (2014) Hadoop online tutorial. Available at: http://hadooptutorial.info/tag/hadoop-pig-architecture-explanation. Accessed 13 Feb 2015
12. Hadoop (2010) Pig latin manual. Available at: https://pig.apache.org/docs/r0.7.0/piglatin_ref2.html. Accessed 20 May 2014
13. Lam C (2010) Hadoop in action. Manning Publications, Greenwich
14. Gates A (2011) Programming Pig. O'Reilly Media Inc, Sebastopol
15. Apache (2007) Getting started-Pig, Apache Software Foundation
16. Apache (2015) When would I use Apache HBase. Available at: Hbase.apache.org. Accessed 10 Feb 2015
17. Grehan R (2014) Review: HBase is massively scalable – and hugely complex. Available at: http://www.infoworld.com/article/2610709/database/review--hbase-is-massively-scalable----and-hugely-complex.html. Accessed 10 July 2015

Chapter 10
Big Data Analytics: Enabling Technologies and Tools

Mohanavadivu Periasamy and Pethuru Raj

Abstract The era of big data is at its full bloom with the data being generated, captured, stocked, polished, and processed in astronomical proportions. This is mainly due to the unprecedented levels of technology adoption and adaption resulting in the connectivity technologies, network topologies, and tools that have enabled seamless connectivity between billions of physical, mechanical, electrical, electronic, and computer systems. This data explosion also challenges the way corporate decisions are being taken. It is for this reason that data is being widely recognized as one of the core strategic assets, and extraction of value from the collected data, using data science approaches, is gaining importance. Apart from the analytics, data management, mining, and visualization are also becoming feverishly articulated tasks on such data. In the knowledge economy and digital world of today, insight-driven decision-making is being tipped as the most important and influential factor. Precisely speaking, competent and cognitive data analytics is capable of bringing forth better and hitherto unforeseen approaches and solutions for a variety of perpetual problems that are constantly hitting hard our society and the global economy. This chapter looks into the big data domain and discusses the relevant analytical methods, platforms, infrastructures, and techniques.

Keywords AaaS • BI • Business intelligence • big data • CSP • data science • Data virtualization • Hadoop • Information visualization • MapReduce • NoSQL • NoSQL databases • SAP HANA

M. Periasamy (✉)
TCL Canada, 1441 rue Carrie-Derick, Montreal H3C 4S9, QC, Canada
e-mail: mohanavadivu@gmail.com

P. Raj
IBM Global Cloud Center of Excellence, Bangalore 560045, India
e-mail: peterindia@gmail.com

Z. Mahmood (ed.), *Data Science and Big Data Computing*,
DOI 10.1007/978-3-319-31861-5_10

10.1 Introduction

There is an infamous adage based on the Parkinson's Law: *Work expands so as to expand the time available for its completion.* So is the case with the generation of more data, which explodes to fill the space available for data storage. We now live in an age where the number of electronic devices has already overtaken the human population worldwide. There are more than seven billion mobile devices on this planet. The data generated by billions of users over billions of devices for human-to-human interaction, human-to-machine interaction, and machine-to-machine interaction is now measured in zettabytes.

Classifying data generated online, these days, is a definite challenge. As an example, for Twitter messages, people try to classify or categorize the message using "hash tags" (#). It is easy to categorize a message with a single # tag into a category, e.g., "I am #happy" can be put into a category named "happy." However, how does one categorize the following message: "When you're #happy, you enjoy the #music. But, when you're #sad, you understand the #lyrics." We can fit this into the category of music, happy, sad, or lyrics. In this case, how do we establish the logic for selecting the category based on the # tag? A seemingly simple statement categorization snowballs into a problem of huge proportion when we take into account the fact that a billion tweets are generated every day, resulting in massive amount of information being uploaded or revised on line and generated through various instant-messaging services. This is where the need for systems for processing the vast amounts of data kicks in.

Big data is one such data framework that aims to manage a vast amount of data very efficiently. A commonly accepted definition of big data can be stated as: *information assets characterized by such a high volume, velocity, and variety to require specific technology and analytical methods for its transformation into value*. Big data has been used successfully by various government departments including the National Security Agency in the USA to process information collected over the Internet. In the UK, the National Health Service uses big data to process information on drug intake; the Indian government uses big data to access the effect of electorate response on various government policies. Corporate agencies use big data to improve the customer response, improve the efficiency in supply chain, etc. Citibank uses big data to assess the credit card spending of customers to recommend discounts. Interestingly, political parties have also taken the big data in a big way. Big data seems to have played a stellar role in the analysis of electorate opinion and tune the successful campaign strategies of President Barrack Obama in the USA in the year 2012 and in the prime ministerial campaign of Narendra Modi in India in the year 2014. Some of the political analysts attribute both these victories to the smart campaigning methods, where big data has played a decisive role. From these examples, we can easily infer that the application of big data has spread into all walks of life.

Organizations face a major challenge in managing and processing big data effectively to work out venerable solutions. The job of data scientists is to discover patterns and relationships and synthesize them into resultant information that addresses the needs of the organizations. Handling, processing, and managing big data is a complex task. These complexities have limited the capabilities of traditional RDBMS and have paved the way for exploring a host of new technologies, approaches, and platforms that are based on the cloud computing paradigm. This has opened up a wide range of possibilities for organizations to use big data network to understand the needs and expectations of customers so as to optimize the use of resources. This requires a rich mixture of skills on the part of data scientists. They need to know how to handle different types of big data and storage mechanisms (e.g., RDBMS, Hadoop HDFS), develop advanced algorithms, develop codes (e.g., Java, Python, R), access data (SQL, Hive, etc.), analyze any inherent constraints, perform regression analysis, communicate to social networks, and finally present the findings to management in business terms through briefings and reports. Invariably, data scientists have established their impact in the advancement of all analytical fields such as statistics, management science, operations research, computer science, logistics, and mathematics.

From the examples mentioned above, one can easily notice two types of requirements or usage categories when it comes to big data application, the first one being real-time operational systems such as capital markets which require instant streaming, analysis, and feedback and the second one being analytical systems such as market research, surveillance, government databases, etc. The concurrency of data users and velocity are critical in real-time operations, whereas data scope and variety are important in analytical systems. Big data handles the operational systems and analytical systems differently as needs of the systems are different. Database management systems for operational systems involve not only SQL (NoSQL) database systems such as MongoDB, whereas analytical systems use DBMS such as MPP-compliant MapReduce.

In the ensuing sections, we aim to focus on big data, its special characteristics, how to capture and crunch big data to extract valuable insights, the technologies and tools available for simplifying and streamlining the process of knowledge discovery and dissemination across a variety of industry verticals, etc. There are some challenges and concerns relating to big data. We also discuss the platforms and products emerging and evolving in order to speed up big data analytics. Finally, we discuss the role of the cloud paradigm in shaping up data analytics and insight-driven enterprises. There are integrated platforms being deployed in cloud environments so that the raging domain of analytics can be taken to cost-effective, elastic, and elegant clouds.

10.2 Characterizing Big Data

Big data typically represents massive amounts of heterogeneous data that are not necessarily stored in the relational form in traditional enterprise-scale databases. Therefore, new-generation database systems are being unearthed in order to store, retrieve, aggregate, filter, mine, and analyze large amount of multi-structured data efficiently. The following are the general characteristics of big data:

- Data storage is defined in the order of petabytes and exabytes in volume as opposed to the current storage limits in terms of gigabytes and terabytes.
- Big data contains structured, semi-structured, and totally unstructured data.
- Different types of data sources (sensors, machines, mobiles, social sites, etc.) and resources for big data.
- Data is time sensitive (near real time as well as real time) which means that big data consists of data collected with relevance to the time zones so that timely insights can be extracted.

The recent developments in big data analytics have generated a lot of interest among industry professionals as well as academicians. Big data has become an inevitable trend, and it has to be solidly and compactly handled in order to derive time-sensitive and actionable insights. There is a dazzling array of tools, techniques, and tips evolving in order to quickly capture data from different distributed resources and processes; analyze and mine the relevant data to extract actionable business-to-insight technology-sponsored business transformation and sustenance. In short, analytics is a flourishing phenomenon in every sphere and segment of the international world today. Especially, with the automated capture, persistence, and processing of tremendous amount of multi-structured data getting generated by men as well as machines, the analytical value, scope, and power of data are bound to blossom further. Precisely, data is a strategic asset for organizations to insightfully plan to sharply enhance their capabilities and competencies and to venture on the appropriate activities that decisively and drastically power up their short- as well as long-term offerings, outputs, and outlooks. Business innovations can happen in plenty and be maintained when there is a seamless and spontaneous connectivity between data-driven and analytics-enabled business insights and business processes.

In the recent past, real-time analytics has brought in much excrescence, and several product vendors have been flooding the market with a number of elastic and state-of-the-art solutions (software as well as hardware) for alleviating on-demand, ad hoc, real-time, and runtime analysis of batch, online transaction, social, machine, operational, and streaming data. There are numerous advancements in this field due to its high potential for worldwide companies in considerably reducing operational expenditures while gaining operational insights. Hadoop-based analytical products are capable of processing and analyzing any data type and quantity across huge volume of commodity server clusters. Stream computing drives continuous and cognitive analysis of massive volumes of streaming data with sub-millisecond response times. There are enterprise data warehouses, analytical platforms, in-memory appliances, etc. Data warehousing delivers deep operational insights with advanced in-database analytics. Here are some examples:

- The EMC Greenplum Data Computing Appliance (DCA) is an integrated analytics platform that accelerates the analysis of big data assets within a single integrated appliance.
- IBM PureData System for Analytics architecturally integrates database, server, and storage into a single, purpose-built, easy-to-manage system.
- SAP HANA is an exemplary platform for efficient big data analytics. Platform vendors are conveniently tied up with infrastructure vendors especially Cloud Service Providers (CSPs) to take analytics to cloud so that the goal of Analytics as a Service (AaaS) sees a neat and nice reality, sooner than later.

There are multiple start-ups with innovative product offerings to speed up and simplify the complex part of big data analysis.

10.3 The Inherent Challenges

There are some uncertainties, limitations, and potential roadblocks that could probably unsettle the expected progress, since big data is still in the emerging era. Let us consider a few of these which are more pertinent.

- *Technology Precedence* – Technologies and tools are very important for creating business value from big data. There are multiple products and platforms from different vendors. However, the technology choice is very important for firms to plan and proceed without any hitch in their pursuit. The tools and technology choices will vary depending on the types of data to be manipulated (e.g., XML documents, social media, sensor data, etc.), business drivers (e.g., sentiment analysis, customer trends, product development, etc.), and data usage (analytic or product development focused).
- *Data Governance* – Any system has to be appropriately governed in order to be strategically beneficial. Due to the sharp increase in data sources, types, channels, formats, and platforms, data governance is an important component in efficiently regulating the data-driven tasks. Other important motivations include data security while in transit and in persistence, data integrity, and data confidentiality. Furthermore, there are governmental regulations and standards from world bodies, and all these have to be fully complied with, in order to avoid any kind of ramifications at a later point in time.
- *Resource Deficit* – It is predicted by MGI that there will be a huge shortage of human talent for organizations providing big data-based services and solutions. There will be requirements for data modelers, scientists, and analysts in order to get all the envisaged benefits of big data. This is a definite concern to be sincerely attended to by companies and governments across the world.
- *Ramification in Framework* – Big data product vendors need to bring forth solutions that simplify the complexities of the big data framework to enable users to extract business value. The operating interfaces need to be intuitive and

informative so that the goal of ease of use can be ensured for people using big data solutions.

Big data's reputation has taken a battering lately, e.g., in the case of the NSA for collecting and storing people's web and phone records, without proper consent. It has led to a wider debate about the appropriateness of such extensive data-gathering activities, but this negative publicity should not detract people from the reality of big data. The fact is that big data is ultimately to benefit society as a whole.

In short, big data applications, platforms, appliances, and infrastructures need to be contrived in a way to facilitate their usage and leverage for everyday purposes. The awareness about the potential of big data needs to be propagated widely, and professionals need to be trained in order to extract better business value out of big data. Competing technologies, enabling methodologies, prescribing patterns, and evaluating metrics, key guidelines, and best practices need to be unearthed and made as reusable assets.

10.4 Big Data Infrastructures, Platforms, and Analytics

The future of business definitely belongs to those enterprises that embrace the big data analytics movement and use it strategically to their own advantages. It is pointed out that business leaders and other decision-makers, who are smart enough to adopt a flexible and futuristic big data strategy, can take their businesses to greater heights. Successful companies are already extending the value of classic and conventional analytics by integrating cutting-edge big data technologies and outsmarting their competitors [8]. There are several forecasts, exhortations, expositions, and trends on the discipline of big data analytics. Market research and analyst groups have come out with positive reports and briefings, detailing its key drivers and differentiators, the future of this brewing idea, its market value, the revenue potentials and application domains, the fresh avenues and areas for renewed focus, the needs for its sustainability, etc.

The cloud computing movement is expediently thriving and trendsetting a host of delectable novelties. A number of tectonic transformations on the business front are being activated and accentuated with faster and easier adaptability of the cloud IT principles. The cloud concepts have opened up a deluge of fresh opportunities for innovators, individuals, and institutions to conceive and formalize new-generation business services and solutions. A dazzling array of pathbreaking and mission-critical business augmentation models and mechanisms have already emerged, and they are consistently evolving toward perfection as the cloud technology grows relentlessly and rewardingly in conjunction with other enterprise-class technologies.

10.4.1 Unified Platforms for Big Data Analytics

Highly synchronized and integrated platforms are prerequisites to automate several tasks enshrined in capturing of data, analysis, and knowledge discovery processes. A converged platform brings out a reliable workbench to empower developers to assist application development and other related tasks such as data security, virtualization, integration, visualization, and dissemination. Special consoles are being used with new-generation platforms for performing other important activities such as management, governance, and enhancement. An example is *Hadoop* [7, 10] that is a disruptive technology for data distribution among hundreds of commodity compute machines for parallel data crunching, and any typical Big Data platform is blessed with Hadoop software suite.

Furthermore, the big data platform enables entrepreneurs; investors; chief executive, information, operation, knowledge, and technology officers (CXOs); and marketing and sales people to explore and perform their experiments on big data, at scale at a fraction of time and cost to what was required previously. That is, platforms are to bestow all kinds of stakeholders and end users with actionable insights that in turn lead to consider and take informed decisions. Knowledge workers such as business analysts and data scientists could be the other main beneficiaries through these empowered platforms. Knowledge discovery is a vital portion, and the platform has to be chipped in with real-time and real-world tips, associations, patterns, trends, risks, alerts, and opportunities. In-memory and in-database analytics are gaining momentum for high-performance and real-time analytics. New advancements in the form of predictive and prescriptive analytics are rapidly emerging with the maturity and stability of big data technologies, platforms, infrastructures, tools and finally a cornucopia of sophisticated data mining and analysis algorithms. Thus, platforms need to be fitted with new features, functionalities, and facilities in order to provide new insights. Here is a simple strategy:

- *Develop Infrastructure* – The infrastructure required to support the acquisition of big data must deliver low and predictable latency both in capturing data and in executing short and simple queries. It should be able to handle very high transaction volumes often in a distributed environment that also support flexible and dynamic data structures. NoSQL databases are the leading infrastructure to acquire and store big data. NoSQL databases are well suited for dynamic data structures and are highly scalable. The data stored in a NoSQL database is typically of a high variety because the systems are intended to simply capture all kinds of data without categorizing and parsing the data. For example, NoSQL databases are often used to collect and store social media data. While customer-facing applications frequently change, underlying storage structures are kept simple. Instead of designing a schema with relationships between entities, these simple structures often just contain a major key to identify the data point and then a content container holding the relevant data. This extremely simple and nimble structure allows changes to take place without any costly reorganization at the storage layer.

- *Integrate Big Data* – In classical data warehousing terms, organizing data is called data integration. Because there is such a huge volume of data, there is a tendency and trend to organize data at the original storage locations. This saves a lot of time and money as there is no movement of data. The brewing need is to have a robust infrastructure that is innately able to organize big data and process and manipulate data in the original storage location. It has to support very high throughput (often in batches) to deal with large data processing steps and handle a large variety of data formats (unstructured, semi-structured, and fully structured).
- *Analyze Big Data* – The data analysis can also happen in a distributed environment. That is, data stored in diverse locations can be accessed from a data warehouse to accomplish the intended analysis. The appropriate infrastructure required for analyzing big data must be able to support in-depth analytics such as statistical analysis and data mining [2] on a wider range of data types stored in those diverse systems to scale the extreme data volumes and to deliver faster response times driven by changes in behavior and also to automate decisions based on analytical models. Most importantly, the infrastructure must be able to integrate analysis on the combination of big data and traditional enterprise data to produce exemplary insights for fresh opportunities and possibilities. For example, analyzing inventory data from a smart vending machine in combination with the event calendar for the venue in which the vending machine is located will dictate the optimal product mix and replenishment schedule for the vending machine.

10.4.2 Newer and Nimbler Applications

The success of any technology is to be squarely decided based on the number of mission-critical applications it can create and sustain. That is, the applicability or employability of the new paradigm to as many application domains as possible is the major deciding factor for its successful journey. As far as the development is concerned, big data applications could differ from other software applications to a larger extent. Web and mobile enablement of big data applications are also important. As big insights are becoming mandatory for multiple industry segments, there is a bigger scope for big data applications. Therefore, there is a big market for big data application development platforms, patterns, metrics, methodology, reusable components, etc.

10.4.3 Tending Toward a Unified Architecture

It is well recognized that an integrated IT environment is a minimum requirement for attaining the expected success out of the big data paradigm. Refer to Fig. 10.1 for an integrated platform. Deploying big data platforms in an IT environment that lacks a unified architecture and does not seamlessly and spontaneously integrate distributed and diverse data sources, metadata, and other essential resources would not produce the desired insights. Such deployments will quickly lead to a torrent of failed big data projects, and in a fragmented setup, achieving the desired results remains a dream forever. Hence, a unified and modular architecture is a need of the hour for taking forward the ideals of the big data discipline. Deployment of big data applications in a synchronized enterprise environment or cloud IT environment makes analytics simpler, faster, cheaper, and accurate while reducing deployment and operational costs.

In the ensuing era of big data, there could be multiple formats for data representation, transmission, and persistence. The related trend is that there are databases without any formal schema. SQL is the standard query language for traditional databases, whereas in the big data era, there are NoSQL databases that do not support the SQL, which is the standard for structured querying. Special file systems such as Hadoop Distributed File System (HDFS) are being produced in order to facilitate big data storage and access. Thus analytics in the big data period is quite different from the analytics on the SQL databases. However, there is a firm place for

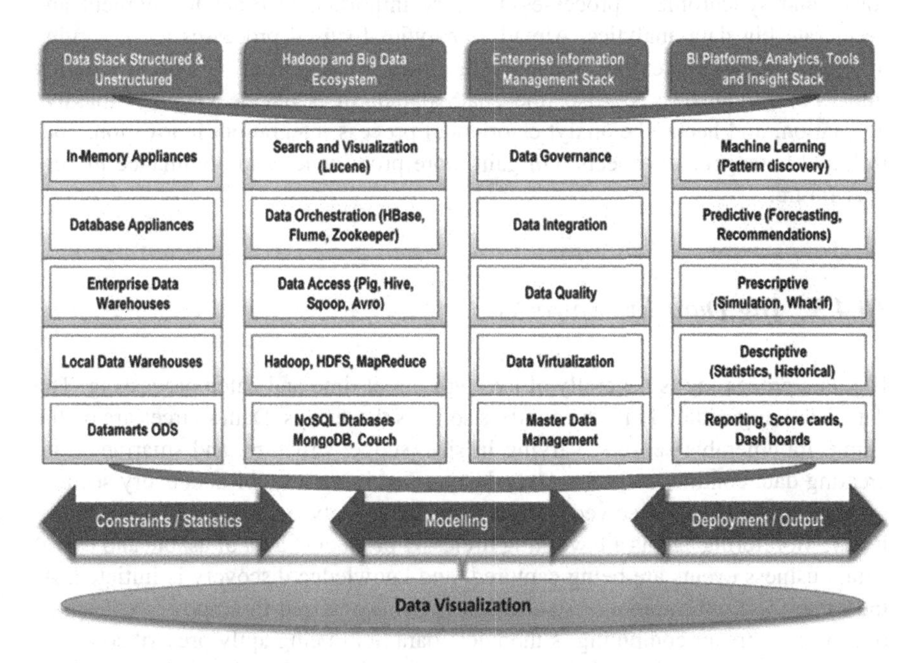

Fig. 10.1 Big Data analytics platforms, appliances, products, and tools

SQL-based analytics, and hence there is an insistence on converging both to fulfill the varying needs of business intelligence (BI). Tools and technologies that provide a native coalesce of classic and new data analytics techniques will have an inherent advantage.

10.4.4 Big Data Appliances and Converged Solutions

Appliances (hardware and virtual) are being prescribed as a viable and value-adding approach for scores of business-critical application infrastructure solutions such as service integration middleware, messaging brokers, security gateways, load balancing, etc. They are fully consolidated and prefabricated with the full software stack so that their deployment and time to operation are quick and simple. There are XML and SOA appliances plenty in the marketplace for eliminating all kinds of performance bottlenecks in business IT solutions. In the recent past, EMC Greenplum and SAP HANA appliances are stealing and securing the attention. SAP HANA is being projected as a game-changing and real-time platform for business analytics and applications. While simplifying the IT stack, it provides powerful features like the significant processing speed, ability to handle big data, predictive capabilities, and text mining capabilities. Thus, the emergence and evolution of appliances represent a distinct trend as far as big data is concerned.

Besides converged architecture, infrastructures, application domains, and platforms and synchronized processes are very important in order to augment and accelerate big data analytics. Already, analytics-focused processes are emerging and evolving consistently. Analytics becomes an important activity to tightly integrate with processes. Also, the AaaS paradigm is on the verge of massive adaptation, and hence the analytics-oriented process integration, innovation, control, and management aspects will gain more prominence and dominance in very near future.

10.4.5 Big Data Frameworks

Big data processing is generally of two types: real-time and batch processing. The data is flowing endlessly from countless sources these days. Data sources are on the climb. Innumerable sensors, varying in size, scope, structure, and smartness, are creating data continuously. Stock markets are emitting a lot of data every second; system logs are being received, stored, processed, analyzed, and acted upon ceaselessly. Monitoring agents are working tirelessly producing a lot of usable and useful data, business events are being captured, and knowledge discovery is initiated. At the same time, information visualization is being realized to empower enterprise operations. Stream computing is the latest paradigm being aptly prescribed as the best course of action for real-time receipt, processing, and analysis of online, live, and continuous data. Real-time data analysis through in-memory and in-database computing models is gaining a lot of ground these days with the sharp reduction in computer memory costs. For the second category of batch processing, the Hadoop technology is being recommended as there is a need for competent products, platforms, and methods for efficiently and expectantly working with both real-time as well as batch data.

As elucidated before, big data analysis is not a simple affair, and there are Hadoop-based software programming frameworks, platforms, and appliances emerging to ease the innate complications. The Hadoop programming model [7, 10] has turned out to be the central and core method to further extend the field of big data analysis. The Hadoop ecosystem is continuously spreading its wings wider, and the enabling modules are being incorporated to make Hadoop-based big data analysis simpler, succinct, and quicker.

10.4.6 The Hadoop Software Family

Apache Hadoop [7, 10] is an open-source framework that allows distributed processing of large-volume data sets across clusters of computers using a simple programming model. Hadoop was originally designed to scale up from a single server to thousands of machines, each offering local computation and storage. Rather than relying on hardware to deliver high availability, the Hadoop software library itself is designed to detect and handle failures at the application layer. Therefore, it delivers a highly available service on top of a cluster of cheap computers, each of which may be prone to failures. Hadoop [7, 10] is based on the modular architecture, and thereby any of its components can be swapped with competent alternatives if such a replacement brings noteworthy advantages.

Despite all the hubbub and hype around Hadoop, few IT professionals know its key drivers, differentiators, and killer applications. Because of the newness and complexity of Hadoop, there are several areas where confusion reigns and restrains its full-fledged assimilation and adoption. The Apache Hadoop product family includes the Hadoop Distributed File System (HDFS), MapReduce, Hive, HBase, Pig, Zookeeper, Flume, Sqoop, Oozie, Hue, and so on. HDFS and MapReduce together constitute the core of Hadoop, which is the foundation for all Hadoop-based applications. For applications in business intelligence (BI), data warehousing (DW), and big data analytics, the core Hadoop is usually augmented with Hive and HBase and sometimes Pig. The Hadoop file system excels with big data that is file based, including files that contain nonstructured data. Hadoop is excellent for storing and searching multi-structured big data, but advanced analytics is possible only with certain combinations of Hadoop products, third-party products, or extensions of Hadoop technologies. The Hadoop family has its own query and database technologies, and these are similar to standard SQL and relational databases. That means BI/DW professionals can learn them quickly.

The HDFS is a distributed file system designed to run on clusters of commodity hardware. HDFS is highly fault-tolerant because it automatically replicates file blocks across multiple machine nodes and is designed to be deployed on low-cost hardware. HDFS provides high-throughput access to application data and is suitable for applications that have large data sets. As a file system, HDFS manages files that contain data. HDFS itself does not offer random access to data since it is file based and provides only limited metadata capabilities when compared to a DBMS. Likewise, HDFS is strongly batch oriented and hence has limited real-time data access functions. To overcome these challenges, it is possible to layer HBase over HDFS to gain some of the mainstream DBMS capabilities. HBase is a product from Apache Hadoop family and is modeled after Google's Bigtable, and hence HBase, like Bigtable, excels with random and real-time access to very large tables containing billions of rows and millions of columns. Currently, HBase is limited to straightforward tables and records with little support for more complex data structures. The Hive meta-store gives Hadoop some DBMS-like metadata capabilities.

When HDFS and MapReduce are combined, Hadoop easily parses and indexes the full range of data types. Furthermore, as a distributed system, HDFS scales well and has a certain amount of fault tolerance based on data replication even when deployed on top commodity hardware. For these reasons, HDFS and MapReduce can complement existing BI/DW systems that focus on structured and relational data. MapReduce is a general-purpose execution engine that works with a variety of storage technologies including HDFS, other file systems, and some DBMSs.

As an execution engine, MapReduce and its underlying data platform handle the complexities of network communication [1], parallel programming, and fault tolerance. In addition, MapReduce controls hand-coded programs and automatically provides multi-threading processes, so they can be executed in parallel for massive scalability. The controlled parallelization of MapReduce can apply to multiple types of distributed applications, not to analytic ones. In a nutshell, Hadoop MapReduce is a software programming framework for easily writing massively parallel applications which process massive amounts of data in parallel on large clusters (thousands of nodes) of commodity hardware in a reliable and fault-tolerant manner. A MapReduce job usually splits the input data set into independent chunks which are processed by the map tasks in a completely parallel manner. The framework sorts the outputs of the maps as input to the reduced tasks which, in turn, assemble one or more result sets.

Hadoop is not just for new analytic applications; it can revamp old ones too. For example, analytics for risk and fraud that is based on statistical analysis or data mining benefits from the much larger data samples that HDFS and MapReduce can wring from diverse Big Data. Furthermore, most 360° customer views include hundreds of customer attributes. Hadoop can provide insight and data to bump up to thousands of attributes, which in turn provide greater detail and precision for customer-based segmentation and other customer analytics. Hadoop is a promising and potential technology that allows large data volumes to be organized and processed while keeping the data on the original data storage cluster. For example,

Weblogs can be turned into browsing behavior (sessions) by running MapReduce programs (Hadoop) on the cluster and generating aggregated results on the same cluster. These aggregated results are then loaded into a relational DBMS system.

HBase is the mainstream Apache Hadoop database. It is an open-source, non-relational (column oriented), scalable, and distributed database management system that supports structured data storage. Apache HBase, which is modeled after Google Bigtable, is the right approach when you need random and real-time read/ write access to big data. It is for hosting of very large tables (billions of rows X millions of columns) on top of clusters of commodity hardware. Just like Google Bigtable leverages the distributed data storage provided by the Google File System, Apache HBase provides Bigtable-like capabilities on top of Hadoop and HDFS. HBase does support writing applications in Avro, REST, and Thrift.

10.5 Databases for Big Data Management and Analytics

10.5.1 NoSQL Databases

Next-generation databases are mandated to be non-relational, distributed, open source, and horizontally scalable. The original inspiration is the modern web-scale databases. Additional characteristics such as schema free, easy replication support, simple API, eventually consistent/BASE (not Atomicity, Consistency, Isolation, Durability (ACID)), etc., are also being demanded. The traditional Relational Database Management Systems (RDBMSs) use Structured Query Language (SQL) for accessing and manipulating data that reside in structured columns of relational tables. However, unstructured data is typically stored in key-value pairs in a data store and therefore cannot be accessed through SQL. Such data are stored in NoSQL [9] data stores and accessed via GET and PUT commands. There are some big advantages of NoSQL databases compared to the relational databases as illustrated in the following web page: http://www.couchbase.com/why-nosql/nosql-database.

- Flexible Data Model – Relational and NoSQL data models are extremely different. The relational model takes data and separates them into many interrelated tables containing rows and columns. Tables reference each other through foreign keys stored in columns. When looking up for data, the desired information needs to be collected from many tables and combined before it can be provided to the application. Similarly, when writing data, the write process needs to be coordinated and performed on many tables.

NoSQL databases [9] follow a different model. For example, a document-oriented NoSQL database takes the data that needs to be stored and aggregates it into documents using the JSON format. Each JSON document can be thought of as an object to be used by the application. A JSON document might, for example, take all the data stored in a row that spans 20 tables of a relational database and aggregate it into a single document/object. The resulting data model is flexible, making it easy to distribute the resulting documents. Another major difference is that relational technologies have rigid schemata, while NoSQL models are schema-less. Changing the schema once data is inserted is a big deal, extremely disruptive, and frequently avoided. However, the exact opposite of the behavior is desired for the big data processing. Application developers need to constantly and rapidly incorporate new types of data to enrich their applications.

- High Performance and Scalability – To handle the large volumes of data (big data) and their concurrent user accesses (big users), applications and underlying databases need to be scaled using one of two choices: scale-up or scale-out. Scaling-up implies a centralized approach that relies on bigger and even bigger servers. Scaling-out implies a distributed approach that leverages many commodity physical or virtual servers. Prior to NoSQL databases, the default scaling approach at the database tier was to scale up, dictated by the fundamentally centralized, shared-everything architecture of relational database technology. To support more concurrent users and/or store more data, here is a requirement for a bigger server with more CPUs, memory, and disk storage to keep all the tables. Big servers are usually highly complex, proprietary, and disproportionately expensive.

NoSQL databases [9] were developed from the ground up to be distributed and scale out databases. They use a cluster of standard, physical, or virtual servers to store data and support database operations. To scale, additional servers are joined to the cluster, and the data and database operations are spread across the larger cluster. Since commodity servers are expected to fail from time to time, NoSQL databases are built to tolerate and recover from such failures to make them highly resilient. NoSQL databases provide a much easier and linear approach to database scaling. If 10,000 new users start using an application, one simply needs to add another database server to a cluster, and there is no need to modify the application as per the scale since the application always sees a single (distributed) database. NoSQL databases share some characteristics, as presented below, with respect to scaling and performance.

- Auto-Sharding –A NoSQL database automatically spreads data across different servers without requiring applications to participate. Servers can be added or removed from the data layer without any application downtime, with data (and I/O) automatically spread across the servers. Most NoSQL databases also support data replication, storing multiple copies of data across the cluster and even across data centers to ensure high availability (HA) and to support disaster recovery (DR). A properly managed NoSQL database system should never be taken offline.
- Distributed Query Support – Sharing a relational database can reduce or eliminate in certain cases the ability to perform complex data queries. NoSQL

database system retains their full query expressive power even when distributed across hundreds of servers.

- Integrated Caching – To reduce latency and increase sustained data throughput, advanced NoSQL database technologies transparently cache data in system memory. This behavior is transparent to the application developer and the operations team, compared to relational technology where a caching tier is usually a separate infrastructure tier that must be developed to and deployed on separate servers and explicitly managed by the operations team.

There are some serious flaws with respect to relational databases that come in the way of meeting up the unique requirements in the modern-day social web applications, which gradually move to reside in cloud infrastructures. Another noteworthy factor is that doing data analysis for business intelligence (BI) is increasingly happening in the cloud era. There are some groups in academic and industrial circles striving hard for bringing in the necessary advancements in order to prop up the traditional databases to cope up with the evolving and enigmatic requirements of social networking applications. However NoSQL and NewSQL databases are the new breeds of versatile, vivacious, and venerable solutions capturing the imagination and attention.

The business requirement to leverage complex data can be driven through the adoption of scalable and high-performance NoSQL databases. This new entrant is to sharply enhance the data management capabilities of various businesses. Several variants of NoSQL databases have emerged in the past decade in order to appropriately handle the terabytes, petabytes, and even exabytes of data generated by enterprises and consumers. They are specifically capable of processing multiple data types such as text, audio, video, social network feeds, and Weblogs that are difficult to handle by traditional databases. These data are highly complex and deeply interrelated, and therefore the demand is to unravel the truth hidden behind these huge yet diverse data assets besides understanding the insights and acting on them to enable businesses to plan and surge ahead.

Having understood the changing scenario, web-based businesses have been crafting their own custom NoSQL databases to elegantly manage high volume of data and its diversity. Amazon Dynamo and Google's Bigtable are the shining examples of homegrown databases that can store huge amounts of data. These NoSQL databases were designed for handling highly complex and heterogeneous data. The key differentiation is that they are not built for high-end transactions but for analytic purposes.

10.5.2 Why NoSQL Databases?

Business-to-consumer (B2C) e-commerce and business-to-business (B2B) e-business applications are highly transactional, and the leading enterprise application frameworks and platforms such as Java Enterprise Edition (JEE) can directly and distinctly support a number of transaction types (simple, distributed, nested, etc.). For a trivial example, flight reservation application has to be rigidly transactional otherwise everything is bound to collapse. As enterprise systems are increasingly distributed, the need for transaction feature is being pronounced as a mandatory one.

In the recent past, social applications have grown fast, and especially the younger generation is totally fascinated by a stream of social computing sites which has resulted in an astronomical growth of further data. It is no secret that the popularity, ubiquity, and utility of Facebook, LinkedIn, Twitter, Google +, and other blogging sites are surging incessantly. There is a steady synchronization between enterprise and social applications with the idea of adequately empowering enterprise applications with additional power and value. For example, the online sellers understand that businesses should be more interactive, open, and inclined toward customers' participation to garner and glean their views to reach out to even more people across the globe and to pour in Richer Enterprise Applications (REAs). There are specialized protocols and web 2.0 technologies (e.g., Atom, RSS, AJAX, and mash-ups) to programmatically tag information about people and places and proclivity to dynamically conceive, conceptualize, and concretize more and more people-centric and premium services.

The dormant and dumb database technology has to evolve faster in order to accomplish these new-generation IT abilities. With the modern data being more complicated and connected, the NoSQL databases need to have the implicit and innate strength to handle the multi-structured and massive amounts of data. A NoSQL database should enable high-performance queries on such data. Users should be able to ask questions such as the following: "Who are all my contacts in Europe?" "Which of my contacts ordered from this catalog?" A white paper entitled "NoSQL for the Enterprise" by Neo Technology [9] points to the uniqueness of NoSQL databases for enterprises. Some of the key points have been reproduced below, from that paper:

- A Simplified Data Representation – A NoSQL database should be able to easily represent complex and connected data that makes up today's enterprise applications [3]. Unlike traditional databases, a flexible schema that allows for multiple data types also enables developers to easily change applications without disrupting live systems. Databases must be extensible and adaptable. With the massive adoption of clouds, NoSQL databases ought to be more suitable for clouds.
- End-to-End Transactions – Traditional databases are famous for "all or nothing" transactions, whereas NoSQL databases give a kind of leeway on this crucial property. This is due to the fact that the prime reason for the emergence and evolution of NoSQL databases was to process massive volumes of data quickly and to come up with actionable inputs. In other words, traditional databases are for enterprise applications, whereas NoSQL databases are for social applications. Specifically, the consistency aspect of ACID transactions is not rigidly insisted in NoSQL databases. Sometimes an operation can fail in a social

application and it does not matter much. For instance, there are billions of short messages being tweeted every day, and Twitter will probably survive if a few Tweets do get lost. But online banking applications relying on traditional databases have to ensure a very tight consistency in order to be meaningful. That does not mean that NoSQL databases are off the ACID hook. Instead they are supposed to support ACID transactions including XA-compliant distributed two-phase commit protocol. The connections between data should be stored on a disk in a structure designed for high-performance retrieval of connected data sets – all while enforcing strict transaction management. This design delivers significantly better performance for connecting data than the one offered by relational databases.

- Enterprise-Grade Durability – Every NoSQL database for the enterprise needs to have the enterprise-class quality of durability. That is, any transaction committed to the database will not be lost at any cost under any circumstances. If there is a flight ticket reserved and the system crashes due to an internal or external problem thereafter, then when the system comes back, the allotted seat still has to be there. Predominantly the durability feature is ensured through the use of database backups and transaction logs that facilitate the restoration of committed transactions in spite of any software or hardware hitches. Relational databases have employed the replication method for years successfully to guarantee the enterprise-strength durability.

10.5.3 Classification of NoSQL Databases

There are four categories of NoSQL databases available today: key-value stores, document databases, column-family databases, and graph databases. Each was designed to accommodate the huge volumes of data as well as to have room for future data types. The choice of NoSQL database depends on the type of data one needs to store, its size, and complexity. Here is a brief mention of the four types:

- *Key-Value Stores* – A key-value data model is quite simple. It stores data in key and value pairs where every key maps to a value. It can scale across many machines but cannot support other data types. Key-value data stores use a data model similar to the popular memcached distributed in-memory cache, with a single key-value index for all the data. Unlike memcached, these systems generally provide a persistence mechanism and additional functionality as well including replication, versioning, locking, transactions, sorting, and several other features. The client interface provides inserts, deletes, and index lookups. Like memcached, none of these systems offer secondary indices or keys. A key-value store is ideal for applications that require massive amounts of simple data like sensor data or for rapidly changing data such as stock quotes. Key-value stores support massive data sets of very primitive data. Amazon Dynamo was built as a key-value store.

- *Document Databases (DocumentDB)* – A document database contains a collection of key-value pairs stored in documents. The documentDB support more complex data than the key-value stores. While it is good at storing documents, it was not designed with enterprise-strength transactions and durability in mind. DocumentDB are the most flexible of the key-value style stores, perfect for storing a large collection of unrelated and discrete documents. Unlike the key-value stores, these systems generally support secondary indexes and multiple types of documents (objects) per database and nested documents or lists. A good application would be a product catalog, which can display individual items, but not related items. One can see what is available for purchase, but one cannot connect it to what other products similar customers bought after they viewed it. MongoDB and CouchDB are examples of documentDB systems.
- *Column-Family Databases* – A column-family database can handle semi-structured data because in theory every row can have its own schema. It has few mandatory attributes and few optional attributes. It is a powerful way to capture semi-structured data but often sacrifices consistency for ensuring the availability attribute. Column-family databases can accommodate huge amounts of data, and the key differentiator is that it helps to sift through the data very fast. Writes are really faster than reads so one natural niche is real-time data analysis. Logging real-time events is a perfect use case, and another one is random and real-time read/write access to the Big Data. Google's Bigtable was built on a column-family database. Apache Cassandra [6], the Facebook database, is another well-known example, which was developed to store billions of columns per row. However, it is unable to support unstructured data types or end-to-end query transactions.
- *Graph Databases* – A graph database uses nodes, relationships between nodes, and key-value properties instead of tables to represent information. This model is typically substantially faster for associative data sets and uses a schema-less and bottom-up model that is ideal for capturing ad hoc and rapidly changing data. Much of today's complex and connected data can be easily stored in a graph database where there is great value in the relationships among data sets. A graph database accesses data using traversals. A traversal is how a graph is queried, navigating from starting nodes to related nodes according to an algorithm, finding answers to questions like "what music do my friends like that I don't yet own?" or "if this power supply goes down, what web services are affected?" Using traversals, one can easily conduct end-to-end transactions that represent real user actions.

10.5.4 Cloud-Based Databases for Big Data

Relational database management systems are an integral and indispensable component in enterprise IT, and their importance is all set to grow. However, with the advances of cloud-hosted and managed computing, network, and storage infrastructures, the opportunity to offer a DBMS, as an offloaded and outsourced service, is gradually gaining momentum. Carlo Curino and his team members have introduced a new transactional "database as a service" (DBaaS) [5]. A DBaaS promises to move much of the operational burden of provisioning, configuration, scaling, performance tuning, backup, privacy, and access control from the database users to the service operator, effectively offering lower overall costs to users. However, the DBaaS being provided by leading CSPs do not address three important challenges: efficient multi-tenancy, elastic scalability, and database privacy. The authors argue that before outsourcing database software and management into cloud environments, these three challenges need to be surmounted. The key technical features of DBaaS are:

1. A workload-aware approach to multi-tenancy that identifies the workloads that can be colocated on a database server achieving higher consolidation and better performance over existing approaches
2. The use of a graph-based data partitioning algorithm to achieve near-linear elastic scale-out even for complex transactional workloads
3. An adjustable security schema that enables SQL queries to run over encrypted data including ordering operations, aggregates, and joins

An underlying theme in the design of the components of DBaaS is the notion of workload awareness. By monitoring query patterns and data accesses, the system obtains information useful for various optimization and security functions, reducing the configuration effort for users and operators. By centralizing and automating many database management tasks, a DBaaS can substantially reduce operational costs and perform well. There are myriad of advantages of using cloud-based databases, some of which are as follows:

- Fast and automated recovery from failures to ensure business continuity
- A built-in larger package with nothing to configure or comes with a straightforward GUI-based configuration
- Cheap backups, archival, and restoration
- Automated on-demand scaling with the ability to simply define the scaling rules or facility to manually adjust
- Potentially lower cost, device independence, and better performance
- Scalability and automatic failover/high availability
- Anytime, anywhere, any device, any media, any network discoverable, accessible, and usable
- Less capital expenditure and usage-based payment
- Automated provisioning of physical as well as virtual servers in the cloud

Some of the disadvantages include:

- Security and privacy issues
- Constant Internet connection (bandwidth costs!) requirement
- Loss of controllability over resources
- Loss of visibility on database transactions
- Vendor lock in

Thus, newer realities such as NoSQL and NewSQL database solutions are fast arriving and being adopted eagerly. On the other hand, the traditional database management systems are being accordingly modernized and migrated to cloud environments to substantiate the era of providing everything as a service. Data as a service, insights as a service, etc., are bound to grow considerably in the days to come as their realization technologies are fast maturing.

10.6 Data Science

Data science [4] is an independent discipline that emerged since the beginning of the twenty-first century. It suggests models and frameworks to support the analysis of complex data. Data analysis has a long history; it has been practiced by librarians, statisticians, scientists, sales executives, financial experts, and marketing technicians for years to induce attention of the customers to focus on their products.

In 1994, many companies introduced their marketing strategies based on databases, as marketing prediction to sell the products to overwhelm the competition. The first conference on data science was held in 1996. The terminologies such as data mining, data analysis, and data science were used by researchers and scientists were introduced at that time. Data mining is defined as an application for the extraction of patterns from data using specific algorithm. Statistics plays a major role in processing of data, through unique statistical methods. Data mining refers to extracting, importing, and analyzing data from huge databases to determine the relationship among internal and external factors and the impacts on sales, profits, and customer satisfaction. At the moment, analyzing the data sounds difficult as it requires information from stochastic statistical model and unknown algorithmic model. Statisticians observed uncertainty in their results while analyzing these data models. This has resulted in research in computing in the field of data analysis, enabling the statisticians to merge statistical methods and computing using web/Internet to produce innovations in the field of data science. The view and the idea are to find pace-based decision-making, instead of traditional calculations, and to find predictive model using quantitative analysis as primary object.

Data science brings together expertise from data visualization, hacker mindsets, and various domains to process data that are generated in a specific domain or across domains. Rightly, the journal of data science defines or claims data science as *almost everything that has something to do with data: collecting, analyzing, modeling, its application, and all sorts of application*. For example, data mining – the processing of data to find interesting patterns in data that are not easily observed using basic queries – is widely used for decision-making in diverse domains such as surveillance, market research, and asset management of municipal infrastructure such as pipe lines. This pattern recognition could be performed on multiple types of data such as audio, video, images, and texts. However there are limitations when we apply these techniques to vast amounts of data, and the traditional data processing techniques become inadequate.

This necessitates for appropriate data-handling frameworks and tools that could handle the vast amount of data in a timely and efficient manner. It should be noted that a lot of junk is also being generated online and that can creep into important decision-making tasks. Though we boast of efficient decision-making techniques in place, most of these systems are also fragile and feeble. For example, a fake tweet on the hacked Associated Press Twitter account that President Barrack Obama was injured in a bomb blast in White House caused a momentary decline in the value of S&P 500 index about 0.9 %. This is just enough to wipe out $ 130 billion in a matter of milliseconds. This is a crude reminder of how sensitive the systems are toward the information generated online. A sensible recommendation based on a thorough data analysis for such a highly impacting decision-making would require the process of streaming large quantities of data from various sources; verifying the authenticity of news from various news agency sites that are in audio (podcasts), video (online live telecasts), and text formats (tweets, webpage contents); arriving at a conclusion based on the analysis as and when the information is generated; and instantly feeding it to the capital markets. All the aforementioned processes will need to be performed at the lightning speed. Delays even by a couple of seconds might render this information completely useless or might create negative and irreparable consequences.

10.6.1 Basic Concepts

Data science is an activity to identify the problem (automated), process the requirement, and visualize the solution for the user. The activity involves collection of data, preparation of the file system and storage, and analysis of data and the request, focusing on result visualization and management to satisfy the customers' requirements. It also involves predictive analysis of other aspects such as scalability, storage, safety, and security.

Key factors involved with data science include classification, probability estimation to predict, value estimation by regression analysis, identification of similarities, determining unity between entities, describing attributes characterization, link prediction from the basis, and reduction of data and technology to come up with innovative models and modeling. Data science concepts and operations are extracted from data engineering, programming analysis with statistics, social data warehousing, and through other related processes. Data science [4] is of interest to business by improving the decision-making concepts. Generally, Data science is the function of quantitative and qualitative methods to solve related questions and forecast conclusions. Businesses require real-time analysis using Data science. The necessary skills relating to data science are as follows:

- Technical communication with users, explaining the statistical aspect and problem requests to clear the technicality of the problem.

- Expertise in adopting the application and getting the required data for the given particular context.
- For complex systems, data scientist has to create constraints according to the problem request, to thoroughly understand the issues from the user side.
- Skills in data description, defining the variables, and accordingly collecting those data and making it available for analysis through network sharing.
- For process and data transformation, the data scientist must have the ability to resolve issues with feasible solutions and to interpret data in presentable form.
- Visualization and keeping the presentable results with precision, detail, and charts for a more effective communication with customers.
- Expertise in handling of ethical issues such as security violation, privacy, and communication limitation and keeping the data away from malpractice.

10.6.2 The Role of Data Scientist

A data scientist could be a single person or a group of people working in an organization, whose main role is to perform statistical analysis, data mining, and retrieval processes on a huge amount of data to determine trends and other diagnose-related information. They normally work on data analysis on huge data sets in data warehouse or common access data centers to resolve many issues with respect to optimization of performance and factors relating to business intelligence (BI). This will be useful for the business to act upon the inherent trends, so they can plan and take correct corporate decisions.

In general, the data scientists' role is to analyze big data or data repository that are preserved by an organization or some other sources and to analyze the objectives using mathematics and statistical models to obtain instructions and recommendation to help with excellent business and marketing decision-making. The responsibilities include handling the product development, technology transformation to fit with current trends, and building architecture in the back end for the analysis and front end for demonstration of the results.

10.7 Conclusion

Big data analytics is now moving beyond the realm of intellectual curiosity and propensity to make tangible and trendsetting impacts on business operations, offerings, and outlooks. It is no longer hype or a buzzword and is rapidly becoming a core requirement for every sort of business enterprise to be relevant and rightful to their stakeholders and end users. Being an emerging and evolving technology, it needs a careful and cognitive analysis before its adoption.

In this chapter, we have provided some detail on enabling big data analytics and data science concepts. We have commenced the chapter with the key drivers for big data. Then we proceeded by describing the significance of generating actionable insights out of data sets in huge amounts in order to determine sustainable business value. We have incorporated the details regarding the proven and potential tools and provided tips for simplifying and streamlining big data analytics. We have also discussed the prominent database systems such as NoSQL databases. This is an introductory chapter to expose what is generally needed for readers to start with in the fast-moving big data field.

Acknowledgment Sincere thanks are due to the following individuals for their contribution and support in developing this chapter:

- Ernestasia Siahaan, PhD candidate, Department of Intelligent Systems – Multimedia Computing, Faculty of Electrical Engineering, Mathematics, and Computer Science, Delft University of Technology, the Netherlands
- Mohanasundar Radhakrishnan, PhD fellow, UNESCO-IHE Institute for Water Education, Delft, the Netherlands
- Vadivazhagan Meiyazhagan, lead administrator, Wipro Infotech, India

References

1. Liang C, Yu FR (2014) Wireless network virtualization: a survey, some research issues and challenges, communications surveys & tutorials. IEEE 17(1):358–380. doi:10.1109/COMST.2014.2352118
2. Han J, Kamber M, Pei J (2011) Data mining: concepts and techniques. Elsevier Publications, Amsterdam/Boston. ISBN 978-0-12-381479-1
3. Grolinger K, Higashino WA, Tiwari A, Capretz MAM (2013) Data management in cloud environments: NoSQL and NewSQL data stores. J Cloud Comput: Adv Syst Appl 2:22. doi:10.1186/2192-113X-2-22
4. Lillian Pierson, Wiley, Carl Anderson, Ryan Swanstrom (2015) Data science for dummies. John Wiley & Sons Inc, New York, United States. ISBN 9781118841556
5. Raj P (2012) Cloud enterprise architecture. CRC Press, Boca Raton. ISBN 9781466502321
6. Strickland R (2014) Cassandra high availability. Packt Publishing, Birmingham, Ebook ISBN:978-1-78398-913-3 | ISBN 10:1-78398-913-0
7. White T, Cutting D (2009) Hadoop: the definitive guide. O'Reil-ly Media, Beijing, ISBN 0596521979; ISBN13: 9780596521974
8. McKinsey Global Institute (2011) Big data: the next frontier for innovation, competition, and productivity. Available at: http://www.mckinsey.com/insights/business_technology/big_data_the_next_frontier_for_innovation
9. Neo Technology (2011) NoSQL for the enterprise, white paper. Available at: http://www.neotechnology.com/tag/nosql/
10. Russom P (2011) Hadoop: revealing its true value for business intelligence, white paper. Available at: http://www.tdwi.org

Chapter 11
A Framework for Data Mining and Knowledge Discovery in Cloud Computing

Derya Birant and Pelin Yıldırım

Abstract The massive amounts of data being generated in the current world of information technology have increased from terabytes to petabytes in volume. The fact that extracting knowledge from large-scale data is a challenging issue creates a great demand for cloud computing because of its potential benefits such as scalable storage and processing services. Considering this motivation, this chapter introduces a novel framework, data mining in cloud computing (DMCC), that allows users to apply classification, clustering, and association rule mining methods on huge amounts of data efficiently by combining data mining, cloud computing, and parallel computing technologies. The chapter discusses the main architectural components, interfaces, features, and advantages of the proposed DMCC framework. This study also compares the running times when data mining algorithms are executed in serial and parallel in a cloud environment through DMCC framework. Experimental results show that DMCC greatly decreases the execution times of data mining algorithms.

Keywords DMCC • Data mining • Cloud computing • Knowledge discovery • Classification • Clustering • Association rule mining

11.1 Introduction

As a result of technological innovations and developments, enormous amounts of data are generated every day in a wide range of areas such as business, education, healthcare, government, finance, social media, and many others. The increase in the amount of data that is generated creates the potential to discover valuable

D. Birant (✉)
Department of Computer Engineering, Dokuz Eylul University, Izmir, Turkey
e-mail: derya@cs.deu.edu.tr

P. Yıldırım
Department of Software Engineering, Celal Bayar University, Manisa, Turkey
e-mail: pelin.yildirim@cbu.edu.tr

Z. Mahmood (ed.), *Data Science and Big Data Computing*,
DOI 10.1007/978-3-319-31861-5_11

knowledge from it. However, it also generates a need to deal with data processing in an efficient and low-cost way.

Data mining is the process of extracting interesting (previously unknown and potentially useful) knowledge or patterns from large data repositories such as databases, data warehouses, data marts, extensible markup language (XML) data, files, and so on. Data mining is regarded as one of the main steps of Knowledge Discovery in Databases (KDD).

The main data mining methods are classification, clustering, and association rule mining (ARM). Classification is the process of developing an accurate model according to classes that use the features in data. It enables the categorization of new data into these predefined classes. Clustering is used to identify groups and discover structure in unlabelled data. ARM is useful for finding the rules among different items in large datasets.

The data mining in cloud computing allows to centralize the management of software and data storage. Cloud computing is a new paradigm that relies on sharing computing resources by providing parallel processing and data storage over the Internet on a pay-for-use basis. Cloud systems can be effectively used to handle the parallel mining of large datasets since they provide scalable storage and processing services as well as software platforms for running data mining applications.

This study presents a novel cloud computing-based framework, data mining in cloud computing (DMCC), that has the ability to apply classification, clustering, and ARM methods on huge amounts of data efficiently. We propose this framework to enhance the existing data mining applications to take advantage of features inherent to cloud characteristics that include scalability, efficiency, and ease of use. This study compares the running times when data mining algorithms are executed in serial and parallel in a cloud environment. A performance comparison between the traditional and DMCC frameworks shows that cloud-based applications can run in a parallel manner and that DMCC greatly decreases the execution time of data mining algorithms.

The focus of this study is to present the advantages of combining three core technologies: data mining, cloud computing, and parallel computing. The novelty and main contributions of the DMCC framework proposed in this study are as follows:

- It reduces the running time of data mining algorithms by simultaneously exploiting both the scalable properties of cloud computing and parallel programming methods.
- It enables on-demand pricing options of cloud platforms to utilize knowledge discovery processes with affordable prices.
- It allows data mining applications to be accessible from anywhere.
- It makes data storage and management easier.

Experimental results in this study show that it is possible to decrease the execution time of data mining techniques with the help of cloud computing deployments and implementation of parallel programming structures.

The rest of this chapter is structured as follows: Sect. 11.2 summarizes the related literature and previous studies. While Sect. 11.3 explains the basic concepts and methods of data mining, Sect. 11.4 presents the deployment models and service models of cloud computing. Section 11.5 introduces the proposed DMCC framework and describes its characteristics, promises, and benefits. In Sect. 11.6, the results obtained from experiments are presented, focusing on an analysis of the various data mining techniques with different input parameters on several datasets and also more briefly commenting on the comparison of parallel and serial executions. Section 11.7 presents concluding remarks and possible future studies.

11.2 Related Work

In this section, some related studies that apply data mining techniques on large datasets using cloud computing technology are described.

Using the advantages of cloud computing technology to tackle the performance problems in large-scale data mining applications is an approach that has gained increasing attention [1–4]. The most common way to perform data mining algorithms effectively on high-volume data is to use *MapReduce*, a programming model for parallel data processing that is widely used in cloud environments [5, 6]. An open-source implementation of MapReduce, Hadoop, can be used for distributed data processing [7]. However, many scientific applications, which have complex communication patterns, still require low-latency communication mechanisms. For this reason, in this study, we introduce a novel framework to help fill this critical void.

The *Sector/Sphere* framework is another way to mine large-scale datasets using computer clusters connected with wide-area high-performance networks [8–10]. While high-level data parallel frameworks like Sector/Sphere simplify the design and implementation of large-scale data processing systems, they do not naturally support many important data mining algorithms and can lead to inefficient learning systems [11]. Thus, in this study, we introduce a novel framework that supports many different data mining algorithms.

In addition to MapReduce, Hadoop, and Sector/Sphere frameworks, other large-scale computation frameworks have been proposed: the GraphLab, "Data Mining Cloud App", and Define-Ingest-Preprocess-Analyze-Report (DIPAR) frameworks. The *GraphLab* abstraction was proposed for machine learning and data mining in the cloud, and it directly targets dynamic, graph-parallel computation in the shared memory setting [11]. The *Data Mining Cloud App* framework performs parameter sweeping operations on clustering and classification algorithms [12]. The *DIPAR* framework, which consists of five stages, define, ingest, preprocess, analyse, and report, was proposed as a way to implement Big Data Science (BDS) in organizations [13].

Association rule mining (ARM) is a major data mining task in the application of large-scale and parallel data mining in cloud computing environments

[14, 15]. Apiletti et al. [16] proposed a novel cloud-based service, SEARUM (SErvice for Association RUle Mining), that applies ARM on the large volume of data using MapReduce jobs that are run in the cloud. In another study [17], the Apriori algorithm, which is a well-known ARM algorithm, was selected as a case study and a pipelined MapReduce framework was utilized.

In addition to ARM, other well-known data mining tasks have also been investigated in cloud environments such as classification [5, 6], clustering [12, 18, 19], sequential pattern mining [20], XML data mining [21], web mining [22], semantic web mining [23], and text mining [24].

In contrast to these studies, the work presented in this contribution (the DMCC framework) focuses on the implementation of different classification, clustering, and ARM algorithms on large-scale data using the Microsoft Azure cloud platform to reduce local workload. In contrast to existing systems, the DMCC is off the shelf, and thus (1) it is ready to use from anywhere, (2) there is no need to move data, (3) there is no cost for managing the underlying hardware/software layers, and (4) the installation of software is not required.

11.3 Data Mining

This section presents most commonly used data mining tasks and the algorithms which DMCC framework offers.

11.3.1 Classification

Classification is the most studied and commonly applied data mining task. It develops a model to assign new patterns into one of several predefined classes. Classification uses a training set $D = (R_1, R_2, \ldots, R_n)$ that has some records R, which consist of a number of attributes $R = (a_1, a_2, \ldots, a_m)$ of which one (a_j) is a target outcome. Classification algorithms try to determine relationships between attributes in a training set to classify new observations.

In this study, four classification algorithms are developed as "Platform as a Service (PaaS)" for cloud end users: Naive Bayes, Decision Tree (C4.5), Random Forest, and AdaBoost.

11.3.1.1 Naive Bayes Classifier

The Naive Bayes classifier is a well-known statistical classifier that uses Bayes theorem to calculate unknown conditional probabilities to identify classes of input samples [25]. The algorithm performs well and returns successful results in many different areas such as medical diagnosis, pattern recognition, document categorization, banking, and marketing.

The Naive Bayes classifier is generally applied to data that consists of discrete-valued attributes. However, in this study, we used a Gaussian distribution to deal with continuous-valued attributes. The Gaussian distribution, sometimes called the normal distribution, uses mean μ and variance σ2 values to calculate its probability density function as follows:

$$f(x,\mu,\sigma) = \frac{1}{\sigma\sqrt{2\pi}} e^{-\frac{(x-\mu)^2}{2\sigma^2}} \tag{11.1}$$

11.3.1.2 Decision Tree (C4.5)

A Decision Tree is one of the commonly used supervised classification methods that aims to develop a tree to identify unknown target attribute values based on input parameters. A Decision Tree consists of internal nodes for attributes, branches to represent attributes values, and leaf nodes to assign classification labels. In this study, we used the C4.5 algorithm, which computes information gain to construct the tree based on entropy and the fraction (f_i) of items labelled with value i in the set as follows:

$$I_E(f) = -\sum_{i=1}^{m} f_i \log_2 f_i \tag{11.2}$$

11.3.1.3 Random Forest

A Random Forest is an ensemble learning method that grows multivalued Decision Trees in different training sets. To classify a new sample, input parameters are given to each tree in the forest. The trees return predicted outcomes and the result is selected by taking the majority vote over all the trees in the forest. The Random Forest is an ideal learning model with the goal of reducing variance and increasing ease of use.

11.3.1.4 AdaBoost

AdaBoost, an acronym of Adaptive Boosting, combines a number of weak learners to form a weighted voting machine to achieve better separation between classes. It is the most appropriate solution for binary classification problems in particular. AdaBoost specifies a boost classifier F_T for each weak learner f_t that takes an object

x as input, and after T layers, it returns a real value indicating the class of the object as follows:

$$F_T(x) = \sum_{t=1}^{T} f_t(x) \tag{11.3}$$

11.3.2 Clustering

Clustering, an unsupervised learning technique, is the process of grouping a set of objects into meaningful clusters in such a way that similar objects are placed within a cluster. Clustering analysis is currently used in many areas such as image processing, pattern recognition, segmentation, machine learning, and information retrieval. The main task of clustering is to compare, measure, and identify the resemblance of objects based on their features by using a similarity measure such as the Manhattan, Euclidean, or Minkowski distance for numerical attributes and Jaccard's distance for categorical values. Clustering algorithms are categorized into five major types, as listed in Table 11.1. One of these, the K-Means, is briefly described below.

11.3.2.1 K-Means

In this study, the K-Means clustering algorithm is developed as a PaaS for the cloud end users because of its popularity, computational simplicity, efficiency, and empirical success. K-Means is a well-known and simple unsupervised learning algorithm that groups a given dataset into k clusters. Let $X = \{x_1, x_2, \ldots, x_n\}$ be a set of observations. The K-Means algorithm divides n observations into k cluster sets $C = \{c_1, c_2, \ldots, c_k\}$ by minimizing the Sum of Squared Errors (SSE) of the k clusters. The algorithm involves the following steps:

1. The k initial centroids are chosen randomly.
2. Each object is assigned to the cluster associated with the nearest centroid.

Table 11.1 Categorization of well-known clustering algorithms

Cluster models	Clustering algorithms
Partitioning methods	K-Means, C-Means, K-Medoids, CLARANS, etc.
Hierarchical methods	Single/complete/average link, BIRCH, ROCK, CAMELEON, etc.
Density-based methods	DBSCAN, DENCLUE, OPTICS, etc.
Grid-based methods	STING, CLIQUE, WaveCluster, etc.
Model-based methods	SOM (Self-Organizing Maps), COBWEB, EM (Expectation Maximization), etc.

3. Centroids are updated to the mean of each segment.
4. Steps 2 and 3 are repeated until no object switches clusters.

11.3.3 Association Rule Mining

Association rule mining (ARM), one of the most important and well-researched techniques of data mining, is the extraction of interesting correlations, relationships, frequent patterns or associations, or general structures among sets of items in the transactions.

Let $I = \{i_1, i_2, \ldots, i_m\}$ be a set of m distinct literals called items, T be a transaction that contains a set of items such that $T \subseteq$ I, and D be a dataset $D = \{t_1, t_2, \ldots, t_n\}$ that has n transaction records T. An association rule is an implication of the form $X \Rightarrow Y$, where $X \subset I$ and $Y \subset I$ are sets of items called frequent itemsets, and $X \cap Y = \emptyset$. The rule $X \Rightarrow Y$ can be interpreted as "if itemset X occurs in a transaction, then itemset Y will also likely occur in the same transaction".

There are two important basic measures for association rules: support and confidence. Usually thresholds of support and confidence are predefined by users to drop those rules that are not particularly interesting or useful. In addition to these measures, additional constraints can also be specified by the users such as time, item, dimensional, or interestingness constraints.

Support of an association rule in the form of $X \Rightarrow Y$ is defined as the percentage of records that contain both X and Y itemsets to the total number of transactions in the dataset D. Support is calculated by the following formula:

$$Support(X \Rightarrow Y) = \frac{\text{Number of transactions contain both } X \text{ and } Y}{\text{Total number of transactions in } D} \quad (11.4)$$

Suppose the support of rule $X \Rightarrow Y$ is 1 %. This means that 1% of the transactions contain X and Y items together.

The *Confidence* of an association rule in the form of $X \Rightarrow Y$ is defined as the percentage of the number of records that contain both X and Y itemsets with respect to the total number of transactions that contain X, as follows:

$$Confidence(X \Rightarrow Y) = \frac{\text{Support}(X \cup Y)}{\text{Support}(X)} \quad (11.5)$$

Suppose the confidence of the association rule $X \Rightarrow Y$ is 85 %. This means that 85 % of the transactions that contain X also contain Y.

In this study, the Apriori algorithm is developed as a PaaS for cloud end users. Apriori is a well-known ARM algorithm that requires many passes over the database, generating many candidate itemsets, pruning those itemsets whose supports are below the predefined threshold, and storing frequent itemsets. In each pass, k itemsets are found. For example, the rule X, Y $\Rightarrow$ Z can be produced as a frequent three itemset, where X, Y, and Z represent in any items in the dataset.

11.4 Cloud Computing

Cloud computing offers resources, software applications, and infrastructures over the Internet to users [26]. This section explains the basic deployment models and service models of cloud computing.

11.4.1 Deployment Models

Cloud computing broadly breaks down into three different deployment models: public, private, and hybrid cloud [27].

11.4.1.1 Public Cloud

The public cloud is the common cloud computing model in which a service provider hosts resources, such as applications and data storage, available to the general public over the Internet.

11.4.1.2 Private Cloud

The private cloud (also called the internal cloud) consists of cloud infrastructure that is operated solely for a single organization, whether managed internally or by a third party.

11.4.1.3 Hybrid Cloud

A hybrid cloud comprises both private (internal) and public (external) cloud services, offering the benefits of multiple deployment models.

11.4.2 Service Models

Cloud computing is divided into three broad service categories: Infrastructure as a Service (IaaS), Platform as a Service (PaaS), and Software as a Service (SaaS) [28].

11.4.2.1 Infrastructure as a Service (IaaS)

Infrastructure as a Service (IaaS) is a cloud computing model that provides virtualized computing resources over the Internet, including some virtual machines and hardware resource units. IaaS can also be classified into two categories: Computation as a Service (CaaS) and Data as a Service (DaaS). CaaS refers to the ability to buy or rent specific resources—mainly processor, RAM, operating system, or deployed software—for a period of time to perform difficult computation calculations. DaaS refers to the ability to buy or rent storage space to store the user's data, regardless of its type. IaaS clouds often offer additional resources such as a virtual-machine disc image library, block- and file-based storage, load balancers, IP addresses, firewalls, and virtual local area networks. Typical examples are Amazon EC2 (Elastic Cloud Computing), GoGrid, Rackspace Cloud, Amazon S3 (Simple Storage Service), and Google Compute Engine.

11.4.2.2 Platform as a Service (PaaS)

Platform as a Service (PaaS) is a form of cloud computing in which cloud providers offer a computing platform that typically includes an operating system, database, web server, and environments for building, testing, and deploying custom applications. Application developers can develop and run their software solutions on a cloud platform without the cost of managing the underlying hardware and software layers. Some common examples of PaaS include Microsoft Azure, Heroku, Force.com, Google App Engine, and Apache Stratos.

11.4.2.3 Software as a Service (SaaS)

Software as a Service (SaaS) is a cloud computing model in which users are provided access to application software and databases. It allows users to run existing applications remotely from the cloud to meet their requirements. Instead of cloud users, cloud providers manage the infrastructure and platforms that run the applications. Some common examples of SaaS include Google Calendar, Microsoft Office 365, Yahoo Maps, Flickr, and Gmail.

This study focuses on developing and deploying a framework in the public cloud using a PaaS approach.

11.5 The Proposed DMCC Framework

DMCC (data mining in cloud computing) is a framework that uses current algorithms to support data mining in a cloud environment. This section describes the design and implementation of the DMCC framework that we developed to archive, analyse, and mine large datasets in the cloud environment. The DMCC framework offers the accessibility to increase computational power for users who need to perform large-scale data/compute intensive data mining applications. However, to perform such computations, two major conditions need to be satisfied: (i) the data mining application should be deployed on the cloud, and (ii) an appropriate parallel runtime support should be implemented.

11.5.1 Overview

In the DMCC framework, data mining techniques are implemented as a dynamic library, and data mining applications are stored on the cloud. The aim is to develop a framework suitable for executing data analysis and knowledge discovery tasks over a wide range of different applications within different sectors such as medicine, education, entertainment, transportation, and accounting. It allows the user to discover useful knowledge from virtually stored data that reduces the costs of infrastructure and storage.

The DMCC framework provides a user interface that facilitates the utilization of data mining applications in a cloud environment. Figure 11.1 shows a screenshot from a data mining application developed in Visual Studio. Weka open-source data mining library is used to perform classification, clustering, and ARM on large-scale datasets. It is hosted on the Microsoft Azure cloud platform. Users can access this

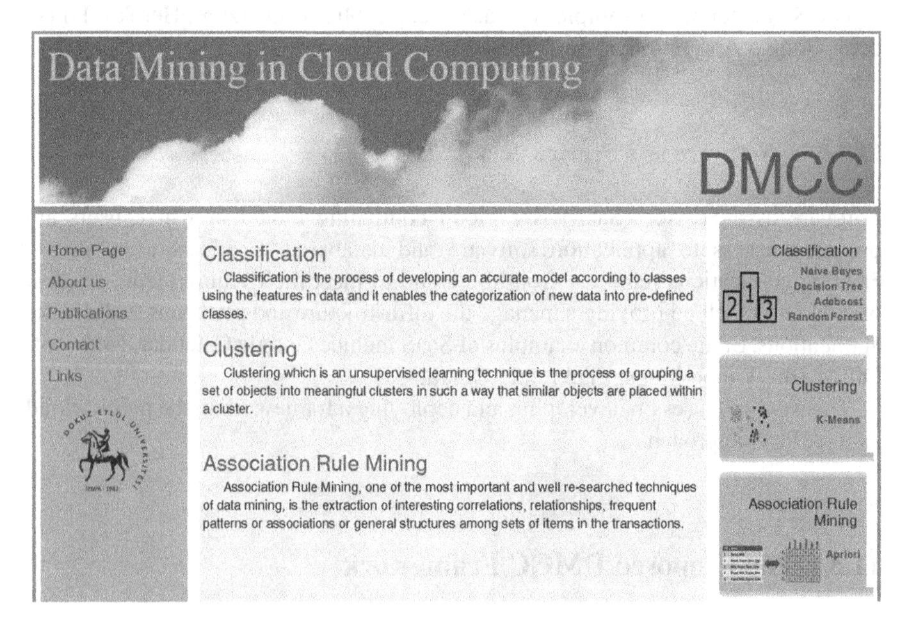

Fig. 11.1 Screenshot from a data mining application deployed on the cloud

data mining application using a web browser, regardless of their location. They perform mining algorithms on selected massive amounts of data, independently from the installed operating system.

11.5.2 DMCC Framework Architecture

When designing a data mining system based on cloud computing, Lin [29] divided the platform, bottom-up, into three layers, the algorithm layer, task layer, and user layer. In contrast to previous studies, the DMCC framework is typically built using a four-layer architecture model consisting of an application layer, data mining engine layer (DMEL), data access layer (DAL), and data layer. The application layer provides the application-user interactions, the DMEL executes the data mining algorithms, the DAL encapsulates the data access functionality, and the data layer is responsible for data persistence and data manipulation. The DMCC framework architecture is a multilayered, parallel processing-enabled, cloud-based architecture, as shown in Fig. 11.2.

The framework is designed so that parallel computing can be done very simply. In particular, if a user defines a function *f* on the parts of a dataset *d*, then he invokes the command

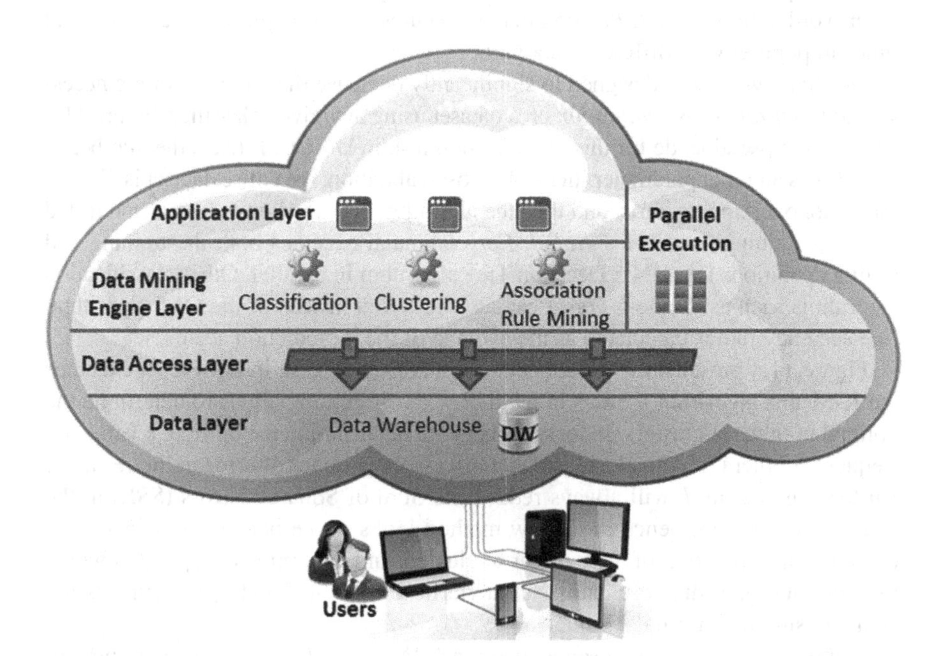

Fig. 11.2 DMCC architecture

```
Classification(int n) {
   Dataset d = new Dataset(new FileReader("data.dat"))
   d.setClassIndex(d.numAttributes() - 1)
   Classifier c = new Classifier.C45()
   Paralel.For (1, n, i => {
      int s = d.size() / n
      Dataset test  = new Dataset(d, s * (i-1), s * i)
      Dataset train = d - test
      c.buildClassifier(train)
      double total = 0, numCorrect = 0
      Paralel.Foreach (test, inst => {
         if (c.classifyInstance(inst) == inst.classValue())
            numCorrect++ ) })
      total=total+(double)numCorrect/(double)test.size()*100.0 })
   print total / n
}
```

Fig. 11.3 Pseudocode for the classification method in DMCC

```
Parallel.For(0, d.parts.count, i =>
{
f(d.part[i]);
});
```

This applies the user-defined function *f* to each part of the dataset *d* in parallel. In other words, the *Parallel. For* command executes the same pieces of code several times in parallel with different data parts.

Our framework was designed to significantly decrease the amount of time necessary to run *n-fold cross-validation* on a dataset using any given classifier. Figure 11.3 presents the pseudocode for this classification task in DMCC. It takes the number of folds (*n*) as an input parameter. In n-fold cross-validation, the entire dataset is divided into *n* nonoverlapping parts, and then the algorithm is trained on *n-1* parts and tested on the remaining one. The *Parallel. For* command executes both the training and testing operations for a C4.5 Decision Tree algorithm in parallel. Other classification algorithms such as Random Forest can also be called instead of the C4.5 algorithm. The accuracy rate is calculated as the average of the *n* prediction accuracies.

Figure 11.4 presents the pseudocode for a clustering task in DMCC. It executes the K-Means algorithm *n* times in parallel with different *k* values to determine the optimal number of clusters. It does this because determining the optimal k value is a frequent problem in data clustering that affects the amount of error in the resulting clusters. Increasing *k* will always reduce the Sum of Squared Errors (SSE) in the resulting clustering; hence the elbow method looks at the percentage of variance explained as a function of the number of clusters and determines the point at which the marginal gain drops, creating an angle in the graph. Finally, the clusters are created using the optimal *k* value.

Figure 11.5 presents the pseudocode for ARM in DMCC. It takes the minimum support and confidence values as input parameters and executes the Apriori algorithm with these parameters to output the rules.

```
Clustering(int n) {
   Dataset d = new Dataset(new FileReader("data.dat"))
   Clusterers.SimpleKMeans skm = new Clusterers.SimpleKMeans()
   skm.setSeed(10)
   Paralel.For (1, n, k => {
      skm.setNumClusters(k)
      skm.buildClusterer(d)
      store skm.getSquaredError() in SSEs
   })
   int optimalK = ElbowMethod(SSEs)
   skm.setNumClusters(optimalK)
   skm.buildClusterer(d)
   print skm.getSquaredError()
}
```

Fig. 11.4 Pseudocode for the clustering method in DMCC

```
AssociationRuleMining(double minSupport, double minConfidence)
{
   Dataset d = new Dataset(new FileReader("data.dat"))
   Associations.Apriori ap = new Associations.Apriori()
   ap.setLowerBoundMinSupport(minSupport)
   ap.setLowerBoundMinConfidence(minConfidence)
   ap.buildAssociations(d)
   print ap.getRules()
}
```

Fig. 11.5 Pseudocode for ARM in DMCC

11.5.3 DMCC Framework Features

The DMCC framework has many advantages over traditional grid-based, distributed, or parallel computing techniques.

Grid computing is a collection of computer resources from multiple locations, acting in concert to perform very large tasks using distributed and parallel computation. However, cloud computing provides the technologies to compute parallel data mining applications at much more affordable prices than traditional parallel computing techniques.

Distributed computing is an environment in which the application and data is broken into pieces and executed on individual machines. However, cloud computing stores the data at a data centre and does not distribute it to individual machines for processing.

The DMCC framework deploys data mining applications in cloud environments for the following reasons:

- Cloud resources are dynamically reallocated per demand; hence the cloud approach maximizes the use of computing power and thus significantly reduces the running time of data mining algorithms.
- Cloud computing eliminates the investment needed for stand-alone software or servers. It allows developers to avoid costs for items such as servers, air conditioning, and rack space, allowing many data mining developers to reduce their costs.
- Cloud computing allows developers to get their data mining applications operational within a short time with improved manageability and less maintenance.
- Developers focus on data mining projects instead of infrastructure.
- Cloud computing offers on-demand pricing so that commercial data mining software can be provided at affordable prices [30].
- In cloud environments, providers are able to devote resources to solving security issues, so security can be improved by data centralization, increased security-focused resources, and so on.
- Cloud computing enables users to access systems using a web browser, regardless of their location or which device they use (e.g. PC, laptop, or mobile phone) so that data mining applications can be accessed via the Internet and users can connect from anywhere.
- Cloud computing offers cost-effective and policy-driven solutions to store big data so that large amount of data can be processed in place without needing to move it.

As a result, the major features of the DMCC framework include high performance, scalability, simplified access interfaces, quality of service, ease of use, cost-effectiveness, and conceptual simplicity in a large-scale and data-massive environment. The DMCC framework is a practical alternative to traditional distributed or grid-based data mining solutions. Data mining in cloud computing can be considered as the future of data mining because of the advantages listed above.

11.6 Experimental Results

Because of explosive data growth and the amount of computation involved in data mining, a high-performance cloud computing is an excellent resource that is necessary for efficient data mining. This study presents the importance of parallel data analysis for a data mining application in a cloud environment. The aim of this study is to show that the execution time of data mining algorithms can be significantly reduced by combining cloud computing and parallel computing technologies.

11.6.1 Dataset Description

We performed experimental studies on four different large datasets obtained from the UCI Machine Learning Repository: the EEG (electroencephalography) Eye State, Skin Segmentation, KDD Cup 1999, and Census Income datasets.

11.6.1.1 EEG Eye State Dataset

This dataset is used to classify an eye state as eye closed (1) or eye opened (0). There are 15 attributes in the dataset that were detected via a camera during the EEG measurement. The dataset consists of 14,980 samples. A brief description of the dataset is presented in Table 11.2.

11.6.1.2 Skin Segmentation Dataset

The Skin Segmentation dataset consists of R (red), G (green), and B (blue) colour values from face images of various age groups (young, middle, or old), race groups (white, black, or Asian), and gender. RGB attributes have continuous values ranging from 0 to 255. The dataset contains 245,057 labelled samples, where 50,859 are skin/face samples (class label 1) and 194,198 are non-skin/not-face samples (class label 2). Table 11.3 shows statistical information about the data.

Table 11.2 Description of the EEG Eye State dataset

Attributes	Min value	Max value	Mean	Std. dev.
AF3	1030.77	309,231	4321.92	2492.07
F7	2830.77	7804.62	4009.77	45.94
F3	1040	6880.51	4264.02	44.29
FC5	2453.33	642,564	4164.95	5216.40
T7	2089.74	6474.36	4341.74	34.74
P7	2768.21	362,564	4644.02	2924.79
O1	2086.15	567,179	4110.4	4600.93
O2	4567.18	7264.10	4616.06	29.29
P8	1357.95	265,641	4218.83	2136.41
T8	1816.41	6674.36	4231.32	38.05
FC6	3273.33	6823.08	4202.46	37.79
F4	2257.95	7002.56	4279.23	41.54
F8	86.67	152,308	4615.20	1208.37
AF4	1366.15	715,897	4416.44	5891.29
Class (eye detection)	0 (eye open) 1 (eye closed)			

Table 11.3 Description of the Skin Segmentation dataset

Attributes	Min value	Max value	Mean	Std. dev.
B (blue)	0	255	125.07	62.26
G (green)	0	255	132.51	59.94
R (red)	0	255	123.18	72.56
Class (skin detection)	1 (skin) 2 (non-skin)			

11.6.1.3 KDD Cup 1999 Dataset

The purpose of this dataset is to build a network intrusion detector capable of distinguishing between attacks and normal connections. In this study, we used the 10 % KDD Cup 1999 dataset, which contains 494,021 connection records, each with 38 numeric features (i.e. duration, number of failed login attempts, number of data bytes, number of urgent packets, and so on), except three categorical attributes (protocol type, service, and status of the connection flag). It contains one normal and 22 different attack types (class labels): normal, back, buffer_overflow, ftp_write, guess_passwd, imap, ipsweep, land, loadmodule, multihop, neptune, nmap, perl, phf, pod, portsweep, rootkit, satan, smurf, spy, teardrop, warezclient, and warezmaster.

11.6.1.4 Census Income Dataset

The Census Income dataset has 48,842 individual records and contains demographic information about people. Each record has 15 attributes for each person: age, working class, final weight, education, years of education, marital status, occupation, type of relationship, race, sex, capital gain, capital loss, weekly working hours, native country, and salary.

11.6.2 Classification Results

We compared the performances of different classification algorithms in the cloud environment in terms of both their accuracy rates and execution times when running in parallel and serial. The average times shown in the figures and tables were calculated using the results of five repeated runs.

The accuracy rate of a classification algorithm is generally defined as the closeness of the predicted values to the actual values and shows the success of the algorithm on the selected dataset. To evaluate the accuracy rates, we used a fivefold cross-validation technique. Thus, the dataset is divided into five parts, 4/5 for training and 1/5 for testing, and the accuracy rates of the algorithms were evaluated according to five experiments that were acquired independently on different test sets.

Tables 11.4, 11.5, and 11.6 show the execution times and accuracy rates of three classification algorithms, C4.5 Decision Tree, AdaBoost, and Random Forest, on three datasets, EEG Eye State, Skin Segmentation, and KDD Cup 1999. According to the results, the Random Forest algorithm has the highest accuracy rate, but performs poorly on the datasets in terms of execution times. The AdaBoost algorithm shows the best performance for the Eye State dataset in the case of parallel execution, but the accuracy rate of this classification algorithm is the worst. Classification algorithms show similar accuracy rates for the KDD Cup 1999 dataset, highlighting the benefit of using a large training set to obtain a better learning mechanism.

Comparing the execution times, we find that the running times of different data mining algorithms on the same dataset are different. However, from these results, it is clearly evident that the parallel run times in the cloud environment perform competitively well for all classification algorithms. Although we used different datasets with different sizes, the Random Forest algorithm shows considerably high accuracy rates compared to the other algorithms; however we still need to compare it with other algorithms or use alternative approaches for different types of problems and datasets.

We measured the performance of the Naive Bayes classification algorithm in a cloud environment. The classifier created from the training set used a Gaussian distribution to deal with continuous data. Figure 11.6 compares the parallel and

Table 11.4 Comparison of classification algorithms on EEG Eye State data

EEG Eye State data			
Classification algorithms	Serial (sec.)	Parallel (sec.)	Accuracy rate (%)
C4.5 Decision Tree	4.184	0.696	84.52
AdaBoost	2.020	0.345	67.16
Random Forest	6.078	0.610	90.24

Table 11.5 Comparison of classification algorithms on Skin Segmentation data

Skin Segmentation data			
Classification algorithms	Serial (sec.)	Parallel (sec.)	Accuracy rate (%)
C4.5 Decision Tree	13.298	3.250	99.92
AdaBoost	16.778	3.423	89.62
Random Forest	46.070	5.890	99.94

Table 11.6 Comparison of classification algorithms on KDD Cup 1999 data

KDD Cup 1999 data			
Classification algorithms	Serial (sec.)	Parallel (sec.)	Accuracy rate (%)
C4.5 Decision Tree	232.822	73.876	99.94
AdaBoost	285.734	64.085	97.78
Random Forest	262.862	48.297	99.97

serial execution of the Naive Bayes algorithm on different datasets in a cloud environment. From these results, it is clearly evident that it is possible to decrease long-term consumption fairly well using many computing resources concurrently. Hence, this study confirms the importance of parallel data analysis in data mining applications to provide good performance in a cloud environment.

Figures 11.7 and 11.8 show the average running times of the Naive Bayes algorithm for varying input sizes, 5000, 10,000, and 14,980 records for the EEG Eye State dataset and 50,000, 100,000, 150,000, 200,000, and 245,057 records for the Skin Segmentation dataset. The results show that the running time increases more sharply when the data size increases. When the amount of data is increased, the gap between serial and parallel executions increases significantly.

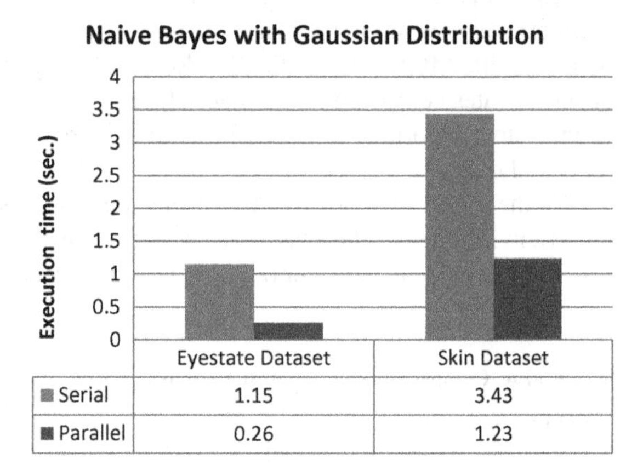

Fig. 11.6 Parallel and serial executions of the same data mining algorithm in a cloud environment

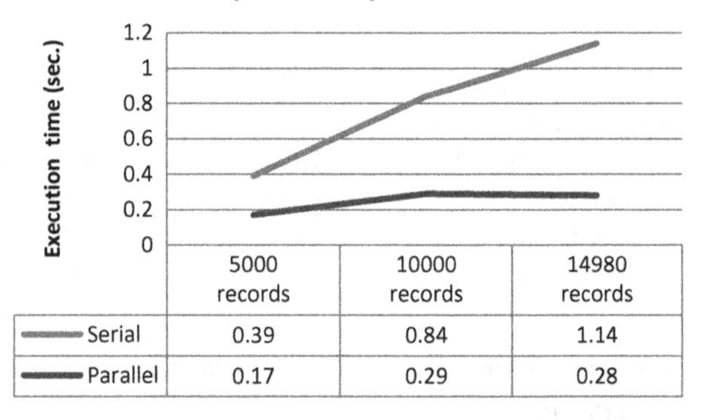

Fig. 11.7 Average running times for varying input sizes of the EEG Eye State dataset

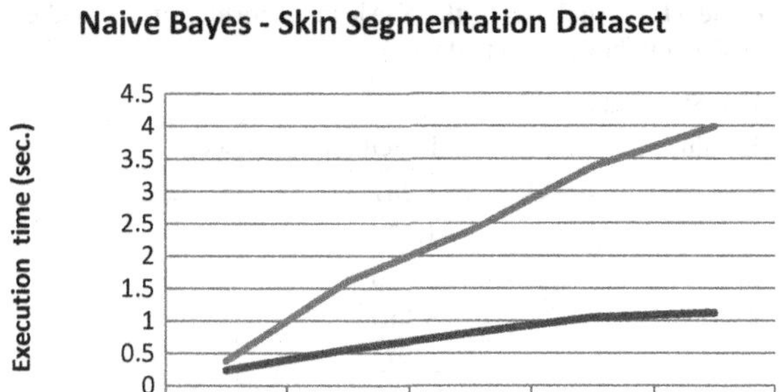

Fig. 11.8 Average running times for varying input sizes of the Skin Segmentation dataset

Naive Bayes - Skin Segmentation Dataset

Execution time (sec.)

	50000 records	100000 records	150000 records	200000 records	245057 records
Serial	0.38	1.61	2.38	3.38	3.99
Parallel	0.24	0.56	0.81	1.05	1.12

11.6.3 Clustering Results

We specifically analysed the performance of clustering applications deployed in a cloud environment to determine the overhead of virtualized resources and understand how clustering algorithms with different numbers of clusters perform on cloud resources. We also evaluated the K-Means clustering algorithm for different numbers of clusters in order to understand clustering errors.

Tables 11.7, 11.8, and 11.9 show the execution times and SSE for the K-Means clustering algorithm on three different datasets. The K-Means algorithm was executed with varying k input values from two to ten in increments of two. These results allow us to predict how the execution time changes as the number of clusters increases. Results show that even as the amount of data grows and the number of clusters increases, the execution time increases within an acceptable range (a clearly normal way), and the application completes in reasonable time. In other words, even when the data size and number of clusters increases rapidly, the execution time increases in a much slower manner than the predicted time.

11.6.4 A Study of Association Rule Mining

We also specifically analysed the number of rules generated by the ARM application deployed in a cloud environment to determine the overhead of virtualized resources and understand how ARM applications with different numbers of support values perform on cloud resources. We also evaluated the Apriori algorithm in a cloud environment for different numbers of confidence values.

The graphs in Figs. 11.9 and 11.10 show the number of rules generated by the Apriori algorithm for varying minimum support and confidence values, respectively. When the minimum support value is kept constant and confidence values are changed, the number of rules decreases almost linearly. However, in the other case,

Table 11.7 Analyses of the K-Means algorithm in a cloud environment for varying numbers of clusters on the Eye State dataset

Eye State data		
K value	Execution time (sec.)	Sum of Squared Error (SSE)
2	1.267	14.45
4	1.330	13.17
6	3.020	10.29
8	5.710	6.94
10	5.675	6.79

Table 11.8 Analyses of K-Means algorithm in a cloud environment for varying numbers of clusters on the Skin Segmentation dataset

Skin Segmentation data		
K value	Execution time (sec.)	Sum of Squared Error (SSE)
2	3.120	20336.13
4	4.700	9842.09
6	6.990	6568.03
8	6.630	5063.96
10	6.285	4055.97

Table 11.9 Analyses of the K-Means algorithm in a cloud environment for varying numbers of clusters on the KDD Cup 1999 dataset

KDD Cup 1999 data		
K value	Execution time (sec.)	Sum of Squared Error (SSE)
2	51.130	551115.68
4	84.530	260735.65
6	89.500	83524.36
8	102.530	57807.90
10	147.620	61061.86

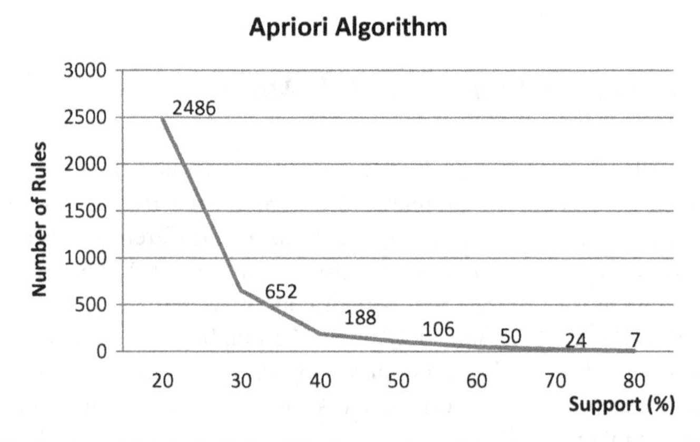

Fig. 11.9 Evaluation of the Apriori algorithm for varying minimum support values and a constant confidence value of 90

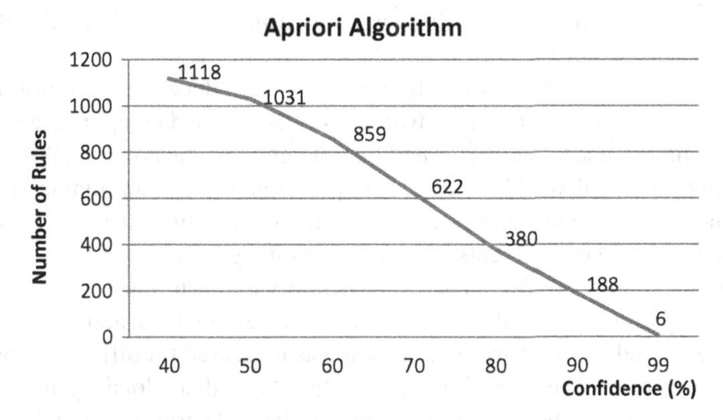

Fig. 11.10 Evaluation of the Apriori algorithm for varying minimum confidence values and a constant support value of 40

the number of rules decreases exponentially. Results show that deploying an ARM application on a cloud environment is more critical when the application is executed with small support values.

Cloud infrastructures allow users to access the hardware nodes in which the virtual machines are deployed. With a cloud infrastructure, we have complete access to both virtual-machine instances and the underlying nodes; as a result, we can deploy different virtual-machine configurations, allocating different CPU cores to each virtual machine. In shared scale mode, six instances are served to an application, while in standard mode, ten instances that have four cores and 7 GB of memory are offered for usage. We increased the core count and memory capacity of each instance by changing the settings to provide better throughput and performance.

As a result, it is clear that the proposed DMCC framework can effectively process very large data in a cloud environment. Based on our observations, we conclude that the DMCC framework can improve the efficiency of data mining applications.

11.7 Conclusion and Suggestions for Future Work

This study proposes DMCC, a cloud-based framework, developed to perform the main data mining tasks (classification, clustering, and ARM) over large datasets. We discussed the main architectural components, interfaces, and features of the DMCC framework. The DMCC framework has many advantages over traditional grid-based, distributed, or parallel computing techniques. The important advantages provided by DMCC framework are that a large amount of data can be processed from anywhere, without moving it, and with on-demand pricing options, increased security, improved manageability, less maintenance, and increased computing power. Furthermore, there is no investment in stand-alone software or servers.

The general aim of the DMCC framework is to reduce the execution times of data mining algorithms using the advantages of both cloud computing and parallel programming methods. We performed experimental studies on the DMCC framework using three real-world datasets. The performance of data mining algorithms within the DMCC framework was measured by executing them in parallel and serial modes. In the experiments, (i) classification algorithms (C4.5 Decision Tree, Random Forest, and AdaBoost) were compared with each other, (ii) the K-Means clustering algorithm was evaluated in terms of cluster numbers, and (iii) the number of rules generated by the Apriori algorithm was measured for different support and confidence values. The experimental results show that cloud systems can be effectively used to handle the parallel mining of large datasets because they provide virtualized, dynamically-scalable computing power as well as storage, platforms, and services.

As future work, this approach could be specialized for the private cloud computing deployment model to enable data mining applications for a specific organization. In this way, corporate companies could benefit from the proposed framework on their private network, providing only authorized access to the system. In addition, it would also be possible to expand this framework by integrating different data mining algorithms.

References

1. Geng X, Yang Z (2013) Data mining in cloud computing. International conference on information science and computer applications, Atlantis Press, 1–7
2. Petre R (2012) Data mining in cloud computing. Database Syst J 3(3):67–71
3. Kamala B (2013) A study on integrated approach of data mining and cloud mining. Int J Adv Comput Sci Cloud Comput 1(2):35–38
4. Hu T, Chen H, Huang L, Zhu X (2012) A survey of mass data mining based on cloud-computing. IEEE conference on anti-counterfeiting, security and identification, 1–4
5. Zhou L, Wang H, Wang W (2012) Parallel implementation of classification algorithms based on cloud computing environment. Indonesian J Electr Eng 10(5):1087–1092
6. Tan A X, Liu VL, Kantarcioglu M, Thuraisingham B (2010) A comparison of approaches for large-scale data mining – utilizing MapReduce in large-scale data mining, Technical Report
7. Nappinna V, Revathi N (2013) Data mining over large datasets using hadoop in cloud environment. Int J Comput Sci Commun Netw 3(2):73–78
8. Grossman RL, Gu Y (2008) Data mining using high performance data clouds: experimental studies using sector and sphere. In: Proceedings of the 14th ACM SIGKDD international conference on knowledge discovery and data mining, 920–927
9. Mishra N, Sharma S, Pandey A (2013) High performance cloud data mining algorithm and data mining in clouds. IOSR J Comput Eng 8(4):54–61
10. Mahendra TV, Deepika N, Rao NK (2012) Data mining for high performance data cloud using association rule mining. Int J Adv Res Comput Sci Softw Eng 2(1)

11. Low Y, Gonzalez J, Kyrola A, Bickson D, Guestrin C, Hellerstein JM (2012) Distributed graphLab: a framework for machine learning and data mining in the cloud. Proc Very Large Data Bases (VLDB) Endowment 5(8):716–727
12. Marozzo F, Talia D, Trunfio P (2011) A cloud framework for parameter sweeping data mining applications. In: Proceedings of the IEEE 3th international conference on cloud computing technology and science, 367–374
13. Villalpando LEV, April A, Abran A (2014) DIPAR: a framework for implementing big data science in organizations. In: Mahmood Z (ed) Continued rise of the cloud: advances and trends in cloud computing. Springer, London
14. Qureshi Z, Bansal J, Bansal S (2013) A survey on association rule mining in cloud computing. Int J Emerg Tech Adv Eng 3(4):318–321
15. Kamalraj R, Kannan AR, Vaishnavi S, Suganya V (2012) A data mining based approach for introducing products in SaaS (Software as a service). Int J Eng Innov Res 1(2):210–214
16. Apiletti D, Baralis E, Cerquitelli T, Chiusano S, Grimaudo L (2013) SeARuM: a cloud-based service for association rule mining. In: Proceedings of the 12th IEEE international conference on trust, security and privacy in computing and communications, 2013, 1283–1290
17. Wu Z, Cao J, Fang C (2012) Data cloud for distributed data mining via pipelined mapreduce, vol 7103, Lecture notes in computer science. Springer, Berlin/Heidelberg, pp 316–330
18. Ismail L, Masud MM, Khan L (2014) FSBD: a framework for scheduling of big data mining in cloud computing. In: Proceedings of the IEEE international congress on big data, 514–521
19. Masih S, Tanwani S (2014) Distributed framework for data mining as a service on private cloud. Int J Eng Res Appl 4(11):65–70
20. Huang JW, Lin SC, Chen MS (2010) DPSP: Distributed progressive sequential pattern mining on the cloud, vol 6119, Lecture notes in computer science. Springer, Berlin/Heidelberg, pp 27–34
21. Li Z (2014) Massive XML data mining in cloud computing environment. J Multimed 9 (8):1011–1016
22. Ruan S (2012) Based on cloud-computing's web data mining, vol 289, Communications in computer and information science. Springer, Berlin/Heidelberg, pp 241–248
23. Lal K, Mahanti NC (2010) A novel data mining algorithm for semantic web based data cloud. Int J Comput Sci Secur 4(2):160–175
24. Ioannou ZM, Nodarakis N, Sioutas S, Tsakalidis A, Tzimas G (2014) Mining biological data on the cloud – a MapReduce approach, IFIP advances in information and communication technology, vol. 437. Springer, 96–105
25. Yıldırım P, Birant D (2014) Naive bayes classifier for continuous variables using novel method (NBC4D) and distributions. In: Proceedings of the IEEE international symposium on innovations in intelligent systems and applications, 110–115
26. Erl T, Puttini R, Mahmood Z (2013) Cloud computing: concepts, technology, & architecture. Prentice Hall, Upper Saddle River
27. Mahmood Z (2011) Cloud computing for enterprise architectures: concepts, principles and approaches. In: Mahmood Z, Hill R (eds) Cloud computing for enterprise architectures. Springer, London/New York
28. Fernandez A, Rio S, Herrera F, Benitez JM (2013) An overview on the structure and applications for business intelligence and data mining in cloud computing, vol 172, Advances in intelligent systems and computing. Springer, Berlin/Heidelberg, pp 559–570
29. Lin Y (2012) Study of layers construct for data mining platform based on cloud computing, vol 345, Communications in computer and information science. Springer, Berlin/Heidelberg
30. Wu X, Hou J, Zhuo S, Zhang W (2013) Dynamic pricing strategy for cloud computing with data mining method, vol 207, Communications in computer and information science. Springer, Berlin/Heidelberg, pp 40–54

Chapter 12
Feature Selection for Adaptive Decision Making in Big Data Analytics

Jaya Sil and Asit Kumar Das

Abstract Rapid growth in technology and its accessibility by general public produce voluminous, heterogeneous and unstructured data resulted in the emergence of new concepts, viz. Big Data and Big Data Analytics. High dimensionality, variability, uncertainty and speed of generating such data pose new challenges in data analysis using standard statistical methods, especially when Big Data consists of redundant as well as important information. Devising intelligent methods is the need of the hour to extract meaningful information from Big Data. Different computational tools such as rough-set theory, fuzzy-set theory, fuzzy-rough-set and genetic algorithm that are often applied to analyse such kind of data are the focus of this chapter. But sometimes local optimal solution is achieved due to premature convergence, so hybridization of genetic algorithm with local search methods has been discussed here. Genetic algorithm, a well-proven global optimization algorithm, has been extended to search the fitness space more efficiently in order to select global optimum feature subset. Real-life data is often vague, so fuzzy logic and rough-set theory are applied to handle uncertainty and maintain consistency in the data sets. The aim of the fuzzy-rough-based method is to generate optimum variation in the range of membership functions of linguistic variables. As a next step, dimensionality reduction is performed to search the selected features for discovering knowledge from the given data set. The searching of most informative features may terminate at local optimum, whereas the global optimum may lie elsewhere in the search space. To remove local minima, an algorithm is proposed using fuzzy-rough-set concept and genetic algorithm. The proposed algorithm searches the most informative attribute set by utilising the optimal range of membership values used to design the objective function. Finally, a case study is given where the dimension reduction techniques are applied in the field of agricultural science, a real-life application domain. Rice plants diseases infect leaves, stems, roots and other parts, which cause degradation of production. Disease identification and taking precaution is very important data analytic task in the field of agriculture. Here, it is demonstrated in the case study to show how the images are collected from

J. Sil (✉) • A.K. Das
Department of Computer Science and Technology, Indian Institute of Engineering Science and Technology, Shibpur, Howrah, West Bengal, India
e-mail: jayasil@hotmail.com

Z. Mahmood (ed.), *Data Science and Big Data Computing*,
DOI 10.1007/978-3-319-31861-5_12

the fields, diseased features are extracted and preprocessed and finally important features are selected using genetic algorithm-based local searching technique and fuzzy-rough-set theory. These features are important to develop a decision support system to predict the diseases and accordingly devise methods to protect the most important crops.

Keywords High-dimensional data • Adaptive decision making • Optimization • Local search • Uncertainty • Big Data Analytics

12.1 Introduction

Due to the advancement of technology and its accessibility to general public, data is generated by leaps and bounds demanding the need for data analysis. The concept of Big Data [1] opens up different areas of research including social networking, bioinformatics, e-health and becomes a big challenge in devising new methods for extracting knowledge. Big Data is characterised by massive sample size, high dimensionality and intrinsic heterogeneity. Moreover, noise accumulation, spurious correlation and incidental endogeneity are the common features in high-dimensional data set. Due to the accumulation of noise, true signals are often dominated and handled by sparsity assumption in high-dimensional data.

Big Data processing is difficult using standard statistical methods due to its volume, variability, rate of growth and unstructured form [2]. In traditional statistical data analysis, we think of observations of instances of a particular phenomenon. We assumed many observations and a few selected variables are well chosen to explain the phenomenon. Today the observations could be curves, images or movies resulting dimension in the range of billions for only hundreds of observations considered for study. The problems include collection, storage, searching, sharing, transfer, visualisation and analysis. A hybrid model may be appropriate to model Big Data. Analysing the highly dynamic Big Data requires simultaneous estimation and testing of several parameters. Errors associated with estimation of these parameters help to design a decision rule or prediction rule depending on the parameters. Big Data can make important contributions in real-life decision making by developing cost-effective methods with minimum risk factor.

The basic approach of processing Big Data is to divide and analyse the subproblems, and finally results from subproblems are combined to obtain the final solution. Dimensionality reduction plays an increasingly important role in the analytics of Big Data and is applied before partitioning of the problem. Dimensionality reduction techniques aim at finding and exploiting low-dimensional structures in high-dimensional data. The methods overcome the curse of dimensionality and reduce computation and storage burden of the decision-making systems. Designing very large-scale, highly adaptive and fault-tolerant computing systems is the solution to manage and extract information from Big Data.

The objective of the chapter is to integrate different fields of machine learning for devising novel “dimensionality reduction” techniques. In this chapter, two different concepts on dimensionality reduction techniques have been proposed to improve decision making. Finally, we discuss how dimension reduction is important for Big data analytics and present a case study in the field of agricultural science.

In the rest of this chapter, we first discuss dimension reduction techniques in Sect. 12.2. This section explores how the important features are selected using genetic algorithm-based optimization technique. The vagueness and uncertainty is also handled using fuzzy-rough-set theory in this section. A case study is provided, based on agricultural data, in Sect. 12.3. This section gives a precise idea about how a decision support system is designed in case of a Big data environment. The section also includes acquisition of images, filtering and extraction of features, selection of important diseased features, etc. The methodology explained in Sect. 12.2 is used for important diseased feature selection. Finally, Sect. 12.4 presents our conclusions.

12.2 Dimension Reductions

Importance of features may vary from one application to another, and often it is difficult to know exactly which features are relevant for a particular task. The irrelevant features not only increase complexity of the systems but affect performance of the system. Dimension reduction plays a vital role in discovering knowledge from high-dimensional data set by removing redundant and irrelevant attributes or features.

The goal of dimensionality reduction is to avoid selecting too many or too few features than is necessary. Too few features result loss of information, while too many irrelevant features dominate important features that overshadow the information present in the data set. The aim of dimension reduction is to select a minimum set of relevant features by preserving essential information content of the data set. This minimum set of features is used instead of the entire attribute set for decision making.

Feature selection [3] is basically searching of a feature space to select the optimal subset of features that can fully characterise the whole data set. There are various optimised searching techniques such as ant colony-based search, Tabu search and simulated annealing which are used frequently for feature subset selection. Memetic algorithms [4] are quite popular for successful searching in a large search space where traditional exhaustive search methods often fail.

Genetic algorithms [5, 6] have great potential for finding the best approximate solutions in the evolutionary optimization community. However, genetic algorithms (GA) in some specific applications may provide local optimal solution due to premature convergence. To overcome this problem, several hybridization of genetic algorithms with local search methods are proposed [6–8]. In [9], the GA incorporates two new operations, namely, cleaning operator and proposition improvement operator to maintain sustained diversity among the population. The work in [10] proposes a multi-objective genetic algorithm integrating local search for finding a set of solutions of a flow shop scheduling problem. In [11], a multi-objective time-dependent route planning problem is solved by integrating NSGA-II [12] with local search.

In this section, two different feature selection methods are proposed. Firstly, a novel neighbourhood-based local search strategy in genetic algorithm for obtaining optimal set of important features is described. Kullback-Leibler (KL) divergence method is used as fitness function with binary-coded chromosomes as population. Steady-state selection, single-point crossover and jumping gene mutation [13] scheme are applied stepwise to implement the proposed algorithm. Secondly, dimension reduction is performed utilising fuzzy-rough concept and genetic algorithm. Optimised range of value associated with each linguistic label of the attributes is evaluated using rough-fuzzy concept. We build an efficient classifier using genetic algorithm (GA) to obtain optimal subset of attributes considering optimal membership values of the linguistic variables. The proposed algorithm reduces dimensionality to a great extent without degrading the accuracy of classifier and avoids being trapped at the local minima.

12.2.1 Hybrid Genetic Search Model (HGSM)

This method deals with selecting relevant features by hybridising genetic algorithm with local neighbourhood-based searching method. Each chromosome is encoded as a binary string where 1/0 in *i*-th position of the string depicts the presence/absence of the corresponding feature. Population size is initialized randomly to explore the search space. Genetic operators are described below.

Selection
A steady-state selection mechanism is implemented where each member of population (i.e. chromosome) is equal probable to get into mating pool.

Crossover
Every pair of parents is made sure to take part in the crossover operation to produce offspring using single-point crossover scheme. Offspring are kept if only if they fit better than the least good individual of the population.

Mutation
Mutation plays a vital role in preserving diversity among the population over generations. Single-bit mutation is most popular in genetic algorithm, but it lacks diversity in population since the most significant bit of a binary string generally does not change. To overcome the demerit, jumping gene or mutation methodology is used for mutating the genes. Jumping genes are a set of genes which can jump dynamically within the chromosome. These genes have a great potential for

maintaining diversity throughout the entire population, crucial for evolutionary searching algorithm.

Let a chromosome in population be $(a_1, a_2, \ldots, a_n)$. Jumping genes of length q ($q << n$), say $(b_1, b_2, \ldots, b_q)$, are selected randomly and replace a portion of the chromosome, decided randomly. Let k be the starting position, so after mutation the muted chromosome is $(a_1, a_2, \ldots, a_{k-1}, b_1, b_2, \ldots, b_q, a_{k+q}, \ldots, a_n)$.

New Population Generation

At each generation, new population is obtained using local search strategy. In a particular generation, proximity of an individual is computed with respect to other members of the population. Let D be the snapshot of current population with four members {1 1 0 1}, {1 0 0 1}, {1 1 1 1} and {1 0 0 0}. Proximity of first member {1 1 0 1} is evaluated using Hamming distance considering the rest of the members in the population. Integer mean of the Hamming distances is taken for selecting the neighbouring region of the first member. In this example, the mean is 1, so assuming the member {1 1 0 1} as schema, specified number of neighbours is generated each with Hamming distance 1, as illustrated in Fig. 12.1. If any binary string in the neighbours is already present in the current population, then it is discarded and another neighbour is created in its place. Now search is performed within the neighbours, and original member {1 1 0 1} is replaced by a neighbour of better fitness value. Similarly, neighbours of the entire population are built and the population is modified by the local search strategy. In this way, searching is fine-tuned resulting better performance in the output.

Fitness Function

Fitness function determines the quality of a chromosome in the population. So, a strong fitness function is imperative for giving good result. The method uses Kullback-Leibler (KL) divergence [14] for computing fitness of the chromosomes. The KL divergence is a primary equation of information theory that quantifies the proximity of two probability distributions. It is a measure in statistics that quantifies in bits how close a probability distribution $p = \{p_i\}$ is to a model (or candidate) distribution $q = \{q_i\}$. The KL divergence is measured using Eq. (12.1).

$$D_{KL}(p||q) = \sum_{i=1}^{\text{no.of feature}} p_i \log_2\left(\frac{p_i}{q_i}\right) \tag{12.1}$$

D_{KL} is non-negative (≥ 0) and not symmetric in p and q. Its value is zero if the distributions match exactly and can potentially equal perpetuity. In the binary string

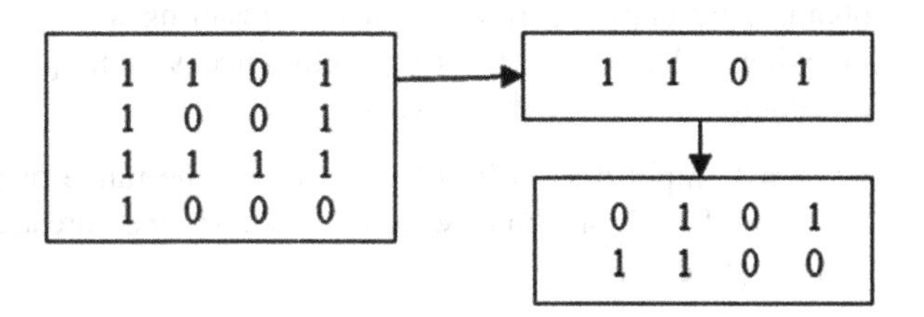

Fig. 12.1 Neighbourhood generation method

population, candidate feature subset is created and associated probability distribution (q_i) is computed. Then the probability distribution (p_i) based on decision attribute over all samples is calculated. We evaluate pairwise D_{KL} value and mean of these D_{KL} value is used as fitness function of the genetic algorithm. The workflow diagram of the methodology is described in Fig. 12.2.

12.2.2 Fuzzy-Rough-Set Approach

Real-world data sets are often vague and redundant, creating problem to take decision accurately. Very recently, rough-set theory [15, 16] has been used successfully with the goal of removing vagueness and extracting knowledge after dimension reduction of the data set. However, rough-set theory is applied on discrete data set, but in real life most often the attributes are continuous. Discretisation of data leads to information loss and may add inconsistency in the data sets. Fuzzy-rough concept [17, 18] has been applied to overcome the above limitations. After removing inconsistency, rough-set theory [19] is applied to obtain the reduced attribute set called *reduct*. However, handling of non-discretised values increases computational complexity of the system. Therefore, to build an efficient classifier, genetic algorithm (GA) has been applied to obtain optimal subset of attributes, sufficient to classify the objects. The proposed algorithm reduces dimensionality to a great extent without degrading the accuracy of classifier and avoids being trapped at local minima. As a next step genetic algorithm has been applied to obtain global optima in the search space for most informative attributes.

The following subsection describes how to generate fuzzy rule set with optimum variation in the range of linguistic labels representing antecedent of the rules. Fuzzy inference system is invoked to obtain membership value of linguistic variables. In the next subsection, we present the dimension reduction algorithm for obtaining reduced attribute set or *reduct*.

12.2.2.1 Data Preparation

To design the fitness function of GA, two parameters are first evaluated, i.e. the membership values of each instances or objects in different clusters and the membership value of the objects in different classes. The former parameters are obtained by applying fuzzy c-means clustering algorithm [20] and the latter by invoking a Mamdani-type fuzzy inference system [19]. Different steps of the procedure are described below:

- Step 1: Input data is fuzzified based on the range of minimum and maximum value of each attribute, which determines the spread of membership value of respective attribute.

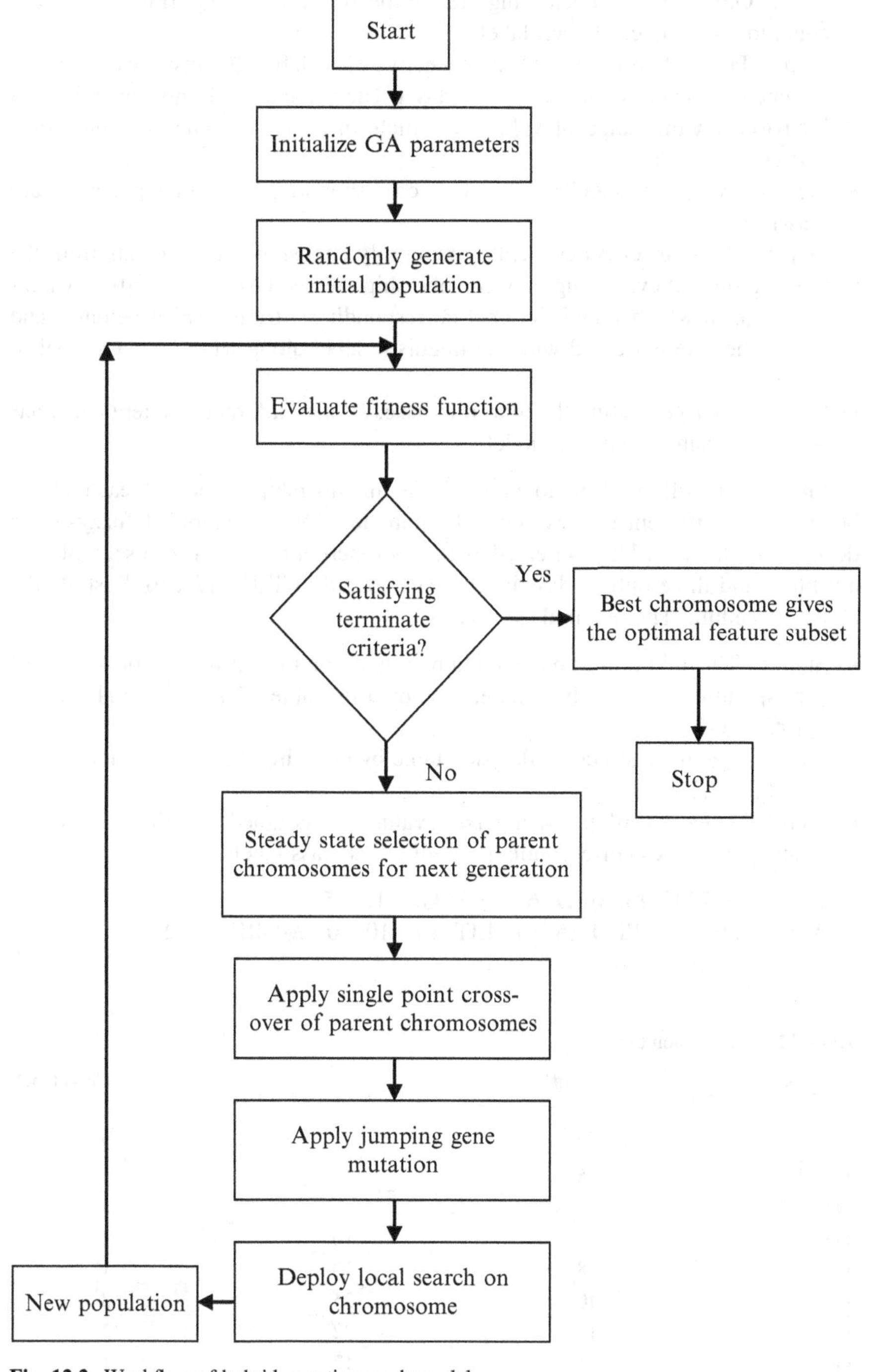

Fig. 12.2 Workflow of hybrid genetic search model

- Step 2: Objects are grouped using fuzzy c-means clustering algorithm, where c is equal to the number of class labels.
- Step 3: For each attribute and its respective class label, linguistic variables are assigned following semantic of the dataset. The value of the linguistic variable is set based on the range of values to which an attribute value is spread for a particular class label.
- Step 4: For each linguistic variable, corresponding membership curves are drawn.
- Step 5: The rule set is designed by randomly choosing data elements from the training set and evaluating their membership values. Using the attribute values we judge in which linguistic label corresponding attribute value belongs, and finally the rule is framed with the linguistic label along with the class label of the rule.
- Step 6: After generating the total rule base, a fuzzy inference system (FIS) has been built using Mamdani model.

Finally, the FIS is used to produce the membership values of each object belonging to different classes (second parameter) for evaluation of fuzzy-rough dependency factor, which is treated as fitness function of GA. A data set with two attributes and three output class label is considered in Table 12.1 to illustrate the above procedure. The required steps are:

- Step 1: The minimum and maximum ranges of the attributes for spread of corresponding membership curve, as shown in Table 12.1, are Attr#1 = 2–15, Attr#2 = 5–25.
- Step 2: Now reconstruct the decision table by grouping the class label as shown in Table 12.2.
- Step 3: For each attribute, membership values are assigned by utilising the range of attribute values corresponding to individual class labels:

  Attr#1: LOW (2–8), MED (6–12), HIGH (10–15)
  Attr#2: VERY LITTLE (5–16), LITTLE (10–20), MORE (21–25)

Table 12.1 A decision table

Objects	Attr#1	Attr#2	Class label
O1	2	10	1
O2	7	5	2
O3	5	15	1
O4	6	8	2
O5	12	16	2
O6	8	20	1
O7	10	25	3
O8	15	22	3
O9	1	17	1
O10	13	21	3

Table 12.2 Output of step 2

Objects	Attr#1	Attr#2	Class label
O1	2	10	1
O3	5	15	1
O6	8	20	1
O9	7	5	2
O2	6	8	2
O4	12	16	2
O5	10	25	3
O7	10	25	3
O8	15	22	3
O10	13	21	3

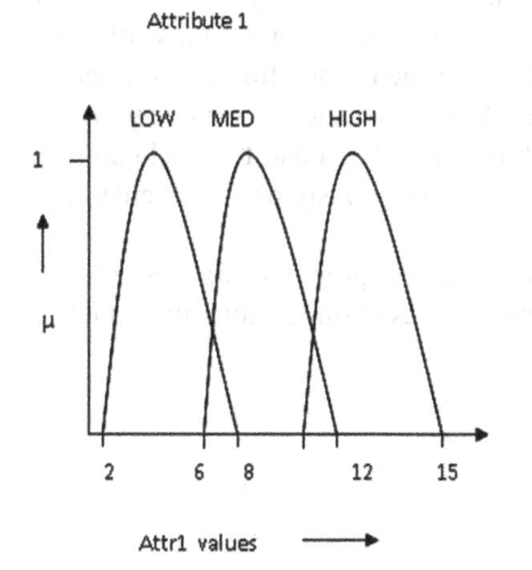

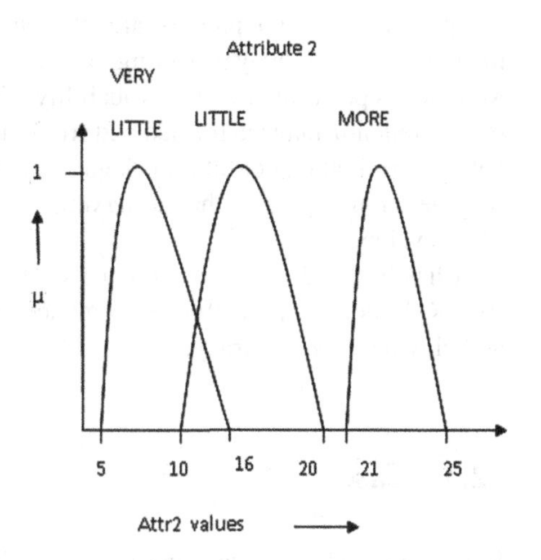

Fig. 12.3 Membership curve of two attributes

- Step 4: The membership curves for the attributes are shown in Fig. 12.3.
- Step 5: The rule set corresponding to the decision system is given below:

1. IF Attr#1 is LOW and Attr#2 is VERY LITTLE THEN class is 1.
2. IF Attr#1 is LOW and Attr#2 is LITTLE THEN class is 1.
3. IF Attr#1 is LOW and Attr#2 is MORE THEN class is 1.
4. IF Attr#1 is MED and Attr#2 is LITTLE THEN class is 1.
5. IF Attr#1 is MED and Attr#2 is MORE THEN class is 1.
6. IF Attr#1 is LOW and Attr#2 is VERY LITTLE THEN class is 2.
7. IF Attr#1 is MED and Attr#2 is VERY LITTLE THEN class is 2.
8. IF Attr#1 is MED and Attr#2 is LITTLE THEN class is 2.
9. IF Attr#1 is HIGH and Attr#2 is VERY LITTLE THEN class is 2.
10. IF Attr#1 is HIGH and Attr#2 is LITTLE THEN class is 2.
11. IF Attr#1 is MED and Attr#2 is MORE THEN class is 3.
12. IF Attr#1 is HIGH and Attr#2 is MORE THEN class is 3.

Finally, Mamdani model has been applied for evaluating degree of membership value of each attribute in different classes.

12.2.2.2 DIM-RED-GA Algorithm

In the proposed method, the dependency of the decision attribute or class label on different set of condition attribute is calculated, and attributes with highest dependency factor are selected as optimum set of *reduct* by applying genetic algorithm.

In the proposed DIM-RED-GA, population size is considered as the size of the total data set, and fuzzy-rough dependency factor is designed as the fitness function. The length of each chromosome is equal to the number of attributes, and each component represents corresponding attribute value of different objects, i.e. population. Two parents are chosen randomly and crossing over between them is performed by selecting a crossover point with a probability of 0.10. Mutation is performed with a probability of 0.02. In each generation, a combination of different attributes is formed and from them few are selected based on the fitness value. Termination condition is kept as combination of two conditions: (1) number of generation is greater than a maximum value set previously or (2) no change in fitness value.

It has been observed that for the UCI database, the proposed algorithm DIM-RED-GA shows appreciable performance both in terms of dimensionality reduction and classification accuracy.

12.3 Case Study

Rice is cropped worldwide and over half of the world population depends on rice as the main food. To cope with the increasing demand for rice, a key element is the development and implementation of effective rice insect management strategies. The use of high-yielding varieties and greater nitrogen fertiliser increased rice yields. Still disease and pests cause 25 % loss of rice in India, which need to be controlled for more production of rice. Losses may be reduced using pesticides. However, the use of pesticides increases the cost of production, and in addition it reduces food quality creating adverse effect on the environment. So instead of using pesticides, researchers try to develop sustainable farming practice that depends on many factors including effective and timely detection of diseases and pest management to protect the crops. Timely diagnosis of crop diseases in the fields is critical for precision on farm disease management.

Rapid development of digital devices for acquiring and storing of data attracts researchers to apply analytical methods for automatic identification of rice diseases using computational frameworks. However, due to the diversity of crops and their associated diseases, application of technology in agriculture is still in research stage. Therefore, the challenge is elaborate investigation and analysis of the big data set for developing automated rice disease classification system. To design an efficient classification system, different types of features are acquired from the rice plant images generating huge volume of data set. Using the proposed feature selection methods, most important features are obtained based on which the rice diseases are classified.

12.3.1 Rice Diseases

Three different types of attacks, namely, weeds, insect/pest and diseases, cause substantial loss in crops. In the present study, five rice plant diseases are considered that are grouped as follows:

- Diseases caused by fungus
- Diseases caused by bacteria
- Diseases caused by virus

To describe the diseases, locations of infection in different parts of the rice plant are to be detected accurately. The rice plant consists of roots, stem, leaves and panicle, as illustrated in Fig. 12.4.

12.3.1.1 Leaf Brown Spot

Brown spot symptoms are observed at the tillering stage and beyond. Small foliar lesions are found and the shapes of the lesions vary from circular to oval with light brown to grey at the centre and reddish brown margin, as shown in Fig. 12.5.

12.3.1.2 Rice Blast

Rice blast is observed in both lowland and upland. Lesions on the leaves are, generally, brown to dark brown colour and may enlarge and coalesce to kill the

Fig. 12.4 Mature rice plant

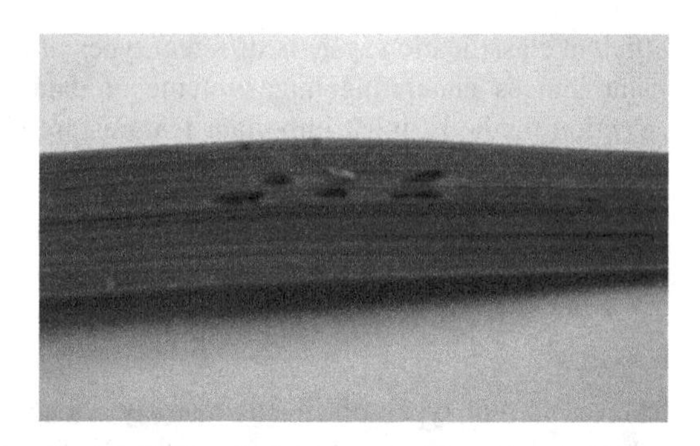

Fig. 12.5 Leaf infected by brown spot disease

Fig. 12.6 Rice leaf infected by blast disease

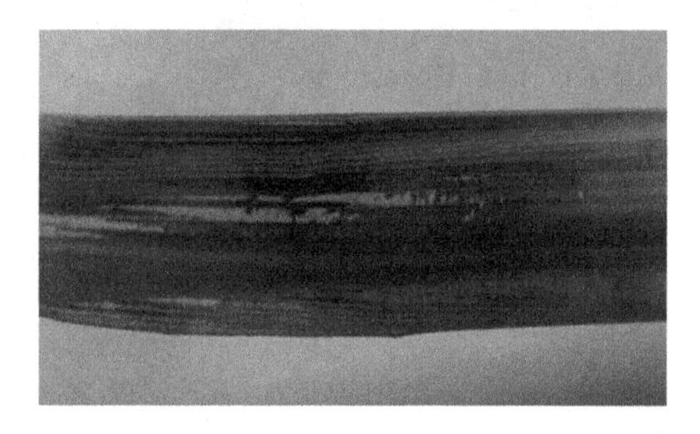

Fig. 12.7 Rice stem infected by sheath rot

entire leaves. Initially white to greyish green circular lesions or spots with dark green borders are found on the leaves. Older lesions on the leaves are elliptical or spindle shaped with more or less pointed ends. A leaf infected by *blast* is shown in Fig. 12.6.

12.3.1.3 Sheath Rot

Sheath rot occurs on the leaf sheath that encloses the young panicles. The lesions start as oblong or irregular spots with variation in colour from grey to light brown at centres, surrounded by distinct dark reddish brown margins, shown in Fig. 12.7. As the disease progresses, the lesions enlarge and coalesce and may cover most of the leaf sheath. Lesions may also consist of diffuse reddish brown discolorations in the sheath. An abundant whitish powdery growth may be found inside the affected sheaths, although the leaf sheath looks normal from outside.

12.3.1.4 Bacterial Blight

Bacterial blight disease appears on old plants, usually at or after maximum tillage or at rooting stage. Water-soaked lesions usually start at leaf margins, a few cm from the tip and spreading towards the leaf base. Length and width of the affected areas increase, while colour changes from yellowish to light brown due to drying. Yellowish border between dead and green areas are found on the leaf, as shown in Fig. 12.8.

12.3.1.5 Rice Hispa

Rice hispa is a defoliator during the vegetative stage of the rice plant. Scraping of the upper surface of the leaf blade leaving only the lower epidermis as white streaks parallel to the midrib is observed, while tunnelling of larvae through leaf tissue causes irregular translucent white patches that are parallel to the leaf veins, as shown in Fig. 12.9.

Fig. 12.8 Rice leaf infected by bacterial blight disease

Fig. 12.9 Rice leaf infected by pest rice hispa disease

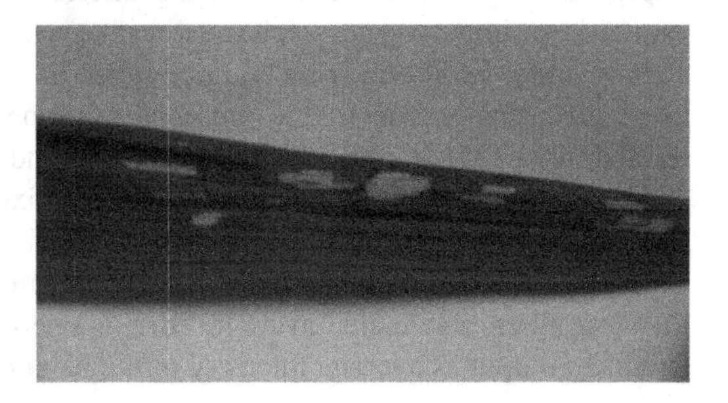

12.3.2 Methodology

Image processing has wide range of applications in biometric identification, defect detection, quality measurement, medical imaging and many more fields. However, in agriculture domain, it has limited applications. One potential area of application in agriculture is rice disease classification where image processing and computational methodologies can be applied to develop an autonomous system. Many researchers have attempted to develop such systems with the capability to understand and reason from observations using intelligent techniques [21, 22]. Real-life implementation has very limited scope due to variety of diseases and variation of symptoms that increase complexity of the system. Presently, rice diseases are detected manually depending on the visual symptoms and opinion of field experts. Visual symptom-based common features are change of colour of the plant, shape of the infected region, colour of different subregions with respect to the background region, texture of infected regions and location of infection. However, evaluation of manually identified visual symptoms are subjective and observer dependent, and lack in timely diagnose often fail to classify diseases accurately. Different types of features are identified and extracted from the infected regions using geometric property of the affected region, statistical characteristics, intensity and texture values. All such features are not significant, and the presence of redundant features affect classification accuracy and increases complexity of the systems. Therefore, features are selected using evolutionary algorithms to reduce complexity of the decision-making system.

12.3.2.1 Image Filtering

The basic task of image processing is to extract pixel information from the image and provide suitable interpretation of such information for understanding of human observers. However, when images are captured optically or electronically, the original image is degraded by the environment and image acquisition system. So the very first step is to filter noise content from the acquired image with an objective to recover the image from degraded observations. Filtering is a low-level image-processing task, which removes high-frequency noise component from the degraded image. Noise removal using different types of low-pass filtering techniques is an important research area in image and signal processing. Linear filter [23] is sufficient in removing additive noise like Gaussian noise [24] but delivers poor performance due to the presence of impulse noise. Non-linear filter like median filter [25] is used to de-noise the salt and pepper noise from the degraded images. In linear filter, the value of the output pixel is obtained using the weighted sum of the input pixels, while non-linear filtering method assigns a value to the output pixel, based on the neighbouring pixel information of the input pixel.

Mean filter, also called average filter, is one kind of low-pass linear filter that produces output as average intensity of the neighbouring pixels. The average filter smoothes local variation in an image, and thus noise is reduced due to blurring effect in the image. In average filtering, if the template size is larger, the blurring effect is more in the output image. Median filtering is a non-linear low-pass filter that is adequate for removing the outliers present in an image. It uses neighbourhood operations to process an image around the input pixel. The pixel values in a neighbourhood region are ordered, and the median value is used to substitute the pixel at the same location of the output image.

12.3.2.2 Feature Extraction

Feature extraction is an important step towards disease identification on which the success of the system depends. A visual feature is defined to capture certain visual properties of an image, either globally for the entire image or locally for a small group of pixels. Most of the common features observed by the farmers in the diseased plants are the visual symptoms in the spatial domain such as change of colour, shape of the infected portion, location of infection and texture. Spots created by the diseases have colour difference between its boundary region, centre or core region and uninfected region. Change of colour of the spot at boundary and core regions with respect to the uninfected portion have also been used as colour features to address the age and nutrient level of the plant. Another set of colour features based on distribution of hue are obtained since hue represents the true colour recognised by the human visual system. Distribution of hue value from the centre towards the boundary of the spot along the radius in eight different directions has been considered as colour features to distinguish the diseases.

Different types of features are extracted from the infected plant images to classify the rice diseases. However, it has been observed that classification accuracy to detect diseases is not proportional to the number of features, rather less number of features but more significant ones. Therefore, feature selection is an important step for accurately classifying rice diseases based on the extracted features. The irrelevant and redundant features should be removed prior to building the classifier in order to achieve better performance and reducing complexity of the systems.

Most common features used for building image-based recognition/classification system are colour features, shape-based features and texture features.

12.3.2.3 Colour Features

Colour is a major distinctive attribute used to distinguish diseases. Colour features are geometric transformation invariance and generally global so is considered to extract colour features. Consumers initially accept or reject a food based on its colour and other visual attributes, which is simulated using machine vision method. To identify different types of diseased plant images in remote-sensing techniques, different types of vegetative indices are used to represent quantitative measurements indicating the vigour of vegetation. Bannari et al. [26] suggest that over forty vegetation indices have been developed during the last two decades and used for classification of different diseases. Vegetative index based on the visual images are also proposed in literature where normalised difference greenness index (NDGI) and redness index (RI) are defined as

$$NDGI = \frac{g - r}{g + r}$$

$$RI = \frac{r + g}{r - g}$$

where r and g represent the red and green value of a pixel, used to identify stress of the images.

Colour is used not only for disease identification but also as a major feature for quality, composition and standards of identity of food. Colour features are widely used for grain quality measurement. Wang et al. [27] developed a real-time grain inspection instrument to classify durum wheat kernels based on their virtuousness. To develop the system, acquired images are normalised into 70×200 size and converted to HSI colour plane. Here it has been assumed that an artificial neural network (ANN) would prefer the HSI model over the RGB model because ANNs are designed based on an analogy to the human neural system. Using various artificial neural network models, they got accuracy in the range of 58–90 %.

12.3.2.4 Shape-Based Features

Shape of the spots created by infection plays an important role for disease identification. Some simple geometric features are used to describe the shapes of the infected regions. Usually, the simple geometric features can only discriminate the shapes with large differences, therefore usually used as filters to eliminate false hits or combined with other shape descriptors to discriminate shapes. A shape can be described by different aspects. Shape parameters available in literature are centre of gravity, axis of least inertia, digital bending energy, eccentricity, circularity ratio, elliptic variance, rectangularity, convexity, solidity, Euler number, profiles and hole-area ratio. Different shape parameters are described below:

- The position of centre of gravity should be fixed in relation to the shape [28] and so used to recognise the shape.
- The axis of least inertia is unique to the shape and serves as a unique reference line to preserve the orientation of the shape. The axis of least inertia (ALI) of a shape is defined as the line for which the integral of the square of the distances to points on the shape boundary is a minimum [29].
- Average bending energy is defined by the average of squared curvature of the shape and is a useful property of shape [30].
- Eccentricity is the measure of aspect ratio, defined as the ratio of the length of major axis to the length of minor axis [28].

- Minimum bounding rectangle is also called minimum bounding box, the smallest rectangle that contains every point in the shape.
- Solidity describes the extent to which the shape is convex or concave [31] and defined by the ratio of area of the shape region where *H* is the convex hull area of the shape.
- Euler number describes the relation between the number of contiguous parts and the number of holes on a shape [32].
- Circularity ratio represents how a shape is similar to a circle [33]. Three different definitions are available in literature to measure the circularity. Circularity ratio is (i) ratio of the area of a shape to the area of a circle having the same perimeters, (ii) ratio of the area of a shape to the shape's perimeter square and (iii) circle variance, ratio of standard deviation and mean of the radial distance from the centroid of the shape to the boundary points.
- Ellipse variance is a mapping error of a shape to fit an ellipse that has an equal covariance matrix as the shape [28].
- Rectangularity represents how rectangular a shape is, i.e. how much it fills its minimum bounding rectangle and defined as a ratio of area of the shape and the area of the minimum bounding rectangle [28].
- Convexity is defined as the ratio of perimeters of the convex hull over that of the original contour [34].
- Hole-area ratio (*HAR*) is defined as the ratio of area of a shape and the total area of all holes present in the shape. Hole-area is the most effective in discriminating between symbols that have big holes and symbols with small holes [32].
- Shape-based features for rice disease classification include area, perimeter length and minimum enclosing rectangle of infected spot as reported in [35]. The shape features used for grain classification and grading system include perimeter, area and shape descriptors (kernel circularity and shape compactness), selected to classify group of kernels which are placed in random direction. Area and perimeter of the spot are calculated using chain code.

12.3.2.5 Texture Features

Texture is one of the most important characteristics of images in identifying defects or flaws that provides important information for recognition and interpretation of images. The term texture may be used to characterise the surface of a given object or region. Success of texture classification depends on the appropriate method of texture analysis for feature extraction, and selection of such method is a critical one. Texture features based on the extraction methods are broadly divided into four categories:

- Statistical features
- Model-based features
- Structural features
- Filter-based features

Statistical Features

Statistical features are determined by applying statistical procedures. Most common statistical texture features are based on the co-occurrence matrix. A variety of concurrence matrix is found in [36, 37]. Most common is the grey label occurrence matrix (GLCM) [36]. It is a two-dimensional histograms of the occurrence of pairs of grey levels for a given displacement vector. Based on which 14 features are calculated such as angular second moment, contrast, correlation, variance, inverse difference moment, sum average, sum variance, sum entropy, entropy, difference variance, difference entropy, information measures of co-relation and maximal correlation coefficient using grey-level co-occurrence matrices.

Model-Based Features

A number of random field models (i.e. models of two-dimensional random processes) have been used for modelling and synthesis of texture. If a model is shown to be capable of representing and synthesising a range of textures, then its parameters may provide a suitable feature set for classification and/or segmentation of the textures. For a model-based approach to be successful, there must exist a reasonably efficient and appropriate parameter estimation scheme, and the model itself should be parsimonious, i.e. use the minimum number of parameters. Popular random field models used for texture analysis include fractals, autoregressive models, fractional differencing models and Markov random fields.

Autoregressive models have been used to model images as random fields (two-dimensional random processes) by a number of researchers [38–40]. In the two-dimensional spatial case, the "previous values" of the time series process are replaced by the grey values of local neighbourhood pixels. Unlike the temporal case, there is normally no preferred direction in a lattice, and neighbourhoods are therefore normally defined to consist noncausal (two-sided) variables. Again the parameters may be estimated either by using least square error or maximum likelihood techniques.

Markov random fields (MRFs) have been popular for modelling images. They are able to capture the local (spatial) contextual information in an image. These models assume that the intensity at each pixel in the image depends on the intensities of only the neighbouring pixels. MRF models have been applied to various image-processing applications such as texture synthesis, texture classification, image segmentation, image restoration and image compression.

Structural Features

Structural approaches generally model a texture as the deterministic or stochastic placement of texture primitives called textons with the emphasis on texton characterisation such as size and shape or both local properties. Structural methods are based on regular or semi-regular placements of textural primitives. In the case of observable or visual textures, it is usually quite difficult to extract the primitives and their placements. Therefore, these approaches are more appropriate only for highly regular deterministic textures.

Filtered-Based Features

Filter-based texture features are gaining popularity for different classification purposes due to its ability to represent texture precisely and overcome the drawbacks of statistical and structural texture features. Filter-based methods can be grouped into three categories: spatial domain filtering, frequency domain filtering and spatial-frequency domain filtering.

Texture-based approach can fail where texton primitives are not readily identifiable. Statistical approaches focus on the global spatial relationships between intensity variations and often fail to capture local properties of the texture. Texture characterisation requires both local texton primitive and global spatial organisation description. Neither structural nor statistical methods satisfy this requirement fully. Statistical methods have been widely accepted over the past two decades.

12.3.3 Position Detection

Another important feature to classify diseases is the position of infection (PSI) with respect to the top boundary of the leaf. A tree structure method has been described here to identify the *PSI* [41]. The background of a segmented image is represented by grey value, and noninfected region is marked by black, while the infected region is represented by white as shown in Fig. 12.10. A rectangle is drawn across the leaf so that it covers the infected region and boundary of the leaf, called image block, as shown in Fig. 12.10. Image blocks of different infected leaves are standardised to 600×900 image size. A vertical line from the middle of the length and a horizontal line from the middle of the width of the rectangle are drawn, which split the image block into four equal subblocks. Thus, the main image block, say B, is divided into four subblocks, named as B_{00}, B_{01}, B_{10} and B_{11}, and a tree is constructed as follows:

- A vertex is considered for each block.
- An edge is drawn from parent block B to subblock B_{ij} with edge label $\{ij\}$, where ij is marked as combination of 0 and 1.

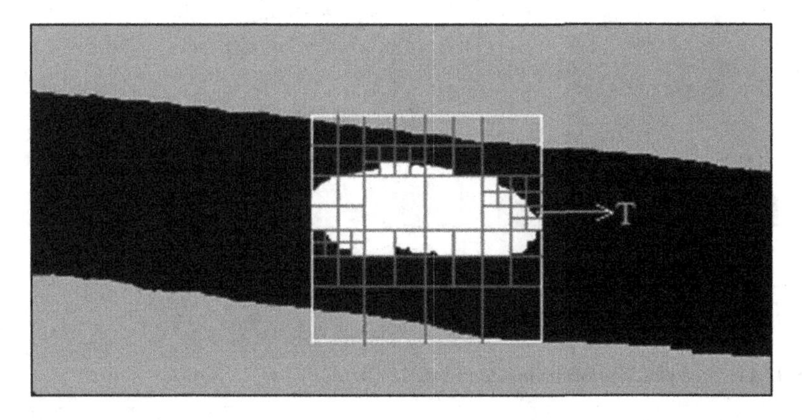

Fig. 12.10 Splitting of regions

The subblocks are either non-uniform or uniform. The non-uniform block contains the mixture of black and white pixels, while uniform block contains either black or white pixels.

Next, all subblocks are checked for uniformity. The subblocks of a parent block are considered as leaf nodes of the tree if at most one subblock is the non-uniform block. Otherwise, each non-uniform subblock of the parent block is divided in the same way, and the tree is expanded until all leaf nodes are generated. Figure 12.11 shows the tree structure of the block, corresponding to the image shown in Fig. 12.10. The block *B* representing the whole image is the root (*R*) of the tree, while the non-uniform subblocks are the leaf nodes represented by grey colour. The leaf nodes with grey colour carry boundary information of the infected region, while white-coloured leaf nodes, i.e. uniform subblocks, have no importance. The intermediate nodes including the root node are used only to measure the distance of the infected region from the top boundary of the leaf. To measure the distance, a binary string is associated to each node of the graph according to Eq. (12.2).

$$
\begin{aligned}
BS(R) &= \{0\} \\
BS(C) &= BS(p)\{ij\};\text{if } P \text{ is parent node (other than root node) of child node } C \\
&= \{ij\} \qquad ; \text{ other wise}
\end{aligned}
\tag{12.2}
$$

where $\{ij\}$ represent the edge label from P to C.

The binary string associated with the non-uniform leaf node (*NL*) ie *BS* (*NL*) is converted to its decimal equivalent and provides distance in terms of blocks from the top of the image. The width of the subblock corresponding to non-uniform node is considered as unit block distance.

Let, for a node *T*, binary string contain *n*-components, i.e. $BS\ (T) = \{i_1j_1\}\{i_2j_2\}\{i_3j_3\}.....\{i_nj_n\}$. The binary code *BC*(*T*) is given as $\{i_1\}\{i_2\}\{i_3\}....\{i_n\} =$

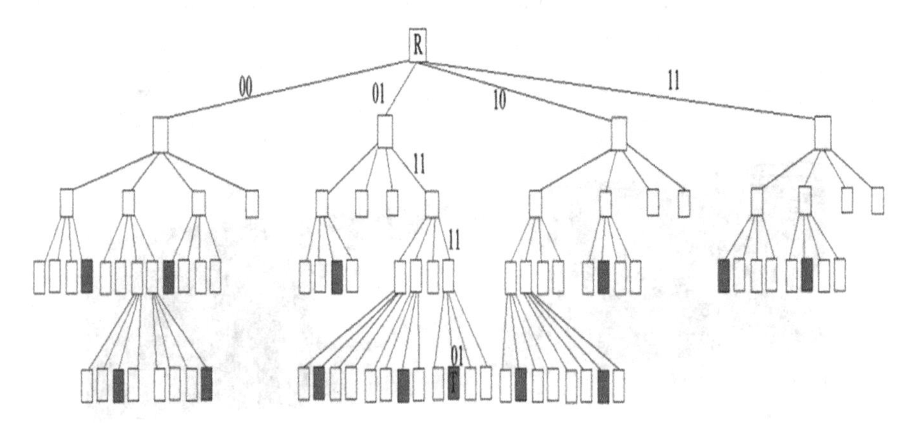

Fig. 12.11 Tree structure of the image in Fig. 12.10

$\{i_1 i_2 i_3 \ldots i_n\}$, whose decimal equivalent is *DC*(*T*). The left most bit of each component of the string is considered for binary code generation because most significant bit carries maximum weight.

For example, in Fig. 12.11, binary string for non-uniform leaf node T is {01}{11}{11}{01}, and its corresponding binary code is {0}{1}{1}{0} = {0110}. It is 6 in decimal code, which implies that subblock *T* is six block distance apart from the top of the main block (i.e. the image).

In reality, infected region created by different diseases vary in size, and as a result a global block with minimum width is used for distance measurement. The width *W*(*T*) of a subblock associated with non-uniform leaf *T* is given by Eq. (12.3).

$$W(T) = \frac{w}{2^l} \quad (12.3)$$

where w is equal to 900 representing standard width of the image block and l is the label of the node T in the quad tree.

The procedure is applied on large number of infected leaves; corresponding trees and width of non-uniform leaf nodes are calculated using Eq. (12.3). Therefore, all the non-uniform nodes in the same label of the tree have associated blocks with same width. Similarly, a node at highest label of the tree has associated block with minimum width. Say, $l_{\max}$ is the highest label among all trees, so the minimum width of the block associated to non-uniform node T' is obtained using Eq. (12.4).

$$w_{\min}\left(T'\right) = \frac{w}{2^{l_{\max}}} \quad (12.4)$$

Thus, the actual block distance *D*(*T*) of the subblock *T* from the top of the image block is given by Eq. (12.5) considering minimum width subblock as the unit block distance.

$$\begin{aligned} D(T) &= DC(T)*\frac{W(T)}{W_{\min}(T')} \\ &= DC(T) \times \frac{2^{l_{\max}}}{2^l} \\ &= DC(T) \times 2^{(l_{\max} - l)} \end{aligned} \quad (12.5)$$

So, if a tree contains b non-uniform leaf nodes $\{T^1, T^2, \ldots, T^b\}$, then the position of the spot in the image from the top is given by Eq. (12.6).

$$pos = \frac{1}{b}\sum_{i=1}^{b} D\left(T^i\right) \quad (12.6)$$

where $D(T^i)$ represents block distance for ith non-uniform leaf node.

For rice disease identification, a wide variation of features are obtained and values are acquired over a period of time. The data set represent characteristics of the images when attacked by different types of diseases. However, the presence of large number of feature may increase redundancy, and dominance of less important features reduces classification accuracy. Therefore, feature selection is an important step for accurately classifying rice diseases; here we have considered five different diseases. The irrelevant and redundant features are removed using feature selection methods prior to classify the diseases.

12.4 Conclusions

Systematic and unbiased approach to classification is of great importance, and therefore, automated discovery of this small and good feature subset is highly desirable. In this chapter, a novel local search based genetic algorithm approach has been proposed to select informative feature subset, which classify the dataset effectively and efficiently. In the second feature selection method, vagueness and continuous domain dataset is managed using fuzzy-rough approach, and optimum features are selected using conventional genetic algorithm.

Finally a real-life case study in agricultural domain is illustrated with sufficient features and large data set. The features are extracted from the objects which are robust, accurate and independent of transformation variance. Colour-based features are robust, transform invariant and represent most important visual changes recognised by human eye. Shape features are used for disease classification, since depending on the type and severity of infection shape of the infected region varies considerably. The presence of strips on the plant parts provides texture information, which are affected by diseases, and so different statistical texture features are also used for classifications. For disease detection, variability of features plays vital role, and at the same time complexity of the decision-making system is important for real-life problem. Genetic algorithm-based optimal feature selection and cleaning the data set using fuzzy-rough approaches are applied for minimising misclassification rate.

Feature dependency and context-based feature selection methods can be taken as future work for dimensionality reduction. More number of classes or diseases can be considered with an aim to enhance its performance and ability with respect to all categories of crops at an early stage. Time-series data can be analysed to understand progression of the disease, and preventive measures could be taken providing appropriate suggestions to control them.

References

1. Hu H, Wen Y, Chua TS, Li X (2014) Toward scalable systems for big data analytics: a technology tutorial. IEEE Access 2:652–686
2. Jagadish HV, Gehrke J, Labrinidis A, Papakonstantinou Y, Patel JM, Ramakrishnan R, Shahabi C (2014) Big data and its technical challenges. Commun ACM 57:86–94
3. Blum A, Langley P (1997) Selection of relevant features and examples in machine learning. Artif Intell 97(1.2):245–271
4. Knowles JD, David WC (2000) M-PAES: a memetic algorithm for multiobjective optimization. In: Proceedings of the 2000 congress on evolutionary computation, 2000, vol. 1. IEEE
5. Goldberg DE, Holland JH (1988) Genetic algorithms and machine learning. Mach Learn 3.2:95–99
6. Deb K et al (2002) A fast and elitist multiobjective genetic algorithm: NSGA-II. IEEE Trans Evol Comput 6.2:182–197
7. Coello CA et al (2007) Evolutionary algorithms for solving multi-objective problems. Springer, New York
8. Kim KW, Yun YS, Yoon JM, Gen M, Yamazaki G (2005) Hybrid genetic algorithm with adaptive abilities for resource constrained multiple project scheduling. Comput Ind 56 (2):143–160
9. Diaz CA, Muro AG, Pérez RB, Morales EV (2014) A hybrid model of genetic algorithm with local search to discover linguistic data summaries from creep data. In: Proceedings of the expert system with appllications, 2014, pp 2035–2042
10. Ishibuchi H, Murata T (1998) A multi-objective genetic local search algorithm and its application to flowshop scheduling. IEEE Trans Syst Man Cybern 28(3):392–403
11. Sharma S, Mathew TV (2011) Multiobjective network design for emission and travel-time trade-off for a sustainable large urban transportation network. Environ Plan B: Plan Des 38.3:520–538
12. Kalyanmoy D et al (2002) A fast and elitist multiobjective genetic algorithm: NSGA-II. IEEE Trans Evol Comput 6.2:182–197
13. Pati SK et al (2013) Gene selection using multiobjective genetic algorithm integrating cellular automata and rough set theory, swarm, evolutionary, and memetic computing. Springer, Cham, pp 144–155
14. Kullback S, Leibler RA (1951) On information and sufficiency. Ann Math Stat 22(1):79–86
15. Pawlak Z (1998) Rough set theory and its applications to data analysis. Cybern Syst 29:661–688
16. Pawlak Z (1991) Rough sets – theoretical aspects of reasoning about data. Kluwer Academic Publishers, Boston/London/Dordrecht, p 229
17. Jensen R, Shen Q (2002) Fuzzy-rough sets for descriptive dimensionality reduction. In: Procceding of the 11th international conference on fuzzy systems, pp 29–34
18. Moumita S, Sil J (2011) Dimensionality reduction using genetic algorithm and fuzzy-rough concepts. In: 2011 world congress on information and communication technologies, IEEE, pp 379–384
19. Qinghua H, Daren Y, Zongxia X (2006) Information-preserving hybrid data reduction based on fuzzy-rough techniques. Pattern Recogn Lett 27(5):414–423
20. Jain A, Murty M, Flynn P (1999) Data clustering: a review. ACM Comput Surv 31(3):264–323
21. Zhigang L, Zetian F, Yan S, Tiehua X (2003) Prototype system of automatic identification of cotton insect pest and intelligent decision based on machine vision. American Society of Agricultural and Biological Engineers
22. Qin Z, Zhang M, Christensen T, Li W, Tang H (2003) Remote sensing analysis of rice disease stresses for farm pest management using wide band airborne data. IEEE 4:2215–2217
23. Gonzalez RC, Woods RE (2007) Digital image processing. Pearson Education, New Delhi
24. Pratt WK (2010) Digital image processing. Wiley, New York

25. Chan RH, Ho CW, Nikolova M (2005) Salt-and-pepper noise removal by median-type noise detectors and detail-preserving regularization. IEEE Trans Image Process 14(10):1479–1485
26. Bannari A, Morin D, Bonn F, Huete AR (1995) A review of vegetation indices. Remote Sens Rev 13(1):95–120
27. Wang N, Dowell FE, Zhang N (2003) Determining wheat vitreousness using image processing and a neural network. Trans Am Soc Agric Eng 46(4):1143–1150
28. Mingqiang Y, Kidiyo K, Joseph R (2008) A survey of shape feature extraction techniques. In: Pattern recognition techniques, technology and applications. I-Tech, Vienna
29. Horn B (1986) Robot vision. MIT Press, Cambridge
30. Loncaric S (1998) A survey of shape analysis techniques. Pattern Recogn 31(8):983–1001
31. Cheng C, Liu W, Zhang H (2001) Image retrieval based on region shape similarity. In: Proceedings of the 13th SPIE symposium on electronic imaging, storage and retrieval for image and video databases
32. Soffer A (1997) Negative shape features for image databases consisting of geographic symbols. In: Proceedings of the 3rd international workshop on visual form
33. Zhang D, Lu G (2002) A comparative study of fourier descriptors for shape representation and retrieval. In: Proceedings of the 5th asian conference on computer vision
34. Mukundan R (2004) A new class of rotational invariants using discrete orthogonal moments. Sixth IASTED international conference on signal and image processing, pp 80–84
35. Yao Q, Guan Z, Zhou Y, Tang J, Hu Y, Yang B (2009) Application of support vector machine for detecting rice diseases using shape and colour texture features. IEEE International Conference on Engineering Computation, pp 79–83
36. Robert M, Haralick KS, Itshak D (1973) Texture features for image classification. IEEE Trans Syst Man Cybern SMC-3(6):610–621
37. Weszka JS, Dyer CR, Rosenfeld A (1976) A comparative study of texture measures for terrain classification. IEEE SMC-6:269–285
38. Kashyap RL, Chellappa R (1983) Estimation and choice of neighbors in spatial interaction models of images. IT V29:60–72
39. Kartikeyan B, Sarkar A (1991) An identification approach for 2-D autoregressive models in describing textures. CVGIP Graph Model Image Process 53:121–131
40. Mao J, Jain AK (1992) Texture classification and segmentation using multi resolution simultaneous autoregressive models. Pattern Recogn 25(2):173–188
41. Phadikar S, Sil J, Das AK (2012) Feature selection and rule generation to classify rice diseases by extracting features using empirical methods. J Comput Electron Agric 75:304–312

Chapter 13
Social Impact and Social Media Analysis Relating to Big Data

Nirmala Dorasamy and Nataša Pomazalová

Abstract Social media is a component of a larger dynamic and complex media and information domain. As the connection with Big Data grows, its impact in the social media domain cannot be avoided. It is vital that while the positive impact needs to be recognized, the negative impact emerging from Big Data analysis as a social computational tool needs to be recognized and responded to by various agencies. There have been major investments in the development of more powerful digital infrastructure and tools to tackle new and more complex and interdisciplinary research challenges. While there is a need to size the opportunities offered by continuing advances in computational techniques for analyzing social media, the effective use of human expertise cannot be ignored. Using the right data, in the right way and for the right reasons, can change lives for the better, especially if Big Data is used discriminately and transparently. This chapter analyzes the impact of Big Data from social media platforms in the social, political, and economic spheres. Further, the discriminate use of Big Data analysis from social media platforms is explored, within the context of ethical conduct by potential users and proposes important imperatives to minimize, if not control, the negative impact of Big Data analysis from a social perspective.

Keywords Social media • Digital technology • Social networking sites • Database tools • Big Data • Big Data analysis • Social media services

---

N. Dorasamy (✉)
Department of Public management and Economics, Durban University of Technology, Durban, South Africa
e-mail: nirmala@dut.ac.za

N. Pomazalová
Department of Regional Development, Mendel University, Brno, Czech Republic

Durban University of Technology, Durban, South Africa
e-mail: natta@atlas.cz

Z. Mahmood (ed.), *Data Science and Big Data Computing*,
DOI 10.1007/978-3-319-31861-5_13

13.1 Introduction

Data, or segments of information, have been collected and used throughout history. However, the potential to collect, store, and analyze data has significantly increased with the advancement in digital technology. The emergence of Big Data has escalated its usefulness for decision-making at various levels of analysis, including individual, group, organizational, and national systems. Organizations have moved from using data stored in relational databases to using data from data mining in general ledger packages, weblogs, social media, e-mail, sensors, photographs, corporate enterprise resource planning (ERP) systems, custom relationship management (CRM) programs, and social networks. While the growth of Big Data has accelerated in the last few years, the ability to find useful information within the Big Data is of crucial importance and requires careful consideration. Managing such data has to be underpinned by quality, protecting privacy and ethical use.

The rapid growth of the web as a publishing tool and the recent explosion of social media and social networking sites have generated opportunities and challenges to social researchers. Currently, there are many types of social media services (SMS). The Personal SMS like Facebook allows users to create online profiles and connect with other users, focusing on social relationships and information sharing such as one's gender, age, interests, and job profile. The Status SMS like Twitter allows users to post short status updates to broadcast information quickly and publicly with other users. The Location SMS like Foursquare and Google Latitude, using GPS-based networks, broadcasts one's real-time location. The Content-sharing SMS like YouTube and Flickr is designed as platforms for sharing content, such as music, photographs, and videos [4]. The Shared-interest SMS like LinkedIn is more a network for a subset of professional users to share information interests like politics and education. These social media services provide datasets that have expanded in size and complexity to the extent that computer-based methods are now required to analyze mass volumes of information. Data, with datasets whose size can range from a few dozen terabytes (TB) to multiple petabytes (PB), is beyond the ability of typical database software tools to capture, process, and store, manage, and analyze. Big Data technologies are required to economically extract value and meaning from very large volumes of a wide variety of data, by enabling high-velocity capture, discovery, and analysis [21].

Today, social media is a key source for Big Data analysis from platforms such as Twitter, Facebook, and Flickr, especially for businesses that depend on data-driven intelligence [3, 12]. IT companies such as Google, Amazon, Facebook, and IBM are fiercely competing in the Big Data analytics market. Social media systems provide valuable information in terms of its detail, personal nature, and accuracy. Since the data is not totally private, it is exposed to scrutiny within a user's network, which can increase the chances of accuracy when compared to data from other sources. Big Data integration and predictive analytics can help overcome the challenges of managing in an environment with increasing rates of change and innovation [12]. Studies have shown that Big Data analytics has resulted in improvements in retail operating margins, reduction in national healthcare expenditures, and savings in operational efficiencies. It has great potential, in that it can generate significant value across sectors, such as healthcare, retail, manufacturing, and the public sector [12]. However, seeing that data is dynamic, there has to be continuous integration of existing data with "living data," if companies want to reap the optimal benefits of data analytics.

Big Data volume is expanding due to the increase of social media, online data, and location data, often resulting from the accelerated usage of sensor-enabled devices. This has resulted in mobile cloud computing being made possible by focusing on Internet technologies that are built on web-based standards and protocols. The key drivers for cloud computing are bandwidth increase in networks, cost reduction in storage systems, and advances in database [12]. It has created a form of virtualization over the Internet, which involves data outsourcing that can be provisioned with minimal management effort or service provider interaction, with no up-front cost and provides just-in-time services. It is a model for enabling convenient and on-demand network access to a shared pool of computing resources and eliminates the cost for in house infrastructure [5]. An area of distinctive growth in the IT ecosystem is social media with smart devices, such as smartphones and tablets, providing a new communication platform with real-time access for customers. Various agencies have recognized such data as an important constituent of innovation and have developed means to use Big Data to develop solutions to business, technical, and social problems in innovative and collaborative ways. While the use of Big Data is invaluable, businesses are challenged by Big Data because it grows so large that they become awkward to work with using on-hand database management tools [12].

While the value of Big Data is clear for tackling complex technical and business problems, the question is on how well Big Data can solve complex social problems? While, business and science have shown the value of Big Data, the social sector needs to show how they can adopt this type of decision-making potential into their operations. The issues that are being addressed in the social sector are more complex than they are in business or science, making the use of Big Data more challenging. In addition, greater focus must be given to the rights, privacy, and dignity of different stakeholders. The large-scale collection, aggregation, analysis, and disclosure of detailed and triangulated information offer the possibility of powerful computational social science tools, but carries with it the potential for abuse by various entities, especially if datasets are not reliable and representative. In spite of these obstacles, progress continues.

13.2 Using Big Data and Social Media Innovatively

Big Data can encompass information such as transactions, social media, enterprise content, sensors, and mobile devices. Since Big Data refers to datasets that extend beyond single data repositories (databases or data warehouses), they are too large and complex to be processed by traditional database management and processing tools. New Big Data technologies have helped to capture analyze and store data to solve problems. Organizations are employing computing power through hardware and software advances to manipulate huge amounts of data. Some new approaches include Hadoop, a software that takes a different approach to data management, and HANA, a hardware approach that handles data manipulation in raw memory to make real-time analytics of the Big Data a reality [7]. Such advances have enabled the cost reduction and volume capacity increase of digital storage mediums and unique ways of manipulating data from social media.

Big Data offers much potential for innovative use, thereby creating value. Fanning and Grant [7] identified the following ways in which Big Data is valuable:

- Big Data can reveal significant value by making information transparent and usable at a much higher frequency. More people looking at the data will bring different perspectives.
- As the data proliferates, there is more accurate and detailed performance information that may expose variability and issues that need attention.
- Controlled experiments using Big Data analysis can assist to make better management decisions regarding tailored products and services.
- Big Data can be used to improve the development of the next generation of products and services through proactive maintenance and preventive measures to minimize failures.

13.2.1 Social Media as Advertising and Marketing Medium

Social media platform providers provide numerous metrics for data analysis to profile users for advertisers by showing the right advertisements to the right people, using masses of hidden information to model which users are likely to respond to a particular advertisement. Facebook can optimize advertising revenue by targeting advertisements to achieve the greatest possible number of clicks. If used scrupulously, valuable information can reach the right users. An additional example is Amazon, which allows users to exercise some control over their data by allowing them to flag purchases not to be used to form recommendations [1]. This is useful both in cases of gift purchases and in cases where the product is of a sensitive nature. Further, by allowing users to instruct a website not to use a specific piece of personal data for a particular purpose represents a significant improvement on the current personal data free-for-all model used by both social networking companies and their corporate customers [17]. Such platforms can also play a beneficial role in exposing, predicting, and helping to eliminate disruptive behavior. Computational social science can serve a positive role in promoting the interests of the community in a social media platform, provided the ethical considerations are evaluated for various purposes of use.

The impact of Big Data can also be seen in marketing. There is a noticeable move away from large-scale mailings of catalogues and offers to various individuals, based primarily on purchased mailing lists or phone directories. This is due to businesses mining Big Data to target individuals based on knowing the preferences of individuals in the population. For example, Amazon can almost instantaneously offer additional purchase opportunities to individuals based on what others have also purchased given a similar purchase showing up in the cart. It can also target those who in the past purchased a product or those who are searching a particular topic [2]. Further, Google offers marketers the opportunity to provide relevant advertisements to individuals based on their search habits.

It is through the improvement in the tools to analyze and collect Big Data that larger markets can be targeted. According to Fanning and Grant [7], digital marketing uses a combination of push-and-pull Internet technologies to execute marketing campaigns. The use of software vendors such as Adobe allows customers to make each digital transaction layered, thus allowing the organization to see in real time how a particular advertising campaign is performing, in terms of what is being viewed, how often, how long, as well as other actions such as responses rates and purchases made [8]. This information provides key information for marketers to make real-time decisions. In addition, the proliferation of mobile phones, tablets, and other means of accessing the Internet has facilitated multichannel marketing by companies like Usablenet.

Applications such as Facebook, LinkedIn, and Google are driven by information sharing that can be used by business clients, government, other users within the social media platform, and the platform provider itself [7]. For example, business clients draw on this computational social science to target users based on constructs that range from age, gender, and geographical location, sexual preferences, education level, and employer.

13.2.2 Adding Value for Social Well-Being

The rapid use of mobile phones and Internet usage, especially in developing economies, gives people the opportunity to improve the quality of their lives. A mobile phone acts as an individual sensor, collecting pertinent information from its environment, which when aggregated and analyzed with information from thousands of other mobile phones can provide vital information, which can then be disseminated back to people on the ground via the same mobile phones. For example, Cell Life, a South African organization, created a mass messaging mobile service called Communicate, which reminds patients to take their medications, links patients to clinics, and offers peer-to-peer support services such as counseling and monitoring [6].

In addition, most modern mobile phones contain global positioning systems technology, which identifies the geographic location of the phone, and other information relating to social media postings. This is important, for example, when researching migration patterns to understand the spread of infectious diseases like Ebola and to help stop the disease from spreading. Information on the patterns of human travel collected from mobile phone usage can be used to develop predictive models to combat diseases in specific regions [13].

A further consideration is census data collected in different countries, which is an important source of information for governments. In the 1800s, information from the national census was logged by hand, microfilmed, and sent to be stored in state archives, libraries, and universities. It took many years to properly tabulate census data after the initial collection. In recent years, countries have streamlined their data collection methods by adopting emerging technologies like geographic information systems, social media, videos, intelligent character recognition systems, and sophisticated data-processing software and processing tools to survey the populace [6].

Thorough data management during crisis communication can provide major benefits. The study by Proctera et al. [18] on the 2011 riots in England found that mainstream media lagged behind crowdsourced ("citizen journalism") reports appearing in social media. Further, it was evidenced that collaborative efforts by large numbers of "producers" of data can provide competing and, at times, better coverage of events than mainstream media. While evidence cannot always be taken at face value, social media provides a platform for robust mechanisms for authentication, so that false rumors are identified more quickly. This is supported by Mendoza, Poblete, and Castillo [16] who noted that users deal with "true" and "false" rumors differently: the former are affirmed more than 90% of the time, whereas the latter are challenged (i.e., questioned or denied) 50% of the time.

There are several examples of social media being a valuable tool for information gathering, for keeping the public informed, and for providing advice during crisis situations. One such case was the August 2011 riots in England, which began as an isolated incident in Tottenham. A study by Proctera, Visb, and Vossc [18] revealed that Twitter was used overwhelmingly for more positive means, especially for the organization of the riot cleanup. Even the police supported keeping social media sites open during the crisis. However, the study confirmed the conclusions of other studies that the police and government agencies in general still need to use social media platforms like Twitter effectively [18].

13.2.3 Gauging Business Performance

The business community has also been a heavy user of Big Data. For example, Netflix collects billions of hours of user data to analyze the titles, genres, time spent viewing, and video color schemes to gauge customer preferences and to give the customer the best possible experience. Following in the path of e-commerce, the rise of social media, Big Data, and cloud computing has impacted businesses in the following ways [11]:

- The ability to identify at the earliest opportunity those who are in danger of leaving organizations and to action the retention of best talent
- The ability to identify the high performers based on "new" live data highlighting performance and ratings and profitability rather than "old" assumptive data such as university attended and the grade of their degree
- The ability to measure the real drivers of performance within the business, thereby identifying "hidden gems" that make a real difference
- The ability to "fine-tune" businesses based on fact and evidence rather than fiction and emotion
- The ability to understand that datasets can be combined to form a more intricate and accurate picture from several data sources

13.2.4 Adding Value in Financial Markets

Financial markets have also benefitted from Big Data. Social media has played a significant role in growing financial markets with respect to the following [11]:

- Significant means by which crowdfunding helps the underserved small- and mid-sized entities (SMEs) and start-ups to access capital in a cost-effective manner.
- New technologies fund opportunities for growing segments of the economy not reached by traditional outlets and also create new jobs in a dynamic, technology-based business model.
- Algorithms and soft-/hardware technology related to "high-frequency trading" has exploded over the last 20 years, and the primary beneficiary has been the market for existing shares and other financial instruments (secondary market). With new cost-effective mechanisms to raise funds, underserved primary markets can now enjoy lower funding costs.
- Provides investors with protection from fraudsters through both proactive education and appropriate regulation by governments through the creation of a new and dynamic mechanism for allocating capital from traditional, institutional stakeholders (such as banks) to individual-driven operations using current and future technologies to touch millions of individuals looking for investment opportunities.

13.3 Discriminate Use of Social Media Analysis

There are multiple dimensions to Big Data, which include volume (considers the amount of data generated and collected), velocity (refers to the speed at which data are analyzed), variety (indicates the diversity of the types of data that are collected), viscosity (measures the resistance to flow of data), variability (measures the unpredictable rate of flow and types), veracity (measures the biases, noise, abnormality, and reliability in datasets), and volatility (indicates how long data are valid and should be stored) [15]. The multiple dimensions make data searches and retrieval more complex, as organizations have to find economical ways of integrating heterogeneous datasets while allowing for newer sources of data (in origin and type) to be integrated within existing systems [6]. The proliferation of social networks and social media requires much of the data being collected to be thoroughly analyzed before decision-making, as the data can be easily manipulated. In view of the high cost of data collection and management, organizations need to analyze the trade-off between accuracy and the cost of inaccurate data. If the quality of data is poor and users cannot make sense of it, then the data has limited value and use.

13.3.1 Unstructured Data

Unstructured data from multimedia networks cannot be categorized or analyzed numerically, as it uses natural language. The explosive growth of social media implies that the variety and quantity of Big Data is growing. A great deal of the growth can be traced to unstructured data. For example, analyzing words and pictures and then collating everything into meaningful and accurately interpreted information requires diverse methods and can be time-consuming. The challenge is exacerbated when time-sensitive issues like monitoring civil riots require data to be aggregated and analyzed in the shortest time possible [19].

Some authors argue that there is no Big Data in the context of social problems, as data is highly unstructured and generally not limited to numbers. In the case of child pornography, the global industry lures thousands of children annually. Increasingly, the producers of child pornography make use of various Internet platforms like mobile phones, social media, and online classifieds. Although, many initiatives exist to curb the problem, few initiatives have attempted to use Big Data. Data from these technologies could be collected and used to identify, track, and prosecute offenders. The problem is that the illicit nature of child pornography makes it difficult to collect reliable primary data for some of the following reasons: there are no valid indicators to measure antipornographic success and information collected often meets organizational needs and not global needs [6]. In addition, because of data privacy and security issues, data held by various organizations are seldom shared in raw form, thereby limiting the creation of global datasets. This is accentuated by agencies combating social evils often competing with each other for scarce resources, therefore not being eager to share data.

Different service providers have their own network management systems. Organizations cannot connect their datasets across other organizations, if such data is immersed in their administrative systems for operational purposes. One such case is the US healthcare industry, which is characterized by large volumes of health plans offered by different service providers with their own network management. This invariably results in data being stored in multiple formats in multiple places. If this data was more efficiently managed, then massive savings can accrue.

13.3.2 Gaps in Governance Standards

The lack of adequate data governance standards has failed to define how data is captured, stored, and used for accountability in the social arena. As a result some of the emergent challenges include integrating different datasets which lack good metadata (data that describe data), poor quality data, and difficult-to-manipulate forms such as PDFs or older file formats. As a result, large inconsistencies in the captured data exist, further complicated by the need to transform data for analysis, which is costly. The number of publicly available government datasets has accelerated, but only limited datasets where there is good metadata, ease of accessibility, and manipulability are ever used. Further, integrating information from multiple data sources requires skilled workers. In a report quoted by Fanning and Grant [7], by 2018 there would be a shortage of 140,000–190,000 people with deep analytical skills in the USA. This could be a serious problem for the analysis of Big Data to make effective decisions, if there are no initiatives to develop data analytics skills.

Without clarifying ethical issues on data storage, access, and use by different stakeholders, advancements in computational social science may put the public at increased risk. If society is to be protected, then there should be legal and ethical limitations on how social media as a computational social science tool can be used. Citing the sentiments of the CEO of Nasdaq, Mark Zuckerberg, to take risks is to "move fast and break things." Oboler et al. [17] argue for external constraints to protect society from the cost of mistakes by social media innovators.

The risk posed by the capacity of computational social science tools and the explosion in the corpus of data, free of the ethical constraints placed on researchers, raises serious questions about the impact that those who control the data and the tools can have on society. Many social media companies are driven by online advertising revenue which places the individual's interest in privacy in conflict with the interest of advertisers in extensive customer profiling. This is facilitated by Web 2.0 sites which have higher advertising rates for advertisers to target selected users. The increase in advertisers targeting specific users highlights unexpected consequences. For example, Target analyzed purchasing patterns to identify potential customers of baby paraphernalia. The analysis, based on purchasing history of unrelated items, highlighted potential pregnancies with a high degree of accuracy. Target sent advertising material to its target market, creating an angry response from a father whose teenage daughter received the advertising, but not knowing at the time that she was really pregnant. The daughter was forced to confirm her pregnancy because the retailer targeted her in the marketing as a result of the data analysis. This does raise ethical concerns regarding privacy of information [10].

Further, social media sites can scrape a great deal of data from users' age and sexual preference to target advertisements for adult products, which can cause distress or unauthorized disclosure of sensitive personal information. This poses both technical and ethical questions like: "Is any technically possible use of personal data ethically acceptable?" [17]. Social networking companies and advertisers need to consider such critical questions. By limiting data acquisition, sharing, and use and by raising public awareness of the implications of its availability through ethical considerations, the risk of abuse can be controlled.

13.3.3 Serving Self-Interest

Large volumes of data are not necessarily representative and reliable to solve problems relating to public interest. Big Data users can exploit Big Data with no regard for data quality, legality, data meaning, and process quality. For example, in 2011, the Rainforest Action Network in the USA discovered that the American Petroleum Institute and its oil lobby allies were able to manipulate social media opinion to show support for a pipeline project to carry oil from Canada to Texas by using fake Twitter accounts to send large numbers of tweets to show support for the project, which falsely represented public opinion. The Rainforest Action Network (RAN) discovered that 14 of 15 accounts were faked and the tweets were generated by an automated process [6].

Desouza and Smith [6] cite the case when public agencies and a newspaper in New York released information about gun owners after the Connecticut school mass shooting. Published information on the names and addresses of licensed gun owners living in the neighborhood can be used by the wrong people like criminals to target vulnerable homeowners who do not own guns or to target homeowners who have guns in order to steal them. According to Oboler, Welsh, and Cruz [17], methodology and ethics, drawn from the underlying fields of computational and social sciences, need to be considered. The authors argue that considerations apply not only to the research context but also to the worlds of government and commerce where philosophical concerns are less likely to counter immediate practical benefits. Most significantly, these concerns need to be considered in the context of social media platforms which have become computational social science tools that are easily accessible to businesses, governments, private citizens, and the platform operators themselves. Governments can exercise their power when they see social media acting against their interests. One such example was when the US government asked Twitter through a court order for data on WikiLeaks founder Julian Assange and those connected to him.

Social media can also promote the agenda of governments. Another example was when the Egyptian government cut off the Internet during the 2011 riots, after it realized that the US government provided training through the Internet to influence social change among Egyptian dissidents. Computational social science tools together with social media data can be used to reconstruct the movements of activists, to locate dissidents, and to map their networks. Governments and their security services have a strong interest in this activity [17].

13.3.4 Moving Away from Comfort Metrics

The analysis of general trends and the profiling of individuals can be investigated through social sciences. In this regard, Kettleborough [11] argues against "comfort metrics," whereby data that is not relevant and focuses merely on the process is collected. There is a need to look at data differently and to be prepared to throw away many old beliefs like forced ranking, which looks at normal distribution to employee performance, as a good employee performance measurement tool. According to Kettleborough [11], there is evidence that these approaches have actually damaged organizations like Microsoft's *lost decade*, which was as a direct result of misplaced or misunderstood data techniques.

Research studies have also shown that many businesses were preselecting and filtering candidates for employment based on social media. If job applicants do not protect their reputation online, then this can compromise their applications, if the interviewer has access to negative attributes posted online. Currently, business and government control large volumes of data used for computational social science analysis. The capacity to collect and analyze datasets on a vast scale provides the magnitude to disclose patterns of individual and group behavior. The potential damage from inappropriate disclosure of information is sometimes obvious. A lack of transparency in the way data is analyzed and aggregated, combined with a difficulty in predicting which pieces of information may later prove damaging, means that many individuals have little knowledge of potential adverse effects of the expansion in computational social science [17].

If data is not correctly understood, then massive mistakes can cause harm. Many employment recruiters are already "looking" at the social media lives of job applicants Although it is seen as justifiable, especially using work-related sites such as LinkedIn, in cases where employers seek data from nonwork-related social media life, there could be some potentially negative consequences. According to the CIPD, "using social media in recruitment or as part of career progression carries the risk of a number of different claims if a candidate is not appointed as a result of information gleaned." These include the following [11]:

- A breach of the Human Rights Act 1998 (incorporating Article 8 of the European Convention on Human Rights) to respect private and family life.
- A breach of the Data Protection Act 1998, which states that data controllers such as prospective employers should not hold excessive information and should process information in a fair way.
- It has been suggested that the over 50s age group will be more cautious with their social media presence than the under 30s, resulting in more potential for negative recruitment decisions for younger people.
- Information about marital status, number of children, and sexual orientation may incorrectly influence a selection decision.
- Information about physical or mental state, such as revealing depression to friends on social media, may be a disadvantage.

If life-changing judgments are to be made about people, then quality and accuracy must be beyond reproach. If employees are hired, promoted or dismissed based on Big Data discrimination, then there can be legal implications [11].

13.3.5 Using Power to Leverage Outcomes

Governments and powerful data-rich companies have the financial support and powerful resources to access data. Such organizations, by their nature, tend at times to assume that the risk of unjustified impacts on individuals is of little consequence when compared with the potential to avert perceived calamities [20]. It is easy to manipulate people, like using computational social science to guide political or product advertising, selling messages that people will favor or withhold information that may compromise support. Google, for example, can sway an election by predicting messages that would engage an individual voter (positively or negatively) and then disseminate content to influence that user's vote. The predictions could be highly accurate by making use of a user's e-mail in their Google-provided Gmail account, their search history, and social network connections. The dissemination of information could include "recommended" videos on YouTube to highlight where one political party agrees with the user's views – also articles in Google News could be given higher visibility to help sway voters into making *the right choice* [17]. Further, this can be complemented with negative messages to appear to create a balance, but in reality may have little or no impact. Such manipulation may not appear obvious, yet powerful to achieve the outcomes of the manipulator.

13.3.6 Risks Relating to Social Media Platforms

Social media platforms have added to their data either by acquiring other technology companies as Google did when acquiring YouTube or by moving into new fields as Facebook did when it created "Facebook Places" providing a geolocation service which generates high value information [14]. The value of information can be maximized by using a primary key that connects this data with existing information like a Facebook user ID or a Google account name, where a user is treated as a single user across all products of the company [14]. One account can connect to various types of online interactions, exposing greater breadth of a user's profile. In such a case all the data is immediately related and available to any query companies like Facebook and Google may have. This can be alarming as there is little privacy, since any information can be collected across platforms about users.

Accounts that are identity-verified, frequently updated, and used across multiple aspects of a person's life present the richest data and pose the greatest risk. For example, Facebook's Timeline feature allows users to mine social interactions that had long been buried. Further, since Timeline is not an option in Facebook, masses of personal data can be held. Another challenge was the Beacon software which, developed by Facebook, connected people's purchases to their Facebook account. It indicated what users had purchased, where they got it, and whether they got a discount. It was eventually closed in view of legal, privacy, and ethical considerations. Further, the emergence of massive open online courses under MOOCs is now causing a stir in the world of Big Data with evidence that student details, including performance data, is being sold online. Recruiters looking for highly motivated candidates with wisdom can hunt potential candidates on this platform [11].

13.3.7 Research Methodology Challenges

The use of social media as a source of social research data can present various methodological challenges. There can be sampling bias which can distort findings, as any particular social medium is not representative of the population as a whole. Avoiding sampling bias in social media sources is a great challenge for researchers in the social sciences. If computational tools are to be appropriately used in social research, then it is important that users are aware of the strengths and weaknesses of such tools. Therefore, it is vital that the capacity of social researchers in developing skills relating to computational methods and tools is developed, so that they can decide when and how to apply them responsibly.

Anonymity among employees during surveys is also causing concern. While it is believed that employee responses will be more truthful if they remain anonymous, their identity can be traced from the demographic details in their social media profiles. If used incorrectly, this "honest data" could be turned against the employees [11]. Employees who are aware of this may not necessarily give the "correct picture."

In terms of reliability and validity, decisions cannot be made with incomplete or incomprehensive data. Rational and fair decisions have to be based on representation. For example, if 20 people are happy with the service at a state hospital, this does not exhibit behavior that is statistically significant for the whole population. As the old adage states, "one swallow does not a summer make." Data focusing on a few does not paint a correct picture. Further, Kettleborough [11] contends that correlation and causation should not be analyzed at face value, since if two items correlate, it does not mean that one causes the other.

Data mining or scrapping of social media sites can result in personal data being used against individuals, even if it has been cleaned to remove personal references. One such example is a study by researchers from the Université Catholique de Louvain in Belgium who identified "95% of the unique users by analyzing only four GPS time and location stamps per person." In addition, researchers at Carnegie Mellon University were able to create a system to uncover Social Security numbers from birthday and hometown information listed on social networking sites like Facebook [11]. Large amounts of data can become the target of the unscrupulous.

13.3.8 Securing Big Data

Big Data, while sourcing data from multiple sources, relies on data that is available. Further, such data must be secured. The challenge is how the data is collected and stored. This raises security issues like internal employees adhering to confidentiality policies. Cases of storage abuses have occurred at Facebook sites. Data can also be lost due to hackers and employees. One example is the two Aviva employees who sold details of people who had accidents to claims companies. The fraud flag was raised when the claimants received calls from firms persuading them to take personal injury claims [11]. Information that is not secured can be used for blackmailing or espionage.

13.3.9 Limitations of Addressing Social Problems

In the social arena, a major gap exists between the potential of data-driven information and its actual use in addressing social problems. Certain social problems can be easily solved using Big Data, such as weather forecasting and areas with high disease rates. However, pandemic problems like drug trafficking and unemployment cannot be easily resolved in a sustainable way with Big Data. According to Desouza and Smith [6], these evil problems are more dynamic and complex than their technical counterparts, because of the diversity of stakeholders involved and the numerous feedback loops among the interrelated constituents. Government agencies and nonprofits are involved in tackling these problems but face the following challenges: limited cooperation and data sharing among them; inadequate information technology resources; their counterparts in the hard sciences work on technical problems or in business who have ready access to financial, product, and customer information; missing and incomplete data; and data stored in silos or in forms that are inaccessible to automated processing. In addition, there are regulatory constraints like policy relating to data sharing agreements, privacy and confidentiality of data, and collaboration protocols among various stakeholders tackling the same type of problem. While various agencies may invest in data technologies, the return on investment for solving social problems is yet to be convincing. This impact on the need is to be provided with information and advice via sources that they can trust in a more timely way.

13.4 Imperatives for Big Data Use from Social Media

In decision-making, context is the key; therefore knowledge of the domain such as social media is crucial. Analysis of Big Data is directly linked to decision-making, which has to be supported by very intricate techniques using wide and deep extensive data sources as shown in Table 13.1. Big Data is about massive amounts of different types of observational data, supporting different types of decisions and decision time frames. According to Goes [9], analytics moves from data to information to knowledge and finally to intelligence. The generation of knowledge and intelligence to support decision-making is critical as the Big Data world is moving toward real-time or close to real-time decision-making. Therefore, the need for context-dependent methodologies that strengthen prediction is pivotal for effective data analysis.

Big Data analytics from social media has to consider the tools, software, and the data to ensure quality results. While the technical ability may exist to gather data, the analytical capacity to draw meaning from such data needs to be developed. For example, visualization can be produced from real-time information as datasets emerge from user activity. However, such visualizations can only be considered powerful representations if visualization specialists are aware of which relationships benefit users. This requires an understanding of how meaning can be created through and across various datasets in social media platforms.

13.4.1 Responsibility of Analytic Role Players

Big Data has enormous potential to inform decision-making to help solve the world's toughest social problems. But for this to happen, issues relating to data collection, organization, and analysis must first be resolved. Much of this responsibility lies with the major analytic players, as shown in Table 13.2, who offer valuable services that help users cope with using Big Data effectively.

The aforementioned major analytic players have to ensure effective use of Big Data from social media platforms. This requires prudent use of analytic tools, incorporating the following guidelines [7]:

Table 13.1 Big data analytics

Decisions	Analytics	Techniques
Real time	Visualization	Statistics
Close to RT	Exploration	Econometrics
Hourly	Explanatory	Machine learning
Weekly	Predictive	Computational
Monthly		Linguistics
Yearly		Optimization
		Simulation

Adapted from [9]

Table 13.2 Major analytic players

Firm	Products	Website
SAP	HANA; Applied analytics	http://www.sap.com/solutions/technology/in-memory-computingplatform/hana/overview/index.epx;
		http://www54.sap.com/solutions/analytics/applications/software/overview.html
Splunk	Splunk Hadoop Connect, Splunk enterprise	http://www.splunk.com/
Tibco	Spotfire	http://spotfire.tibco.com/
Google	Big query	https://cloud.google.com/products/big-query
IBM	Pure data system	http://www-01.ibm.com/software/data/puredata
SAS	High-performance analytics	http://www.sas.com/software/high-performance-analytics/index.html
Metamarkets	Data pipes; Druid	http://metamarkets.com/platform
Oracle	Big data appliance; Exadata; Exalytics	http://www.oracle.com/us/technologies/big-data/index.html
Tableau	Tableau	http://www.tableausoftware.com/
Cloudera	Hbase	http://www.cloudera.com/content/cloudera/en/products/clouderaenterprise-core.html
Qliktech	QlikView	http://www.qlikview.com/
GoodData	GoodData bashes	http://www.gooddata.com/what-is-gooddata/

Adapted from [7]

- Use of in-memory database technology that avoids resources swapping databases between the storage medium and the memory, but rather operating within memory with only limited accessing of alternative storage mediums.
- The sheer size and complexity of data cannot be handled by traditional technologies built on relational or multidimensional databases, as there is a need to have flexibility to have questions answered in real time.
- Use caves of data from unstructured data to improve service levels, reduce operations costs, mitigate security risks, and enable compliance.
- Use tools that break down traditional data silos and attain operational intelligence that benefits both IT and the business, which is valuable for capturing machine-generated data into a system that would provide operational real-time data.
- The need to use technology to efficiently store, manage, and analyze unlimited amounts of data that can process any type of data differently than relational databases.

Since information in the various social media platforms is not static, if it is not updated and cleaned, then "dirty data" will arise. Considering the garbage in, garbage out syndrome, poor quality data will produce poor results.

13.4.2 Evidence-Based Decision-Making

In addition, the following four recommendations have the potential to create datasets useful for evidence-based decision-making [6]:

- The global community needs to create large data banks on critical issues like homelessness and malnutrition, which must have the capacity to hold multiple different data types along with metadata that describes the datasets. This requires multi-sector alliances that promote and create data sharing on sectoral issues. At the 2012 G-8 Summit, leaders committed to the New Alliance for Food and Nutrition Security to help 50 million people out of poverty over the next 10 years through sustained agricultural growth. This is supported by a number of databases like Agrilinks.org, Feed the Future Initiative website, and Women's Empowerment in Agriculture Index.
- Citizens and professionals can help create and analyze these datasets. With the growth of data through open data platforms, citizens are creating new ideas and products through what has become known as "citizen science." A bike map and map of the London tube were created by citizens, using the raw data from the London Datastore which is managed by the Greater London Authority.
- Big Data cannot be left to the pure sciences and business, but needs analysts in the social sciences to be statistically equipped to collect data for large-scale datasets. Skills in data organization, preservation, visualization, search, and retrieval, identifying networked relationships among datasets, and how to uncover latent patterns in datasets need to be developed. These are valuable skills that go beyond simply searching the web for information.
- Virtual experimentation platforms which allow individuals to interact with different ideas and work collaboratively to find solutions to problems can create large datasets, develop innovative algorithms to analyze and visualize the data, and develop new knowledge for tackling social challenges. The use of open forums such as wikis and discussion groups can help the community share lessons learned, collaborate, and advance new solutions.

In addition, Oboler et al. [17] argue that social networking has provided a diverse range of datasets covering large sections of the population, granting researchers, governments, and businesses the powerful ability to identify trends in behavior among a large population and to find vast quantities of information on an individual user. As the industry develops, social media computational tools will increase the scope, accuracy, and usefulness of such datasets. In view of the ethical and privacy implications, regulatory barriers restricting the collection, retention, and use of personal information require consideration. While laws protect human rights, there is a need for greater protection of the customer.

13.4.3 Protection of Rights

The rights of users need to be protected, as social media platform providers and various agencies provide innovative services to targeted users. The debate is whether consumers are protected from preventable harm only after proving damage or are rules set by law. In the first approach, advertisers have more freedom to mine data from various social media platforms, data over which the user has no control especially if it is outdated or hacked by third parties. The safeguarding of personal rights and freedoms is more favored through the setting of regulations and laws. This would place the burden on social media to restrict the storage, accessibility, and manipulation of data in ways that limit its usefulness. This can prevent unscrupulous use of data. However, this will require legislators to use multilateral legislating, since websites can freely choose the physical location of their hosting infrastructure where there is least regulation.

The ethical barriers for data use in the social sciences are much higher than pure science research as the data collection of personal information is higher in the social sciences. Oboler et al. [17] illustrate some suggestions to manage ethical use of social media data, as given below:

- In keeping with the code of ethics developed by professional bodies, example for engineers, these should be applied to social media as well. Such guidelines commit members to act in public interest, by not causing harm or violating the privacy of others. Social media platforms are a form of computational social science which requires recognition of the ethical concerns in the social sciences. This can reduce the opportunity for the abuse of a very powerful tool. Users of social media have an ethical responsibility to one another.
- A code of conduct for producers and consumers of online data which can highlight the issues to be considered when publishing information. For example, when a Twitter user uploads photographs, their action may reveal information about others in their network; the impact on those other people should be considered under a producer's code of ethics. A consumer code of ethics is also needed; such a code would cover users viewing information posted by others through a social media platform. A consumer code could raise questions of when it is appropriate to further share information, for example, by retweeting it.
- Guidelines for principles of engagement can help users determine what they are publishing and to create awareness of the potential impact of publishing information. The power of social media can be used to warn the owner when the content may pose a risk, especially when accesses open.
- A cultural mind shift is needed to become more forgiving of information exposed through social media, an acceptance that social media profiles are private and must be locked down with more intricate filters and used only in certain settings.

The aforementioned suggestions would change the nature of social media as a computational social science tool, by filtering what should be included out of the tools field of observation. As an instrument-based discipline, the way the field is understood can be changed either by changing the nature of the tool or by changing the way we allow it to be used.

13.4.4 Knowing the Context

Understanding and knowledge of the context is fundamental. For example, marketing depends, to a large extent, on information technology accruing from social networks. Researchers have to master the collection and analysis of web data and user-generated content, using advanced techniques. This is necessitated by the massive amounts of observational data, of different types, supporting different types of decisions and decision time frames [9]. In this regard, if researchers want to explain the growth in online shopping among teenagers in developing economies using social media networks, then it is imperative to use models like longitudinal models, latent models, and structured models to explain the causes within the context. Since the Big Data environment is targeting real-time decision-making, it is imperative that tools employed to analyze social media networks use context-dependent methodologies that enhance prediction in a valid and reliable way. The reason being that not only are the networks intricate but also require knowledge of the complex models. Consideration of these dynamics can produce valuable information from Big Data, allowing modeling of individuals at a very detailed level with a rich proliferation of the environment surrounding them [9].

Big Data analytics has the ability to yield deeper insights and predictions about the individuals. According to Waterman and Bruening [22], even though data may be processed accurately, the results may have profound effect on personal life choices. The authors argue that understanding the sources and limitations of data is critical to mitigate harm to individuals. This necessitates understanding and responding to the implications of choices about data and data analytic tools, integrity of analytic processes, and the consequences of applying the outcomes of analytic models to information about individuals [22].

13.5 Conclusion

The proliferation of Big Data has emerged as the new frontier in the wide arena of IT-enabled innovations and opportunities. Attention has focused on how to harness and analyze Big Data. Social media is one component of a larger dynamic and complex information domain, and their interrelationships need to be recognized. As the connection with Big Data grows, we cannot avoid its impact. Without being familiar with the data, the benefits of Big Data cannot be reaped. Large volumes of data cannot be analyzed using conventional media research methods and tools. The current Big Data analytics trend has seen the tools used to analyze and visualize data getting continuously better. There has been a major investment in the development of more powerful digital infrastructure and tools to tackle new and more complex and interdisciplinary research challenges.

Current programs have seen companies like Splunk, GoodData, and Tibco providing services to allow their users to benefit from Big Data. Users with the ability to query and manipulate Big Data can achieve actionable information from Big Data to derive growth by making informed decisions. Access to data is critical. However, several issues require attention in order to benefit from the full potential of Big Data. Policies dealing with security, intellectual property, privacy, and even liability will need to be addressed in the Big Data environment. Organizations need to institutionalize the relevant talent, technology, structure workflows, and incentives to maximize the use of Big Data.

It is imperative that apart from the power users in marketing, financial, healthcare, science, and technical fields, those involved in daily decision-making must be empowered to use analytics. As more and more analytical power reaches decision-makers, enhanced and more accurate decision-making will emerge in the future. While there is a need to size the opportunities offered by continuing advances in computational techniques for analyzing social media, the effective use of human expertise cannot be ignored. Using the right data in the right way and for the right reasons to innovate, compete, and capture value from deep and real-time Big Data information can change lives for the better. Big Data has to be used discriminately and transparently.

References

1. Boyd D, Crawford K (2012) Critical questions for big data: provocations for a cultural, technological, and scholarly phenomenon. Inf Commun Soc 15(5):662–679
2. Brown B, Chui M, Manyika J (2011) Are you ready for the era of 'big data'? McKinsey Q 4:24–35
3. Bughin J, Chui M, Manyika J (2010) Clouds, big data, and smart assets: ten tech-enabled business trends to watch. McKinsey Q 56(1):75–86
4. Chen H, Chiang RH, Storey VC (2012) Business intelligence and analytics: from big data to big impact. MIS Q 36(4):1165–1188
5. Colgren D (2014) The rise of crowdfunding: social media, big data, cloud technologies. Strategic Financ 2014:55–57
6. Desouza KC, Smith KL (2014) Big data for social innovation. Stanf Soc Innov Rev 2014:39–43
7. Fanning K, Grant R (2013) Big data: implications for financial managers. J Corporate Account Financ 2013:23–30
8. Gandomi A, Haider M (2014) Beyond the hype: big data concepts, methods, and analytics. Int J Inf Manag 35:137–144
9. Goes PB (2014) Big data and IS research. MIS Q 38(3):iii–viii
10. Hill K (2012) How target figured out a teen girl was pregnant before her father did. Forbes (16 February). http://www.forbes.com/sites/kashmirhill/2012/02/16/how-target-figured-out-a-teen-girl-was-pregnant-before-her-father-did/. Accessed 9 Dec. 2014
11. Kettleborough J (2014) Big data. Train J 2014:14–19
12. Kim HJ, Pelaez A, Winston ER (2013) Experiencing big data analytics: analyzing social media data in financial sector as a case study. Northeast Decision Sciences Institute Annual Meeting Proceedings. Northeast Region Decision Sciences Institute (NEDSI), April 2013, 62–69
13. LaValle S, Lesser E, Shockley R, Hopkins MS, Kruschwitz N (2013) Big data, analytics and the path from insights to value. MIT Sloan Manag Rev 21:40–50

14. McCarthy J (2010) Blended learning environments: using social networking sites to enhance the first year experience. Australas J Educ Technol 26(6):729–740
15. McKelvey K, Rudnick A, Conover MD, Menczer F (2012) Visualizing communication on social media: making big data accessible. arXiv (3):10–14
16. Mendoza M, Poblete B, Castillo C (2010) Twitter under crisis: can we trust what we RT. In: 1st workshop on Social Media Analytics (SOMA '10). ACM Press, Washington, DC
17. Oboler A, Welsh K, Cruz L (2012) The danger of big data: social media as computational social science. First Monday 17(7):60–65
18. Proctera R, Visb F, Vossc A (2013) Reading the riots on Twitter: methodological innovation for the analysis of big data. Int J Soc Res Methodol 16(3):197–214
19. Qualman E (2012) Socialnomics: how social media transforms the way we live and do business. Wiley, New York
20. Wigan MR, Clarke R (2013) Big data's big unintended consequences. IEEE Comput Soc 2013:46–53
21. Young SD (2014) Behavioral insights on big data: using social media for predicting biomedical outcomes. Trends Microbiol 22(11):601–602
22. Waterman KK, Bruening PJ (2014) Big data analytics: risks and responsibilities. Int Data Privacy law 4(2):89–95

Index

Z. Mahmood (ed.), *Data Science and Big Data Computing*,
DOI 10.1007/978-3-319-31861-5

P

Q

R

S

T

Printed in the United States
By Bookmasters